I0043708

Frederic Moore, William Chapman Hewitson

Descriptions of new Indian lepidopterous insects from the collection

of the late Mr. W.S. Atkinson

Frederic Moore, William Chapman Hewitson

Descriptions of new Indian lepidopterous insects from the collection of the late Mr. W.S. Atkinson

ISBN/EAN: 9783337305420

Printed in Europe, USA, Canada, Australia, Japan

Cover: Foto ©berggeist007 / pixelio.de

More available books at **www.hansebooks.com**

DESCRIPTIONS

OF

NEW INDIAN LEPIDOPTEROUS INSECTS

FROM THE

COLLECTION OF THE LATE MR. W. S. ATKINSON, M.A., F.L.S., &c.,

DIRECTOR OF PUBLIC INSTRUCTION, BENGAL.

RHOPALOCERA,

BY

WILLIAM C. HEWITSON, F.L.S.

HETEROCERA
(SPHINGIDÆ—HEPIALIDÆ),

BY

FREDERIC MOORE, F.Z.S. ETC.,

ASSISTANT CURATOR, INDIA MUSEUM, LONDON.

WITH AN INTRODUCTORY NOTICE,

BY

ARTHUR GROTE, F.Z.S. &c.,

HONORARY MEMBER OF THE ASIATIC SOCIETY OF BENGAL.

CALCUTTA:

PUBLISHED BY THE ASIATIC SOCIETY OF BENGAL.

LONDON: PRINTED BY TAYLOR AND FRANCIS, RED LION COURT, FLEET STREET

1879.

SYSTEMATIC LIST OF CONTENTS.

PART I.

RHOPALOCERA. By W. C. Hewitson.

HETEROCERA. By F. Moore.

Note. Parts II. and III. will contain the descriptions of the species of the remaining families.

INTRODUCTION.

At the Monthly General Meeting of the Asiatic Society of Bengal, held on November 7th, 1877, the Chairman, Mr. W. T. Blanford, F.R.S., and Vice-President of the Society, announced as follows:—

"That arrangements had been made for publishing an extra volume, containing the descriptions by Messrs. Moore and Hewitson of the new species of Lepidoptera in the late Mr. Atkinson's collections, as announced at the March Meeting. It was proposed that the volume should be in Four Parts, quarto. The first Part would be put in hand at once; and the remaining three Parts would be completed in the course of the next two years. The work would be illustrated by coloured Plates. Owing to the expense, 225 copies * only would be printed, and would be available to Members by purchase after presentations had been made to the Societies interested in Natural History exchanging with the Society."

The scheme of publication here described was an enlargement of that which, on the suggestion of Mr. F. Moore of the India Museum, I had proposed to the Society early in 1877. The Society's 'Proceedings' for March of that year, to which Mr. Blanford makes reference, record the letter submitting my recommendation, as well as the Council's action thereon.

Briefly, my letter stated that Mr. Atkinson's magnificent collection of Lepidoptera, specially rich as it was known to be in Heterocera, had been purchased by Mr. Hewitson, and had since gone to Germany, but that Mr. F. Moore had made arrangements with the

---

* Afterwards extended to 525 copies, of which 300 were to be published with uncoloured plates. The work, moreover, will consist of Three Parts only.

b

purchaser to examine the novelties among the Nocturnals, and had offered to describe them in the Society's Journal if the Council would find room for his paper. It was estimated that the new species would number 650, thus ranged—

Bombyces	200
Noctuæ	200
Geometræ	200
Pyrales	50

and that they might be described and illustrated by uncoloured figures of selected types at no great cost. Such a paper I observed might be published as an extra number of the Journal, with a brief Introductory Memoir, which I would supply, on the plan followed in the case of the late Mr. Blyth's Burmah Catalogues.

The promptitude with which the Council adopted the above recommendation showed their appreciation of the motives which led me to make it. The Society could not offer a more appropriate tribute to the memory of a distinguished Member and office-bearer than by securing for their own publications the record of his many and valuable contributions to the Lepidopterous fauna of India.

The larger scheme of publication adopted by the Council, and their sanction of the use of coloured Plates, enabled me to enlist the ever-willing services of the late Mr. Hewitson in the undertaking. On learning that a quarto Plate could be placed at his disposal, he at once offered to figure four of the new species of Diurnals which he had obtained from the Atkinson collection, and to republish with the figures the descriptions which he had already given of them in the 'Entomologists' Monthly Magazine' (Dec. 1876).

Plate I. of this work has now a melancholy interest, inasmuch as its proofs were corrected on the dying bed of this enthusiastic Lepidopterist. His last note to me in connexion with the proofs is dated May 23rd, five days only before his death.

I proceed to give a short personal notice of the assiduous Collector whose successful labours in the field will be found recorded and illustrated by such competent hands in the following pages.

WILLIAM STEPHEN ATKINSON was the eldest son of the Rev. Thomas D. Atkinson, of Chesterton, in Suffolk, where he was born in September 1820. He was educated at home; and his boyish tastes for the study of nature seem to have been developed on Cannock Chase, on his father becoming Vicar of Rugeley. It was here that he commenced what gradually became a very complete collection of British Lepidoptera, including the Tineina, which latter were with him favourite objects of study.

He entered Trinity College, Cambridge, in 1839, and obtained a Scholarship, passing out as 26th Wrangler in 1843. After some further stay at Cambridge, during which he occupied himself in tutorial duties, he came to London to study for the profession of Civil Engineer; but on being offered the appointment of Principal of the Martiniere College, he went out to Calcutta in November 1854, having previously married Miss Montford, daughter of the Vicar of East Winch, in Norfolk. He joined the Asiatic Society in July 1855, and towards the close of the same year succeeded me as its Secretary, continuing to hold that office till 1864.

The new and rich fauna which now offered itself at once fascinated him; and as I had then a considerable collection of Bengal Lepidoptera, consisting for the most part of the larger Diurnals and Nocturnals obtained by the late Mr. Robert Frith, he was a frequent visitor at my house. Sharing as he did my interest in the transformations of the many species which could be observed in the neighbourhood of Allipore, the breeding-cages and the progress made by my native artist * in figuring the several stages of the imago had for him a special attraction. His experience as a collector, and consequent familiarity with the larvæ of British species, gave a point to his observations and criticisms which often proved of the greatest assistance to me.

He had not been many months in Calcutta before he reported to Mr. Stainton his first discoveries of Tineina, a group which no Indian entomologist had yet ventured to touch. His practised eyes detected traces of a miner on the leaf of *Bauhinia purpurea*; and *Lithocolletis Bauhiniæ* was, I believe, the first Microlepidopterous species that ever reached this country from India. It was described in 1856, with two other species, by Mr. Stainton in the Trans. Entom. Soc., new ser. vol. iii. p. 301. Mr. Atkinson at the same time became a Member of the Entomological Society.

How assiduously and with what success he pursued his researches in this new and difficult field of Indian Entomological research, may be gathered from the opening remarks of a later paper by Mr. Stainton when describing, in 1858, twenty-five species of Indian Microlepidoptera (Trans. Entom. Soc. new ser. vol. v. p. 111). One of these species, the type of a new genus of Elachistidæ named after its discoverer, *Atkinsonia clerodendronella*, I hope to see reproduced, as this work proceeds, from a drawing in my own possession.

In a third paper of Mr. Stainton's, describing and figuring nine exotic species of the genus *Gracilaria* (Trans. Ent. Soc. 3rd ser. vol. i. p. 291, 1862), I find that five species were

---

* In one of his letters to Mr. Stainton I find him paying the following just tribute to the drawings of this artist, the late Munshi Zynulabdin, who for many successive years worked in Mr. Frith's and my employ:—"His execution is as good as any I ever saw; in fact it could hardly be better; but he has never drawn under a microscope."

contributed by Mr. Atkinson. Zeller's 'Monograph of the Crambidæ,'[*] again, published in 1863, records the following five species obtained from the same source :—

Scirpophaga auriflua.	Schœnobius minutellus.
" gilviberbis.	Calamotropha Atkinsoni.
Schœnobius punctellus.	

It must be borne in mind that all the specimens sent home by Mr. Atkinson were bred by himself, and that for most of them the larvæ were described; and I cannot here refrain from extracting from one of his letters to Mr. Stainton (July 11, 1856) a passage which sets forth some of the difficulties which will always beset the tropical Micro-lepidopterist :—

"The smaller Tineina are especially difficult to manage: one's fingers and every thing else are sticky with damp and heat; and one can hardly live, except under a punkah. You can imagine, therefore, the difficulty of pinning creatures of the size of *Nepticulæ*, and will not be surprised to hear that I have failed entirely in pinning two small species of *Phyllocnistis*, and several other things of the like size. The Indian Tineina are certainly much smaller, on the average, than those of Europe, which is singular, considering the luxuriance of vegetation, and in marked contrast with the large size of the predominant Diurnal Lepidoptera."

Mr. Atkinson generally employed the short autumn vacations in excursions to places within easy reach of Calcutta; and in 1860 he paid his first visit to Darjiling. It was in this year that he was appointed Director of Public Instruction under the Bengal Government, a post which he held till he finally quitted India. I think I may say that the bulk of his fine collection of Moths were captured during his repeated visits to Darjiling. His wife had preceded him thither, and by simply showing a lamp against a white wall had been in the habit every evening of taking numerous novelties. The exceptional richness in Nocturnals of Mr. Atkinson's cabinets is due entirely to the systematic way in which, thus aided by his wife, he attracted and collected his own specimens, instead of relying on the casual captures of native collectors. When at Darjiling myself, in 1864, I witnessed the success of his ambush on more than one occasion.

I regret that his widow has not been able to find any notes of Mr. Atkinson's tour in the Sikkim Hills, made in company with the late Dr. Thomas Anderson, of the Calcutta Botanic Garden. He often spoke to me of this expedition, and of the useful information which it obtained for him in regard to the localities and elevations in which the numerous so-called "Darjiling" Lepidoptera are met with. The vague indications of locality so commonly furnished by museum-labels were highly obnoxious to him; and I have in my

* 'Chilonidarum et Crambidarum Genera et Species.'

possession copy of revised localities which he compiled for the improvement of any further edition of Horsfield and Moore's India-Museum Catalogues. I have now before me a memorandum-book kindly lent to me by Mrs. Atkinson, in which I find noted for Kashmir, which he visited in 1874, just the particulars which I should have expected in his Sikkim diary.

Mr. Atkinson resigned the Secretaryship of the Asiatic Society in 1864, and in the following year was elected one of its Vice-Presidents and nominated a Trustee of the New Indian Museum. His official duties necessarily occupied the greater part of his time; and the want of leisure as well as of books of reference prevented him from venturing on the description of any of the novelties which now filled his drawers. He was, however, in frequent correspondence with Mr. F. Moore, who, in his paper "On the Lepidopterous Insects of Bengal" (P. Z. S. 1865, p. 755), described, but did not figure, three new species of Saturniidæ sent home from Darjiling by Mr. Atkinson. These were *Cricula drepanoides*, *Saturnia anna*, and *Loepa sikkima*—the two last having the provisional names assigned to them by Mr. Atkinson. Another fine species of the same group, *Loepa miranda*, Atkinson, was described by Mr. Moore in Trans. Entom. Soc. ser. 3, vol. ii. p. 424, 1865; but neither of this has any figure yet been published.

Mr. F. Walker's Catalogue of the Heterocerous Lepidoptera in the British Museum contains in its Supplements (Parts 31–35, 1865–66) several new species sent home by Mr. Atkinson to be described and named. As the types of these were, I believe, returned to India by Mr. Walker, it is to be feared they are lost to this country, though duplicates, where they existed, will probably be found among the specimens which he told me he had presented, before leaving Calcutta, to the Indian Museum.

In Hewitson's 'Exotic Butterflies' I find figures only of the three following species of Mr. Atkinson's:—

Vol. iii. 1862–66. ERYCINIDÆ . . .	*Dodona dipoea.*	Darjiling.
Vol. iv. 1867–71. HESPERIDÆ { . .	*Hesperia phœnicis.*	India *.
. .	*Hesperia eltola.*	Darjiling.

It was not until 1865, when Mr. Atkinson came to England on three months' leave, that he made Mr. Hewitson's acquaintance. Of the other novelties from Darjiling described by the latter in this work, I find *Myrina symira* figured in Supplementary Plate III. *b* of Part VIII. (and last) of his 'Illustrations of Diurnal Lepidoptera,' which has been published since his death. The two Hesperidæ (*H. cephala* and *H. cerata*) were doubtless reserved to be figured in another Part of the same work which he proposed to devote to this family.

---

* Identical, Mr. Moore tells me, with a *Hesperia* which both Mr. Atkinson and I had reared on the Date-tree at Allipore.

As Mr. Hewitson mentioned twelve species of Rhopalocera in the Atkinson collection as being hitherto unknown, there yet remain for description four species, to be dealt with by the British Museum authorities, into whose hands Mr. Hewitson's unique collection has passed.

The first original descriptions by Mr. Atkinson were undertaken on the return of Major Sladen's expedition from Yunan. His paper, which was communicated to the Zoological Society by Dr. J. Anderson in 1871 [*], described three new species of Diurnal Lepidoptera from Western Yunan, collected by Dr. Anderson in 1868. These species were *Æmona lena*, *Zophoessa andersoni*, and *Plesioneura liliana*, all of them being well figured.

In 1873 Mr. Atkinson contributed two more papers to the same Society. One of them [†] described and figured a new and beautiful species of Butterfly which had been sent to him by Dr. Lidderdale from Bhutan, and which, in the belief that it would form a new genus intermediate between *Thais* and the Chinese *Sericinus*, he named *Bhutanitis lidderdalii*. He was not aware, nor was I till Mr. Moore called my attention to the fact, that a closely allied species had already been described by Blanchard in the 'Comptes Rendus,' 1871, p. 809, and constituted the type of his new genus *Armandia*, named after the Abbé Armand David, who had sent it home from Moupin. Blanchard's genus will take precedence of *Bhutanitis*, owing to the delay which occurred in Mr. Atkinson's publication of his description of *Armandia lidderdalii*, which seems to have come into his hands in May 1868. A figure of the Abbé David's species, *Armandia thaitina*, will be found in the second livraison of the 'Études d'Entomologie,' printed in November 1876 (but apparently not published), at M. Charles Oberthür's press at Rennes. A comparison of the figures leaves no room to doubt the close affinity of the two species, which are both from high elevations, and will probably be found to meet each other on the slopes of the mountains which divide Assam from Western China. It will be interesting to ascertain the range eastward of *A. lidderdalii*, and to learn whether its larva feeds on the *Aristolochia*, which the Abbé [‡] tells us is the food-plant both of *Sericinus* and *Armandia thaitina*, the two forms which, he adds, replace *Thais* in Europe.

Mr. Atkinson's third paper [§], contributed to the Zoological Society in the same year, described and figured two new species of Butterflies from the Andaman Islands. These were *Papilio mayo*, a fine species of the *Polymnestor* group, and *Euplœa andamanensis*. He confined himself to bare descriptions, though no one was more competent than he was to offer observations on the distribution of allied species, and the isolation of these two

[*] P. Z. S. 1871, p. 215.
[†] P. Z. S. 1873, p. 570.
[‡] 'Journal de mon 3ᵐᵉ Voyage d'Exploration' &c. Hachette, 1875.
[§] P. Z. S. 1873, p. 736.

forms of genera, so abundant in Indian species, is a fact which affords room for much interesting comment.

This is not the place in which to speak of Mr. Atkinson's career as a public officer. His duties of superintending the Education Department in Bengal were highly interesting to him, and such as previous training well qualified him to discharge efficiently. When he left India on furlough in the early part of 1875, he hardly calculated on returning to that country, and brought home with him his entire collections. He intended to remain in England for three years, whether his term of office was renewed * by Government or otherwise, and to devote his furlough to working out the new species which form the subject of this paper. But before settling down to this task, he was advised to recruit his health by an excursion on the Continent, and he resolved to spend his first winter in Italy. He called on me in the autumn to announce his plans, and to inform me of an arrangement which he had fortunately been able to make for leaving undisturbed within Custom-House limits the numerous cases containing his collections. He was then in the best of spirits, and dwelt much on the pleasant prospect of opening his cases on his return to England in the spring. But this prospect was not to be realized. He was attacked by pneumonia at Rome in January 1876, and died, after a few days' illness, on the 15th of that month.

It will be borne in mind that the figures here published represent but a small proportion of the new species of Heterocera described by Mr. Moore, who, being limited to a certain number of Plates, had necessarily to select the species to be reproduced in them, and who has exercised his discretion wisely in figuring the more typical forms in each family. For the Lithosiidæ he has been able to dispense with figures altogether, as he has so recently revised the family in a well-illustrated paper contributed to the Zoological Society in January 1876, in which appear many of the forms of the Atkinson Collection.

I have to acknowledge the assistance rendered to me by Mr. Stainton in kindly placing at my disposal several letters addressed to him by Mr. Atkinson on his first arrival in India.

ARTHUR GROTE.

London, July 1879.

* This term was renewed, I learn from Mrs. Atkinson ; but the orders of Government did not reach this country till after his death.

DESCRIPTIONS

INDIAN LEPIDOPTERA RHOPALOCERA

FROM THE

COLLECTION OF THE LATE MR. W. S. ATKINSON.

BY

W. C. HEWITSON, F.L.S., F.Z.S., ETC.

———————

Family NYMPHALIDÆ.

Genus ADOLIAS.

ADOLIAS SATROPACES.   (Plate I. figs. 6, 7, 8.)

Adolias satropaces, Hewitson, Entomologist's Monthly Magazine, 1876, p. 150.

Male. Upperside dark brown; anterior wing projecting at the apex, as in *A. cocytus*, marked in the cell by a black line and by two large pale spots bordered with black, by a spot and two short black lines below these, and by another pale undefined spot near the apex; the outer margin, except at the apex, rufous-grey; posterior wing with the outer half of the same colour.

Underside ochreous-yellow; anterior wing with the spots in the cell and a linear submarginal band of brown; posterior wing with some scarcely seen spots and bands before and after the middle.

Female pale rufous-brown; anterior wing with the spots in and below the cell as in the male: marked beyond the middle by six transparent spots, five in a transverse band, and one near the apex; crossed near the outer margin from the apex by a dark brown band, which is continued to the middle of the inner margin of the posterior wing. Posterior wing with two spots in the cell and a series of submarginal lunular spots of brown.

B

Underside as above, except that it is orange-yellow, and that the submarginal band has its origin at a different part of the apex, and is scarcely seen on the posterior wing.

Expanse, ♂ 2$\frac{4}{10}$, ♀ 3$\frac{1}{10}$ inches.

Hab. Moulmein. In coll. W. C. Hewitson.

Note.—The female of this species very closely resembles *A. aphidas.*

Family SATYRIDÆ.

Genus DEBIS.

DEBIS SERBONIS. (Plate I. figs. 4, 5.)

Debis serbonis, Hewitson, Entomologist's Monthly Magazine, 1876, p. 151.

Upperside rufous-brown; both wings with two submarginal brown lines; anterior wing with two indistinct pale spots on the costal margin beyond the middle; posterior wing with a series of three black eye-like spots, and an ocellus marked with white near the anal angle.

Underside rufous; anterior wing with a zigzag brown line and a large pale spot, bordered on both sides with brown within the cell; the discocellular nervure brown: crossed beyond the middle by a dark brown band, bounded outwardly, near the costal margin, by a dull white spot; a white spot near the apex, and below two small ocelli, one of which is incomplete, followed by a band of brown, and a submarginal band also brown: posterior wing crossed by two brown bands, one before, the other at the middle; a brown line at the end of the cell; a series of six ocelli, the first and fifth larger and more distinct than the rest; the outer margin and a line near it black.

Expanse 2$\frac{9}{10}$ inches.

Hab. Darjiling. In coll. W. C. Hewitson.

Genus ZOPHOESSA.

ZOPHOESSA ATKINSONIA. (Plate I. figs. 2, 3.)

Zophoessa atkinsonia, Hewitson, Entomologist's Monthly Magazine, 1876, p. 151.

Upperside: male dark brown, rufous towards the base; both wings with a submarginal black line: anterior wing marked by several rufous-orange spots; two in the cell, a quadrifid band beyond these, three (one bifid) near the apex, and five below the middle: posterior wing with a rufous-orange band near the outer margin marked by five black spots; a submarginal rufous line.

Underside rufous brown; a spot in the cell, which is bordered on both sides with brown, and a continuous band beyond the middle, sinuated and bordered inwardly with dark brown, both yellow; a series of four small white spots near the apex, and two submarginal linear brown bands: posterior wing tinted with green near the base, marked by two short bands of yellow, followed by a broader band of the same colour; the outer half of the wing rufous-brown, marked by five black ocelli, with rufous iris and pupil of blue: a submarginal band of white.

Expanse 2$\frac{1}{10}$ inches.

Hab. Darjiling. In coll. W. C. Hewitson.

Family ERYCINIDÆ.

Genus DODONA.

DODONA DEODATA. (Plate I. fig. 1.)

Dodona deodata, Hewitson, Entomologist's Monthly Magazine, 1876, p. 151.

Upperside brown. Both wings crossed near the middle by a broad common band of white. Anterior wing with the base pale rufous-brown crossed by a band of paler colour; the outer half dark brown marked by ten white spots; four in a band from the costal margin to the anal angle, and two bands of three spots each, near the apex. Posterior wing with the inner margin broadly rufous-brown, crossed by a band of paler colour, the outer margin dark brown, traversed by two bands of white spots.

Underside as above, except that several of the small white spots, near the apex of the anterior wing, meet and form a band, that there is a submarginal white line, and two white spots near the anal angle, and a linear band leading to two black spots near the outer margin of the posterior wing, and that the lobe at the anal angle is bordered above with orange-yellow marked by two black spots.

Expanse $1\frac{15}{20}$ inch.

Hab. Moulmein. In coll. W. C. Hewitson.

Family LYCÆNIDÆ.

Genus MYRINA.

MYRINA SYMIRA.

Myrina symira, Hewitson, Entomologist's Monthly Magazine, 1876, p. 152.

Upperside brown, tinted with purple, the outer margin dark brown: posterior wing with the anal angle dark brown: two tails, a long one in continuation of the first median branch and a short one inside of it.

Underside rufous-orange: posterior wing with a black spot crowned with silvery blue at the base of each tail.

Expanse $1\frac{1}{16}$ inch.

Hab. Darjiling. In coll. W. C. Hewitson.

This species will be figured in the 'Illustrations of Diurnal Lepidoptera,' Lycænidæ, part viii.

Family HESPERIDÆ.

Genus HESPERIA.

HESPERIA CEPHALA.

Hesperia cephala, Hewitson, Entomologist's Monthly Magazine, 1876, p. 152.

Upperside dark brown, the fringe brown and white alternately; anterior wing with three transparent white spots and an opaque spot near the inner margin—one at the middle bifid, one at the apex trifid, and one below it; posterior wing with two transparent spots near the middle.

Underside: anterior wing as above, except that the costal margin from the base to the transparent spot, and the outer margin from the apex to the middle, are yellow; posterior wing yellow, with a black spot near the base, a third white spot adjoining the transparent white spots, which are bordered below with rufous-brown, the outer margin rufous-brown.

Expanse $1\frac{7}{10}$ inch.

Hab. Darjiling. In coll. W. C. Hewitson.

HESPERIA CERATA.

Hesperia cerata, Hewitson, Entomologist's Monthly Magazine, 1876, p. 152.

Upperside dark brown: anterior wing with four transparent white spots; one in the cell sinuated on both sides, two below this between the branches of the median nervure, and one near the apex bifid: posterior wing with a central series of four or five indistinct white spots.

Underside as above, except that both wings have a submarginal series of pale spots, that the posterior wing has a white spot near the base, and a transverse central series of six distinct white spots.

Expanse $1\frac{4}{10}$ inch.

Hab. Darjiling. In coll. W. C. Hewitson.

Genus PTERYGOSPIDEA.

PTERYGOSPIDEA BADIA.

Pterygospidea badia, Hewitson, Annals & Mag. of Natural History, 1877, xx. p. 322.

Upperside dark brown: anterior wing with a central narrow band and four white spots near the apex (one of which is very minute and considerably below the rest) transparent white: posterior wing with the fringe yellow.

Underside as above, except that the posterior wing has a central yellow spot. Antennæ with a white ring near the point.

Expanse $2\frac{1}{10}$ inches.

Hab. Darjiling. In coll. W. C. Hewitson.

DESCRIPTIONS

OF

INDIAN LEPIDOPTERA HETEROCERA

FROM THE

COLLECTION OF THE LATE MR. W. S. ATKINSON.

BY

FREDERIC MOORE, F.Z.S. ETC.,

ASSISTANT CURATOR, INDIA MUSEUM, LONDON.

Family SPHINGIDÆ*.

Genus APOCALYPSIS, Butler.

Apocalypsis, Butler, Trans. Zool. Soc. Lond. ix. pt. x. p. 641 (1876).

Allied to *Euryglottis*; similar in pattern, but at once distinguished by the much smoother thorax, shorter and more slender antennæ, more prominent and less crested head.

APOCALYPSIS VELOX.

Apocalypsis velox, Butler, *l. c.* p. 641 (1876).

Fore wing long and pointed, sepia-brown, with the veins, a chain-like discal excavated transverse band, and some oblique lines connecting its outer border with the apices of the radial veins pale brown; an oblique white streak from the apex to the upper radial: hind wing smoky brown, the costal and abdominal areas paler; a basal hairy patch, two diffused abbreviated discal bands distinct towards abdominal area, and a broad outer border deep brown, darkest at the anal angle; margin blackish, the fringe and a diffused narrower streak at anal angle white. Body smoky brown, head and collar darker; a central longitudinal

---

* The new species of Sphingidæ were handed over to Mr. Butler for description, he being then engaged in writing a revision of that family. They therefore appear here as described by that author.—F. M.

streak, the borders of the thorax, and a series of lateral abdominal transverse bands black-brown; lateral margins of head, fringe of tegulæ, back of thorax, and front margins of the abdominal segments white; antennæ whitish, with testaceous serrations. Wings below smoky brown: fore wing with the internal area whitish; base of discoidal cell testaceous; a white apical streak, less distinct than above: hind wing with the base and abdominal area whitish; two diffused ill-defined transverse bars; margin as above. Body below smoky brown; palpi, sides and hinder part of pectus, and centre of venter white.

    Expanse $6\frac{2}{1}$ inches.

    Hab. Darjiling. In coll. Dr. Staudinger.

Genus AMBULYX, *Walker.*

Ambulyx, Walker, Catal. Lep. Het. B. M. viii. p. 120.

AMBULYX FLORALIS.

Ambulyx floralis, Butler, Trans. Zool. Soc. ix. pt. x. p. 639 (1876).

    Male. Shining bronzy clay-colour: fore wing with the apical half of costal area, the central and interno-discal areas washed with green; subbasal area dusky olivaceous, limited externally by an oblique olive line, a second parallel line crossing the wing over the base of the first median branch; three ill-defined oblique waved lines, the outermost undulated, crossing the disk; between the second and third a very indistinct diffused sinuous line; inner margin and the lines as they approach it blackish; a white-pupilled round black spot on the lower discocellular; a tuft of rose-red hairs at the base of inner margin: hind wing with the basal two thirds rose-red; costal area whitish; external third washed with green, especially towards apex, brownish towards the anal border; fringe for the most part white. Head and collar brown; tegulæ and abdomen washed with green; antennæ testaceous, pectinations brown; anterior tibiæ and tarsi above brown. Wings below much paler, testaceous, washed with pale green: fore wing with the interno-discoidal area rose-red; costal area greenish; a transverse brown litura beyond the cell; a transverse oblique nearly straight white-bordered olive discal line; a zigzag line nearer to the outer margin, becoming black towards inner margin; a submarginal series of spots, only distinct and blackish at the external angle: hind wings crossed by three parallel white-bordered olive lines; a squamose brown submarginal spot near anal angle. Body below whity brown; palpi, pectus, and legs slightly dusky.

    Female altogether less lively in colour, the green colouring less perceptible.

    Expanse, $\male$ $4\frac{8}{10}$ inches, $\female$ $4\frac{1}{2}$ inches.

    Hab. Darjiling. In coll. Dr. Staudinger.

    Allied to *A. superba*, but smaller, and clouded with bronze and green

Genus DILUDIA, *Grote.*

Diludia, Grote, Proc. Ent. Soc. Phil. v. p. 188.

DILUDIA TRANQUILLARIS.

Diludia tranquillaris, Butler, Trans. Zool. Soc. ix. pt. x. p. 641 (1876).

Female. Nearly allied to *D. grandis*, slightly smaller, the markings much less strongly defined, the central irregular transverse congregation of parallel bands broader; a black band only visible on costal area; the apical patch more uniformly dark grey, much narrower and longer, oblique behind, more narrowly black-bordered; hind wing with only one abbreviated black zigzag band across the grey anal patch. Body slightly browner in tint : head not varied with white; abdomen with lateral diffused brown longitudinal bands instead of the black spots. Wings below more uniform in colour, the transverse bands less strongly marked, narrower, and nearer to the outer border; the central blackish band of fore wing obsolete.

Expanse $5\frac{3}{12}$ inches.

Hab. Darjiling. In coll. Dr. Staudinger.

Genus DAPHNUSA, *Walker.*

Daphnusa, Walker, Catal. Lep. Het. B. M. viii. p. 237.

DAPHNUSA PORPHYRIA.

Daphnusa porphyria, Butler, Trans. Zool. Soc. ix. pt. x. p. 640 (1876).

Fore wing reddish brown, the basal area transversely marked with an irregularly arched whitish line; external area rather darker, marked from near external angle to a little below apical angle by a diffused whitish curved streak; a subapical sepia-brown excavated quadrangular patch; apex grey, with a large semicircular sepia-brown spot bordered externally by a white lunule on the outer margin; a broad central red-brown band, bordered on each side by a whitish streak, its outer line angular, much broader in front than behind, transversely clouded with grey, its outer third beyond the discoidal cell darker; the base of second median interspace and the discocellulars blackish piceous; two dissimilar whitish-edged black spots on the veins near external angle : hind wings pale brown, with two very indistinct discal streaks, clearly discernible upon the abdominal area; outer border rather broadly smoky brown; costa whitish; anal angle marked with a greyish and ferruginous dash, upon which is a black spot; a nearly marginal grey line. Body pale brown, varied with dark brown; a black spot on the crest of the head. Fore wing below greyish brown : apical half of costa and internal area pale greyish; apical markings as above, but redder : hind wing pale rosy greyish, paler on the abdominal area; three angulated ferruginous diffused discal lines; outer border rather broadly pale ferruginous, fringe dark piceous. Body below pale reddish brown; palpi chocolate-brown.

Expanse $2\frac{3}{12}$ inches.

Hab. Darjiling. In coll. Dr. Staudinger.

Allied to *D. colligata.*

Genus MACROGLOSSA, *Ochs.*

Macroglossa, Ochs. Eur. Schmett. iv. p. 41.

MACROGLOSSA LEPCHA.

Macroglossa lepcha, Butler, Trans. Zool. Soc. ix. pt. x. p. 635 (1876).

Allied to *M. obscuriceps*, from which it differs in having the head and thorax olive-green, the black band across the abdomen feebly developed, the fore wings narrower, the central greyish band, which is scarcely distinguishable in *M. obscuriceps*, quite obsolete, the subbasal lines bounding it internally converted into a black band, which is broad on the inner margin and tapers towards the costa. Wings below with the transverse lines less distinctly marked; the internal orange area brighter.

Expanse 2 inches.

Hab. Calcutta. In coll. Dr. Staudinger.

Allied to *M. avicula* and *M. bombylans*, Boisd. Spec. Gén. Lép. Hét. Sphinges, p. 334.

Family ÆGERIIDÆ.

Genus ÆGERIA, *Fabr.*

Ægeria, Fabr., Illiger's Mag. vi. p. 287 (1807).

ÆGERIA FLAVA, n. sp.

Wings transparent: fore wing with narrow borders and discocellular band and veins at apex cupreous-brown, base and apex yellowish; cilia of hind wing brown. Antennæ yellowish, tip brown; thorax and abdomen above brown, each segment with a yellow band; anal tuft yellow; palpi, collar, streaks on vertex, and sides of thorax yellow; legs yellow, with brown bands.

Expanse $\frac{12}{17}$ inch.

Hab. Darjiling. In coll. Dr. Staudinger.

ÆGERIA TRICINCTA, n sp.

Male. Wings transparent: fore wings with purplish-brown borders and a band at end of cell; transparent parts with steel-blue reflections: hind wing with narrow purple-brown borders and discocellular streak and cilia. Antennæ greyish at base; body and antennæ purple-brown; palpi beneath, collar, streak on sides of thorax, a basal, medial, and anal band on abdomen, and streak on side chrome-yellow; legs purple-brown, fore tarsi white, middle and hind tarsi and hind tibiæ fringed with white.

Expanse $\frac{7}{8}$ inch.

Hab. Darjiling. In coll. Dr. Staudinger.

PRAMILA, nov. gen.

Allied to *Sphecia*; wings shorter: fore wing narrow at base, broad at apex; no recurrent vein: hind wing broad; no subcostal branch; discocellulars slightly oblique, lower the shortest; one radial; upper median veinlets straight. Thorax broad; abdomen tapering to tip; head small; palpi slender, ascending, hairy; antennæ slender at base, biserrate; legs long, femora and tarsi squamous; tibiæ slightly hairy, spurred.

PRAMILA ATKINSONI, n. sp. (Plate II. fig. 1.)

Female. Wings transparent; costa, cilia, and a narrow discocellular streak on hind wing blackish-brown: fore wing with a broad orange-red black-bordered discocellular spot. Body and antennæ bluish-black; legs brown; palpi beneath and tarsal joints fringed with white.

Expanse $1\frac{3}{8}$ inch.

Hab. Darjiling. In coll. Dr. Staudinger.

Genus SCIAPTERON, *Staudinger.*

SCIAPTERON SIKKIMA, n. sp.

Male. Fore wing dark purple-black, with steel-blue reflections: hind wing transparent; a narrow costal and outer border and discocellular streak blackish-purple; cilia cupreous brown. Thorax blackish-purple, with a bright chrome-yellow streak on each side; head, antennæ, abdomen, and legs black above and slightly streaked with yellow beneath; fore tarsi white; tarsi of middle and hind legs spotted with white.

Expanse $1\frac{2}{10}$ inch.

Hab. Darjiling. In coll. Dr. Staudinger.

TRILOCHANA, nov. gen.

Fore wing long, narrow; costa slightly convex at base and apex, exterior margin short, oblique; subcostal vein four-branched, the fourth also forked at half its length; three radials; median vein two-branched, branches starting from angle of lower discocellular: hind wing broad, exterior margin very convex below apex; subcostal vein extending direct to apex; discocellulars very oblique; one radial; median vein three-branched, two arising together at lower end of cell, the third at a distance from its end; one submedian and two interior veins. Body long; head broad; eyes projecting laterally; thorax long, projecting much beyond base of wings; abdomen attenuated towards base, tuft fan-like; fore legs hairy; mid legs squamous, slightly hairy at sides; hind femora squamous, tibiæ clothed with dense and long hairs; antennæ long, minutely serrate.

TRILOCHANA SCOLIOIDES, n. sp. (Plate II. fig. 2.)

Upperside shining olive-green ; a short narrow transparent longitudinal basal streak on fore wing, and two triangular streaks on hind wing. Body dark purple-brown ; abdominal segments edged with grey ; last two segments on sides and beneath and large anal tuft red ; tuft fringed dorsally with black hairs ; antennæ dark brown ; legs black, hind tarsi yellowish ; palpi purple-brown. Underside as above ; costa of fore wing slightly steel-blue.

Expanse 1⅖ inch.

Hab. Darjiling. In coll. Dr. Staudinger.

Genus MELITTIA, *Hübn.*

Melittia, Hübner, Verz. bek. Schmett. p. 128.

MELITTIA NEPCHA, n. sp.

Margins and veins purple-black ; apical border narrow, speckled with a few purple-grey scales ; apical area large, traversed by four veins and partly by the subcostal upper branch ; basal area traversed by median vein only ; abdominal area fringed with golden hairs ; thorax greenish in front, with a collar composed of a few steel-blue scales ; hind part of thorax golden yellow ; abdomen above bluish-black, with narrow golden-yellow segmental bands, yellow beneath ; palpi, pectus, fore legs, mid femora and tibia pure white fringed with black ; hind legs densely clothed with long black hairs, femora above fringed with long red and white hairs ; antennæ black above, pectinated surface red.

Exp. 1¾ inch.

Hab. Darjiling. In coll. Dr. Staudinger.

MELITTIA NEWARA, n. sp.

Male. Costal margins and veins purplish-black ; apical band of fore wing narrow, speckled with pale purple scales ; cilia brown ; apical area traversed by four veins, basal area by median vein and a short recurrent vein from discocellulars ; abdominal area and base of fore wings fringed with yellow hairs. Thorax bright golden-yellow, top slightly purple-black ; abdomen bluish-black ; palpi, pectus, fore legs, and femur and tibia of mid legs yellow fringed with black ; hind legs densely clothed with long purplish-black hairs, femora fringed above with yellow hairs, and a few red hairs radiating at lower joint ; abdomen beneath yellow ; antennæ purple-black above, pectinated portion reddish.

Expanse 1⁴⁄₁₀ inch.

Hab. Darjiling. In coll. Dr. Staudinger.

Family AGARISTIDÆ.

Genus EUSEMIA, *Dalman.*

Eusemia, Dalman, Vet. Akad. Handl. (1824).

EUSEMIA AUSTENI, n. sp.

Male. Upperside black: fore wing with yellowish cream-coloured spots, two being disposed within the cell, a round one below it and two beyond, also a submarginal curved series of small spots; some bluish-grey scales at end of cell: hind wing orange-yellow, a black basal patch, a broad irregular marginal band, a large geminate spot above anal angle, and a small spot at end of cell; some white apical spots within marginal band. Thorax black, with a pale lemon-yellow narrow stripe on each side; abdomen orange-yellow, banded with black.

Underside as above; middle of thorax orange-yellow; legs and antennæ black.

Expanse 2⅛ inches.

Hab. Khasia Hills. In coll. Dr. Staudinger.

The nearest ally to this species is *E. villicoides,* Butler, from Japan.

NIKÆA, nov. gen.

Fore wing long, narrow; costa and hind margin nearly straight; apex and posterior angle slightly pointed; outer margin convex, very oblique; discoidal cell dilated its entire length: discocellulars bent inwards; subcostal vein five-branched, second to fifth on a foot-stalk arising from end of cell; one radial from upper angle of first discocellular; median vein four-branched, upper three arising from lower angles of cell and contiguous at their base, fourth from half the length of cell. Hind wing trigonal, apex and anal angle rounded, abdominal margin short; cell broad; discocellulars slightly bent inwards; subcostal vein two-branched at end of cell; median four-branched, disposed as in fore wing; one submedian and an internal vein. Body short, extending to anal angle; head small; palpi projecting beyond head, ascending, slightly pilose, broad and flat, third joint rather long; legs long, squamose; middle and hind tibiæ with short spurs; antennæ setaceous, thickest at base.

Type *N. longipennis* (*Hypercompa longipennis,* Walk. Cat. Lep. Het. B. M. iii. p. 655). *Hab.* Upper Kumaon.

Family ZYGENIDÆ.

Genus NORTHIA, *Walker.*

Northia, Walker, Cat. Lep. Het. B. M. i. p. 141.

(Syn. ZAMA, *H.-Sch.*)

NORTHIA KHASIANA, n. sp.

Black: wings subhyaline, blackish ; veins black ; margins and cilia black; hind margin of fore wing and anterior margin of hind wing fuliginous: antennæ greenish black ; legs black.

Expanse 1 inch.

Hab. Khasia Hills. In coll. Dr. Staudinger.

Genus SYNTOMIS, *Ochs.*

Syntomis, Ochs. Schmett. Eur. ii. p. 103.

SYNTOMIS NEWARA, n. sp.

Near *S. multiguttata.* Yellowish hyaline ; veins and cilia blackish : fore wing with a brownish-black narrow streak at end of the cell, short curved band on apex, and small square spot on outer margin between lower median veinlets : antennæ black ; body brownish-black ; front of head, collar, streaks on thorax, and rather broad bands on abdomen yellow, tip blackish ; legs brown above, yellowish beneath.

Expanse $1\frac{3}{16}$ inch.

Hab. Darjiling. In coll. Dr. Staudinger.

SYNTOMIS LEPCHA, n. sp.

Near *S. multiguttata.* Yellowish hyaline ; veins and cilia black: fore wings with a broad, black, elongated hexagonal streak at end of cell, and a short apical marginal band of the same width, which stops at and is slightly dentate on the upper median veinlet: hind wing with decreasing apical marginal black band. Body black ; front of head brown ; collar and four narrow stripes down thorax, narrow bands and two lateral stripes on abdomen, yellow, tip of abdomen reddish ; legs brown, yellowish beneath ; coxæ and femora of fore legs red.

Expanse 1 inch.

Hab. Darjiling. In coll. Dr. Staudinger.

SYNTOMIS CHERRA, n. sp.

Nearest allied to *S. mandarina,* Butler. Wings dark cupreous-brown : fore wing with whitish hyaline space in cell, an elongated space below it, and two subapical vein-crossed spots: hind wing with a large hyaline basal spot. Body blackish, spot on middle and band at base of thorax, and broad abdominal bands orange-yellow, paler beneath; front of head white; antennæ tipped with grey; legs brown, tarsi greyish.

Expanse 1⅜ inch.

Hab. Cherra Punji, Assam. In coll. Dr. Staudinger.

SYNTOMIS HYALINA, n. sp.

Wings yellowish hyaline: fore wing with subcostal vein and narrow band on outer margin black; the hind margin with a short dentate point on its middle: hind wing with narrow black anterior and exterior marginal band, dentate between the median and submedian veinlets; abdominal margin orange-yellow. Antennæ black, tip whitish; body black; bands on middle of abdomen orange-yellow; front of head white; legs brown, first joint of tarsi white.

Expanse 1¾ inch.

Hab. Darjiling. In coll. Dr. Staudinger.

SYNTOMIS DISCINOTA, n. sp.

Wings yellowish hyaline; subcostal vein, narrow band on outer margins, and cilia black; a short black streak from middle of discocellular of fore wing, and slight dentation at end of lower median veinlet in both wings. Antennæ black, tip whitish; body greenish black; collar, spot on sides and base of thorax, and narrow lateral bands on abdomen orange-yellow, tip black; front of head and pectus light yellow; legs black, femora and tibiæ streaked with grey, and first joint of tarsi white.

Expanse 1⅜ inch.

Hab. Khasia Hills. In colls. Dr. Staudinger and F. Moore.

Genus ARTONA, Walker.

Artona, Walker, Cat. Lep. Het. B. M. ii. p. 439.

ARTONA POSTVITTA, n. sp.

Male and Female. Upperside brown; hind wing with a narrow medial longitudinal white streak and white cilia. Antennæ and body above blackish; legs blackish, streaks beneath and tarsal joints white; abdomen and palpi whitish beneath.

Underside paler on fore wing at base, and white space on hind wing curved.

Expanse ₁₃⁄₁₀ inch.

Hab. Darjiling. In colls. Dr. Staudinger and F. Moore.

ARTONA FULIGINOSA, n. sp.

Male and female. Fuliginous-brown; fore wing blackish along the costa and middle of wing. Antennæ and palpi black; legs brown.

Expanse, ♂ $\frac{6}{10}$, ♀ $\frac{7}{10}$ inch.

Hab. Darjiling. In colls. Dr. Staudinger and F. Moore.

Family CHALCOSIIDÆ.

ARACHOTIA, nov. gen.

Wings long, narrow: costa and hind margin of fore wing nearly straight; apex slightly pointed; outer margin short, oblique; cell dilated only at half its length; subcostal vein three-branched; discocellulars slightly sinuous, very short; median vein five-branched, each branch starting at end of cell from discocellulars, and thus forming radials; submedian vein contiguous and running parallel to median to end of cell, two-branched, the upper starting at angle of cell, the lower at a short distance before the cell; two lower submedian or internal veins extending to exterior margin, the upper terminating above and the lower at the posterior angle: hind wing produced and acute at the apex, anal angle lobular; cell divided by a recurrent veinlet its entire length, emitted from middle of lower discocellular; lower discocellular slightly angled, twice the length of upper; subcostal vein two-branched; median vein three-branched, first or radial continuous with recurrent veinlet of the cell, second and third at equal distances at their base; submedian and three lower internal veins extending to anal lobe. Body long; antennæ very long, extending to three fourths of the costa, bipectinate in male to near tip, which is slightly thickened; filiform and slightly thickened near tip in female; palpi short, thin; second and third joints naked; legs long, squamous.

ARACHOTIA FLAVIPLAGA, n. sp.

Male. Upperside: fore wing metallic olive-green at base, bluish green at apex; a narrow streak below base of cell, space within the cell, a contiguous triangular space below it, and a rounded subapical space transparent; a short transverse yellowish streak near base of hind margin: hind wing transparent, with black anterior margin and an irregular-shaped streak from anal angle on hind margin. Antennæ dark brown; vertex, thorax, and tip of abdomen bluish green; base of abdomen blackish, two lateral yellow bands on middle of abdomen; front of head and palpi greyish brown.

Underside of wings as above, but anterior margin of hind wing broadly metallic bluish green. Abdomen beneath greyish, banded with blue-green; legs brown.

Female. Differs only in the absence of the basal transparent longitudinal streak, and in having a broader subbasal short transverse yellowish streak on fore wing, and a broad black abdominal margin. Antennæ whitish near apex.

Expanse, ♂ $1\frac{4}{8}$, ♀ $1\frac{5}{8}$ inch.

Hab. Darjiling. In colls. Dr. Staudinger and F. Moore.

Genus TRYPANOPHORA, *Kollar*.

Trypanophora, Kollar, Hügel's Kaschm. iv. p. 457.

TRYPANOPHORA ATKINSONI, n. sp.

Male. Wings black: fore wing with a hyaline rounded upper basal divided spot and a large medial space traversed by the black veins, the black extending as a streak along the lower median veinlet, and also forming a spot at end of the cell: hind wing with a lower basal and subapical hyaline space traversed by the black veins; anterior margin and cell yellow; apex, outer margin, anal angle, and a spot on cell black. Antennæ brown, shaft greenish-black; body black; front of head, collar, spot on side and middle of thorax, and band on five segments of the abdomen orange-yellow; tip of abdomen bluish-black; legs blackish-brown, greyish at sides.

Expanse 1⅛ inch.

Hab. Darjiling. In coll. Dr. Staudinger.

Distinguished from *T. humeralis* by the much larger hyaline area of the fore wing.

Genus CHELURA, *Hope*.

Chelura, Hope, Trans. Linn. Soc. xviii. p. 444.

CHELURA EROXIOIDES, n. sp.

Female. Upperside subhyaline, fuliginous, veins brown, interspaces of base of both wings broadly streaked and apex of fore wing spotted with dull white; a yellow stripe across immediate base of fore wing. Underside as above. Head, thorax, and antennæ fuliginous-black; abdomen and legs yellowish.

Expanse 2⅔ inches.

Hab. Darjiling. In coll. Dr. Staudinger.

Genus ETERUSIA, *Hope*.

Eterusia, Hope, Trans. Linn. Soc. xviii. p. 445.

ETERUSIA LATIVITTA, n. sp.

Female. Upperside—fore wing dark brownish olive-green, with a broad yellow transverse medial band, a small yellow spot at base of costa, a single spot at end of cell, and a subapical series of small spots: hind wing black; a yellowish patch on middle of anterior margin, and a small spot before the apex, the veins at the apex, and streaks between them steel-blue. Lower part of abdomen golden-yellow. Underside black: fore wing marked as above, the veins beyond the band blue-streaked: hind wing with the costal patch and a large yellowish-white spot before and two below the apex, the veins being blue-streaked.

Expanse 3 inches.

Hab. Darjiling. In coll. Dr. Staudinger and F. Moore.

Nearest allied to *E. edocla*, Doubleday, from Silhet, but may be easily distinguished by the different-shaped and much broader medial transverse band on the fore wing.

ETERUSIA ALOMPRA, n. sp.

Male. Upperside—fore wing dark olive-green, crossed by a rather broad oblique medial yellow band, which is bordered without on both sides by black streaks, one between each vein : hind wing black ; streaks at base and a short maculated apical marginal band cobalt-blue ; a transverse discal curved series of bright yellow ill-defined spots. Front of head, thorax, and abdomen dark metallic olive-green ; antennæ, vertex, and base of thorax steel-blue. Underside bluish-black : both wings crossed by a yellow band and a submarginal row of bluish-white spots : base of wings longitudinally streaked with metallic emerald-green and blue. Legs metallic olive-brown ; abdomen beneath brown.

Expanse 2¾ inches.

Hab. Sibsagar, Assam. In coll. Dr. Staudinger.

Genus SORITIA, *Walker.*

Soritia, Walker, Catal. Lep. Het. B. M. ii. p. 435.

SORITIA FUSCESCENS, n. sp.

Male. Upperside dark umber-brown : hind wing pale brown from below the cell to abdominal margin. Antennæ black ; collar light red ; head brown ; thorax and abdomen fuliginous black ; legs brown. Underside as above ; apex of both wings and costal margin of hind wing pale brown.

Female. Same as male, excepting that the hind wing is broadly whitish brown on abdominal margin.

Expanse, ♂ 1⅛ inch, ♀ 1⅘ inch.

Hab. Darjiling. In coll. Dr. Staudinger.

SORITIA OLIVASCENS, n. sp.

Male. Upperside olive-brown : fore wing with a yellow basal streak, oblique discal patch, and a small subapical spot : hind wing with anterior margin and short transverse subapical streak yellow. Body and antennæ brown ; collar crimson ; tegulæ golden bronze. Underside brown : fore wing with a narrow basal streak, broad discal patch, small subapical spot, and apical lunule yellow : hind wing with costal margin and apex yellow ; a row of black subapical spots. Front of head, palpi, legs, and abdomen beneath yellow.

Female. Upperside—fore wing olive-brown, veins yellow : hind wing pale yellow at base, brown at apex, divided by curved series of subapical black spots : head and collar crimson ; tegulæ yellow ; abdomen pale steel-blue with greyish bands. Underside pale buff-yellow ; apex of fore wing olive-brown as above ; apex of hind wing grey, with subapical spots as above. Legs and abdomen beneath yellowish.

Expanse, ♂ 1⅘ inch, ♀ 1⅗ inch.

Hab. Khasia Hills. In colls. Dr. Staudinger and F. Moore.

CODANE, nov. gen.

Male. Fore wing elongate-ovate ; costa convex ; apex rounded, exterior margin oblique, hind margin nearly straight ; cell broad, divided by a recurrent veinlet its entire length ; subcostal vein five-branched, the third, fourth, and fifth on a foot-stalk arising at end of the cell ; upper discocellular angled near subcostal, bent obliquely inward and descending into the cell, second short and bent obliquely outward ; one radial from upper angle of disco-cellular ; median vein four-branched, two lower at equal distances before end of cell, two upper on a short foot-stalk from lower end of second discocellular ; a submedian and an internal vein terminating above posterior angle. Hind wing ovate, apex and outer margin rounded ; cell fusiform, divided by a recurrent veinlet its entire length ; discocellulars sharply angled ; subcostal vein two-branched, starting from upper angle of first discocellular ; median vein four-branched, two upper on short foot-stalk from end of second discocellular, two lower at equal distances ; two submedian veins contiguous at the base, and two internal veins. Antennæ long, extending beyond half length of costa, broadly bipectinate to end ; body short ; head broad ; palpi small, third joint pointed, minute ; legs naked.

Type *C. zenotea* (*Pidorus zenotea*, Walk. Catal. Lep. Het. B. M. p. 425).

Hab. Cherra Punji, Assam.

Genus CAMPYLOTES, *Westwood.*

Campylotes, Westwood, Royle's Illm. Bot. p. 53.

CAMPYLOTES ATKINSONI, n. sp.

Female. Upperside very dark bright olive-green : fore wing with a deep carmine narrow longitudinal streak at base of costa within the cell, and still narrower streaks below ; two subapical oblique series of pure white spots : hind wing with an indistinct blackish carmine streak between the veins, bright crimson only towards the apex. Underside dull indigo-bluish-green ; base and upper discoidal cell of fore wing, basal streaks and anterior spots between veins of hind wing dull deep crimson. Antennæ black : body dark olive-green ; abdomen beneath and legs black.

Expanse 3¼ inches.

Hab. Darjiling. In coll. Dr. Staudinger.

Distinguished from *C. histrionicus* by its much darker colour, deeper crimson, and nar-rower streaks between the veins.

Genus CHALCOSIA, *Hübner.*

Chalcosia, Hübner, Verz. bek. Schmett. p. 173.

CHALCOSIA ARGENTATA, n. sp.

Upperside—fore wing with the veins metallic emerald-green ; the base broadly silvery-white ; a basal costal streak and an oblique partly transverse medial maculate band black,

D

washed with emerald-green; apex broadly black enclosing a series of small white spots; oblique discal intermediate space cream-white: hind wing pale yellow; apical curved black band crossed by green veins. Underside as above, except that in the fore wing the base is black to below the cell. Antennæ steel-blue; front of head metallic brown and white; vertex and collar carmine red; thorax metallic blue; abdomen and legs metallic greyish-white.

Expanse 2 inches.

Hab. Hazar D. [? Hazara District]. In coll. Dr. Staudinger.

Nearest allied to *C. idæoides*, H.-Sch. Lep. Spec. Nov. pl. 1. f. 6.

Genus CYCLOSIA, *Hübner.*
Cyclosia, Hübner, Verz. bek. Schmett. p. 177.

CYCLOSIA CARDINALIS, n. sp.
Allied to *C. fuliginosa,* Walk. Catal. Lep. Het. B. M. ii. p. 418.

Male. Upperside—fore wing with the base and a partly contiguous wavy transverse discal band dark fuliginous; apical half of wing white; veins and outer margin darker fuliginous: hind wing metallic yellowish-fuliginous, yellowest on abdominal margin; discoidal vein streaked with blue: outer margin broadly white bordered and crossed with black. Underside as above, but paler, and veins at the base streaked with blue. Front of head metallic-blue, hind part of head and collar crimson, thorax and abdomen above purplish black; antennæ steel-blue: legs and abdomen beneath greyish-fuliginous.

Expanse 2¼ inch.

Hab. Darjiling. In coll. Dr. Staudinger.

Genus HERPA, *Walker.*
Herpa, Walker, Catal. Lep. Het. B. M. ii. p. 441.

HERPA SUBHYALINA, n. sp.
Male. Wings subhyaline, pale yellowish: fore wing with fuliginous veins. Underside as above, costal border of both wings fuliginous. Antennæ long, jet-black, pectinations very broad; head fuliginous-black; thorax yellowish; abdomen greyish-yellow; legs and abdomen beneath blackish-fuliginous.

Expanse 2¾ inches.

Hab. Lachung Valley, Sikhim. In coll. Dr. Staudinger.

Genus HISTIA, *Hübner.*
Histia, Hübner, Verz. bek. Schmett. p. 198.

HISTIA NILGIRA, n. sp.
Female. Upperside—fore wing fuliginous-black, veins and parallel intervening streaks black: hind wing at base pale metallic emerald-green streaked with fuliginous between the

veins; apex black: intervening discal space white, forming a large transverse oval patch. Underside—fore wing pale fuliginous, streaks darker, washed with indigo-blue : hind wing with base and apex fuliginous-black, washed with indigo-blue ; central patch white as above. Front of head, vertex, and collar crimson ; thorax black ; abdomen black above, crimson at tip and beneath, with a lateral and lower row of black spots ; palpi and legs black ; antennæ steel-blue.

Expanse 3 inches.

Hab. Nilgiri Hills. In coll. Dr. Staudinger.

Allied to *H. vacillans,* Walk. (*libelluloides,* II. Sch. Pl. 2. f. 13 ♀), but is a much larger insect, having longer wings, these being elongated as in *H. papilionaris.*

Genus MILLERIA, *H.-Sch.*

Milleria, H.-Sch. (Syn. *Laurion,* Walker, Catal. Lep. Het. B. M. ii. p. 426, 1854.)

MILLERIA ALBIFASCIA, n. sp.

Male and female. Upperside dark olive-brown : fore wing crossed by a medial curved oblique white band terminated on the costa on the inner side by a yellow spot: hind wing with a rather large apical yellow spot. Underside as above, apex of both wings washed with greyish-blue. Body, antennæ, and legs pale metallic bluish-black.

Expanse, ♂ 1⅝ inch, ♀ 1⅞ inch.

Hab. Cherra Punji, Assam. In coll. Dr. Staudinger.

Differs from *M. circe,* H.-Sch. Lep. Spec. Nov. pl. 1. f. 2, 1853, in being duller-coloured, the oblique band on fore wing being white, and in the larger size of spot on costa of hind wing.

Family NYCTEMERIDÆ.

ARBUDAS, nov. gen.

Fore wing long, rather narrow, costa very slightly arched, apex acute, exterior margin short, posterior margin nearly of the same length as the costa ; costal vein arched at the base ; cell broad, long ; subcostal vein recurved, five-branched, first branch arising from near end of cell, second trifurcate and arising beyond end of the cell, the outer fork starting from below half the length of the middle fork ; discocellulars extending some distance within the cell, angled at the upper and lower end, from the upper angle starts the radial ; median vein recurved, four-branched, the upper branch starting from angle of the lower discocellular, the next from lower end of the cell, the two lower branches at equal distance from near end of the cell ; submedian vein curved upward and running near the median ; internal vein at some distance from posterior margin. Hind wing long, anterior margin of the same length as posterior margin of fore wing, apex acute, exterior margin and anal angle very convex ; costal vein running close to the margin, first branch of the subcostal vein starting before end of

D 2

the cell; upper discocellular short, both bent inward; four branches to median as above; a submedian and two internal veins. Body small, slender; antennæ bipectinated; palpi very small, slender, naked; legs slender, naked, middle and hind tibiæ with very short spurs.

ARBUDAS BICOLOR, n. sp. (Plate II. fig. 19.)

Male. Fore wing greyish fuliginous-brown; hind wing white, with a fuliginous-brown marginal band and white cilia. Underside as above, the fore wing having a white subapical patch, which is slightly distinguishable from the upperside; marginal band on hind wing not extending to anal angle. Body dusky-fuliginous; antennæ black; palpi and legs greyish fuliginous.

Expanse 1 inch.

Hab. Darjiling. In coll. Dr. Staudinger and British Museum.

Family EUSCHEMIDÆ.

Genus EUSCHEMA, *Hübner.*

Euschema, Hübner, Verz. bek. Schmett. p. 175.

EUSCHEMA NIGRESCENS, n. sp.

Male. Upperside black: fore wing with a white streak at base of costa, crossed by three rows of spots, the first row subbasal and creamy white, formed of three spots commencing from within base of cell, the next medial and of six spots terminating above second spot of first row, the third row smallest, submarginal, both bluish-white: hind wing with the base and a medial narrow maculated band cream-white or yellowish; a submarginal row of indistinct small yellowish spots. Underside as above, with all the markings more prominent. Thorax white-streaked; front of head partly black; abdomen yellow, banded with black; legs fuliginous, yellowish beneath; antennæ black.

Expanse 2¼ inches.

Hab. Darjiling. In coll. Dr. Staudinger.

Allied to *E. flavescens,* Walk. Cat. Lep. Het. B.M. ii. p. 406.

Family CALLIDULIDÆ.

HERIMBA, nov. gen.

Fore wing elongated, costa produced; apex acute; exterior margin very oblique, posterior margin short; costal vein short, at some distance from the margin; cell short, broad; subcostal vein five-branched, first branch arising very near the base, second and third contiguous at their base and arising before end of the cell and extending to apex of costa, fourth and fifth also contiguous at their base and arising from end of the cell; discocellulars of equal length, slightly angled in the middle ; one radial starting from middle of the cell ; median vein four-branched, the two upper branches from end of the cell, two lower at equal distances before its end ; submedian vein contiguous to posterior margin. Hind wing broad ; anterior margin short, exterior margin convex, abdominal margin long ; cell short ; subcostal vein two-branched ; median vein four-branched, three upper branches contiguous at their base. Body stout, short ; antennæ slender, filiform ; palpi robust, ascending, projected slightly beyond the head, apical joint conical ; legs long, squamous, femora thickened, clothed with short adpressed hair-like scales.

HERIMBA ATKINSONI, n. sp. (Plate 11. fig. 3.)

Upperside very dark glossy cupreous-brown : fore wing with a pearly-white oblique transverse medial narrow band with waved borders extending from near the costa to near posterior angle ; two minute pearly-white spots within the cell, another below it, and a slightly larger spot near the apex : hind wing with a minute pearly-white spot within the cell, one beyond it, and another before the apex. Underside with markings as above : both wings also with short transverse pale cupreous streaks between the veins. Body beneath and legs brown ; tarsi with white bands ; antennæ blackish.

Expanse $1\frac{1}{8}$ inch.

Hab. Darjiling. In coll. Dr. Staudinger.

DATANGA, nov. gen.

Allied to *Callidula*, Hubn. (*C. evander*, Cram.). Wings shorter, broader, the fore wing having the second, third, and fourth branches of the subcostal vein starting all together at a short distance beyond the cell.

DATANGA MINOR, n. sp.

Male and female Upperside rufous-brown : fore wing with a red oblique narrow straight subapical band, which is of equal width throughout. Underside saffron-yellow ; both wings with numerous dark short strigæ, which are partially red-bordered on the disk ; two small black-bordered white spots within the cell and an elongated spot at its end.

Expanse, ♂ 1, ♀ $1\frac{1}{8}$ inch.

Hab. Moulmein, Burmah. In colls. Dr. Staudinger and F. Moore.

Allied to *D. sakuni* (*Petavia sakuni*, Horsf.), from Java.

DATANGA ATTENUATA, n. sp.

Upperside rufous-brown : fore wing with a pale red oblique subapical band, broadest in female, and attenuated at its lower end in both sexes. Underside dark saffron-yellow ; fore wing with a black-bordered subapical yellow band, as above ; both wings with numerous confluent dark strigæ, which are bordered and much suffused with deep red ; two small black-bordered white spots within the cell, and an elongated spot at its end.

Expanse, ♂ 1⅔, ♀ 1⅘ inch.

Hab. Darjiling. In colls. Dr. Staudinger and F. Moore.

Family LITHOSIIDÆ.

Genus CALPENIA, *Moore.*

Calpenia, Moore, P. Z. S. 1872, p. 571.

CALPENIA KHASIANA.

Calpenia khasiana, Moore, P. Z. S. 1878, p. 5.

Female. Upperside—fore wing brownish fawn-colour, vinous-tinged, paler along the veins ; a broad pale yellow band extending longitudinally from base below the cell and upward to near the costa, and crossed by the veins ; some yellow spots on the costa ; a small spot and streak at base of the cell, and two outer submarginal rows of small dentate spots of chrome-yellow : hind wing chrome-yellow, with four transverse rows of vinous brownish-black spots and a rayed basal streak, the spots broadly lunate, irregular in size, the outer row being marginal and the smallest. Underside the same as above. Antennæ, front of head, and palpi black ; thorax dark greyish-brown, longitudinally streaked with chrome-yellow ; abdomen chrome-yellow, with dorsal and lateral row of spots ; legs dark grey.

Expanse 3¼ inches.

Hab. Khasia Hills. In coll. Dr. Staudinger.

Genus SIDYMA, *Walker.*

Sidyma, Walker, Cat. Lep. Het. B. M. vii. p. 1686.

SIDYMA APICALIS.

Sidyma apicalis, Moore, P. Z. S. 1878, p. 9, pl. i. f. 2.

Male. Upperside purplish-black : fore wing with a narrow triangular white apical patch. Underside as above. Collar, thorax beneath, and anal tuft orange-red.

Expanse 1¾ inch.

Hab. Darjiling. In coll. Dr. Staudinger.

Smaller than *S. albifinis,* Walker, from Masuri, and distinguished from it by having the apical white patch only on the fore wing.

Genus VAMUNA, *Moore.*

Vamuna, Moore, P. Z. S. 1878, p. 10.

Fore wing long, narrow; costa slightly arched beyond the middle; apex acute; exterior margin short and slightly truncate below the apex, oblique hindward, the angle slightly convex; posterior margin convex towards the base in male. Hind wing broad; apex slightly produced; exterior margin convex in the middle. Veins similar to those in *Churinga* (*C. rufifrons*), excepting that in the fore wing the lowest or fifth branch of the subcostal is emitted from upper angle of the cell, the median branches nearer together at their base, and the lowest nearer the end of the cell. Body moderately stout; antennæ in both sexes setose; palpi moderately long, ascending, projecting beyond the head, second joint long, third joint very short; legs long, naked, middle and hind spurred.

Type *V. remelama* (*Lithosia remelana*, Moore, P. Z. S. 1865, p. 798).

Hab. Darjiling.

VAMUNA MACULATA.

Vamuna maculata, Moore, P. Z. S. 1878, p. 10, pl. i. fig. 5.

Male. Upperside—fore wing greyish-ochreous: hind wing pale dull ochreous at base and whitish externally, with three marginal black spots, the two upper large, the lower small and at some distance from anal angle. Underside dull ochreous basally, whitish externally: fore wing with a brownish patch at the apex, a black longitudinal basal streak, and a broad large transverse subapical patch: hind wing with only the lower and middle marginal black spots. Body ochreous.

Female much paler; the apex of fore wing slightly brownish-ochreous: hind wing above and beneath with smaller medial and lower black spots; the fore wing with the basal black streak less defined, and the subapical spot small.

Expanse, ♂ 1⅜, ♀ 2 inches.

Hab. Darjiling. In colls. Dr. Staudinger and F. Moore.

VAMUNA BIPARS.

Vamuna bipars, Moore, P. Z. S. 1878, p. 10, pl. i. fig. 11.

Male. White; costal edge of the base and apical half of fore wing fuliginous black; hind wing with a fuliginous-black submarginal band, the middle portion of which is very broad and blackest. Underside duller white; the black band on fore wing confined to the disk, the apical border being pale brownish-ochreous; upper portion of band on hind wing obsolete. Antennæ, front of head, and tip of palpi black; fore and middle legs above, a terminal spot on hind tibiæ and its tarsi black; tarsal joints with a white band; base of palpi, thorax and abdomen, legs beneath, and anal tuft ochreous.

Expanse 2 inches.

Hab. Darjiling. In colls. Dr. Staudinger and F. Moore.

Genus MAHAVIRA, *Moore*.

Mahavira, Moore, P. Z. S. 1878, p. 11.

Male. Fore wing long, narrow; costa arched, apex acute; exterior margin oblique, posterior margin recurved ; subcostal vein five-branched, first and second branches arising before and near end of the cell, third trifid, the two lower branches at equal distances from end of the cell ; discocellulars long, straight, upper obliquely inward, lower obliquely outward ; radial from upper end of cell ; median vein four-branched, the two upper branches from lower end of cell, third at some distance before its end ; submedian vein curving towards hind margin. Hind wing longer than broad, exterior margin convex ; subcostal vein two-branched beyond end of cell ; discocellulars as in fore wing ; median vein four-branched, the two upper branches from end of cell and contiguous at their base, third immediately before end of the cell. Body slender ; antennæ serrate, with long and delicate pectinations ; legs slender ; femora slightly pilose beneath, middle and hind tibiæ spurred ; palpi slender, basal joint pilose, apex pointed.

MAHAVIRA FLAVICOLLIS.

Mahavira flavicollis, Moore, P. Z. S. 1878, p. 11, pl. i. fig. 3.

Male. White ; costa slightly edged with brown at the base ; collar yellow ; antennæ yellow ; fore femora with a blackish streak on the inner side.

Expanse 1¾ inch.

Hab. Darjiling. In colls. Dr. Staudinger and F. Moore.

Genus KORAWA, *Moore*.

Korawa, Moore, P. Z. S. 1878, p. 11.

Male and female. Fore wing long, narrow ; costa slightly arched, apex somewhat acute, exterior margin very oblique ; subcostal vein five-branched, first branch ascending and touching the costal, but free at its end, second starting before end of the cell, third at end of the cell, trifurcate, lowest branch at one third beyond end of the cell; upper discocellular angled outward at its middle, lower oblique ; radial from angle of upper discocellular ; median vein four-branched, the two upper from lower end of the cell ; submedian slightly recurved. Hind wing moderately short, apex slightly produced and convex, exterior margin rounded ; subcostal two-branched beyond end of the cell ; discocellulars of equal length, bent inward ; median vein four-branched, two upper beyond end of the cell, middle branch from end of the cell ; submedian nearly straight. Body moderate ; palpi small, pilose, porrect ; antennæ minutely and finely bipectinate ; legs pilose, sparsely in male.

Korawa pallida, Moore, P. Z. S. 1878, p. 12.

Male and female. Semidiaphanous: fore wing pale fleshy-yellow ; hind wing white. Thorax slightly ochreous-yellow ; abdomen white ; palpi pale yellow, black-tipped ; legs white ; fore tibiæ with a black longitudinal line in front.

Expanse, ♂ 1½, ♀ 2 inches.

Hab. Darjiling. In colls. Dr. Staudinger and F. Moore.

Genus CHURINGA, *Moore.*

Churinga, Moore, P. Z. S. 1878, p. 9.

Male and female. Wings ample, broad, somewhat short in male. Fore wing with the costa slightly arched towards end, apex pointed ; exterior margin oblique, posterior angle rounded ; first subcostal branch very oblique, free ; second from near end of the cell, trifurcate, fifth from end of the cell, with a loop-spur to the base of third ; radial starting from below the fifth subcostal beyond end of the cell ; discocellulars of equal length, bent inward at the middle ; median vein four-branched, two upper from end of cell, next at some distance before, and lower from, half length of the cell. Hind wing convex at the apex ; exterior margin rounded, abdominal margin short ; two subcostal branches at one-fourth beyond end of the cell ; median four-branched, two upper from end, next before the end, and lower from half length of the cell. Body large ; thorax with long pilose tegulæ ; palpi slender, ascending, basal joint pilose beneath, second and third joints very long, tip blunt ; legs long, femora slightly pilose beneath ; antennæ in male broadly pectinate, setose in female.

Type *C. rufifrons.*

Churinga rufifrons, Moore, P. Z. S. 1878, p. 9, pl. i. fig. 12.

Male and female. Upperside—fore wing pale purplish ochreous-brown ; costal and posterior margins narrowly edged with ochreous-yellow : hind wing and abdomen pale ochreous-yellow. Underside paler. Thorax brown ; tegulæ edged with yellow ; head, collar, and palpi reddish-ochreous ; thorax beneath and legs bright ochreous ; tip of palpi, antennæ, fore legs above, and tarsi brown.

Expanse, ♂ 1$\frac{9}{10}$, ♀ 2$\frac{3}{10}$ inches.

Hab. Darjiling. In colls. Dr. Staudinger and F. Moore.

Genus HESUDRA, *Moore.*

Hesudra, Moore, P. Z. S. 1878, p. 12.

Male. Wings short, rather broad. Fore wing with the costa nearly straight, apex acute ; exterior margin very oblique ; subcostal vein five-branched, first branch very oblique, free, running close to the costal ; second arising before end of the cell, trifurcate ; fifth from end of the cell, looped to third near its base ; radial from upper end of the cell ; discocellulars bent in the middle ; median vein four-branched, two upper from one-fourth beyond the cell, third near its end. Hind wing produced at the apex ; exterior margin very oblique, nearly straight ; abdominal margin long ; two subcostal branches from one-fourth beyond the cell ; median four-branched, two upper from one-third beyond the cell, third close to its end. Antennæ broadly pectinate ; palpi slender, curved upward, slightly pilose at base ; legs slender.

HESUDRIA DIVISA.

Hesudra divisa, Moore, P. Z. S. 1878, p. 12, pl. i. fig. 4.

Male. Upperside pale testaceous : fore wing with a broad dark purplish-grey band occupying the posterior half of the wing ; base of the costa also purplish-grey ; costal border yellowish. Underside—fore wing pale purplish-grey, costal border yellowish ; hind wing narrowly edged along the anterior margin with greyish-black. Thorax, head, side of palpi, and legs above purplish grey ; abdomen pale greyish testaceous beneath ; tip ochreous ; base of palpi and legs beneath pale ochréous ; antennæ purplish-brown.

Expanse 1¹³⁄₁ inch.

Hab. Darjiling. In colls. Dr. Staudinger and F. Moore.

Genus GHORIA, *Moore.*

Ghoria, Moore, P. Z. S. 1878, p. 12.

Fore wing long, narrow ; costa slightly arched beyond the middle ; apex acute ; exterior margin oblique and slightly convex, posterior margin long ; first subcostal branch very oblique, free ; second trifurcate, fifth from end of the cell and bent upward, looped to the third near its base ; radial starting from below the fifth subcostal at one-third beyond the cell ; discocellulars obliquely concave ; median four-branched, the two upper from end of cell, third immediately before its end, fourth at half its length. Hind wing broad, apex slightly produced ; exterior margin convex in the middle ; two subcostal branches at one third beyond the cell ; median four-branched, two upper beyond the cell, third at its end. Body slender, extending beyond hind wing ; antennæ setose ; legs slender, squamous, spurred ; palpi slender, porrect, base pilose beneath.

GHORIA SERICEIPENNIS.

Ghoria sericeipennis, Moore, P. Z. S. 1878, p. 13.

Male. Upperside—fore wing silky-white, with a broad dark cinereous-brown band along posterior margin: hind wing white, pale cinereous-brown at the apex and along anterior margin. Underside—fore wing, anterior border and apex of hind wing brown. Middle of thorax, tegulæ, and front of head dark cinereous-brown; abdomen above and beneath white; anal tuft brown; collar yellowish; palpi ochreous-yellow; legs above cinereous-brown, whitish beneath.

Expanse 1¾ inch.

Hab. Darjiling. In coll. Dr. Staudinger.

GHORIA ALBOCINEREA.

Ghoria albocinerea, Moore, P. Z. S. 1878, p. 13, pl. i. fig. 10.

Male. Upperside—fore wing silky white, with a cinereous-brown band along posterior margin: hind wing cinereous, cilia white. Underside cinereous-brown, palest on hind wing; abdominal margin cinereous-white; costal edge of fore wing yellowish towards the apex. Thorax and abdomen white; middle of thorax and streak on tegulæ cinereous-brown; collar and front of head, and antennæ, yellowish; anal tuft pale brownish-ochreous; legs cinereous-brown above, yellowish beneath; palpi yellowish, tip brown.

Expanse 1$\frac{1}{16}$ inch.

Hab. Darjiling. In colls. Dr. Staudinger and F. Moore.

Genus TARIKA, *Moore.*

Tarika, Moore, P. Z. S. 1878, p. 14.

Fore wing long, broad; costa arched; exterior margin convex, posterior margin rounded at the base; first subcostal branch short, oblique, anastomosing with the costal, second from near the end of the cell, third from near base of second, bifurcate, fifth from end of the cell, bent upward and touching the third at one-fourth its length; discocellulars bent inward, upper angled close to subcostal, lower shortest; radial from angle of upper discocellular; median three-branched, two upper branches from one-third beyond the cell; submedian nearly straight. Hind wing broad, apex slightly produced; subcostal branches at one-fourth, and median branches at one-third from end of the cell. Palpi small, short, slightly decumbent; legs stoutish, naked; antennæ minutely pectinate; body slender, not extending beyond hind wing.

Type *T. varana* (*Lithosia varana,* Moore, P. Z. S. 1865, p. 797).

TARIKA NIVEA.

Tarika nivea, Moore, P. Z. S. 1878, p. 15.

Male and female. Fore wing silky-white, costal edge pale yellow. Head, front of thorax, palpi, body, and legs beneath yellow; tip of palpi brown; fore and middle legs brown above.

Expanse, ♂ 1¾, ♀ 1⅘ inch.

Hab. Darjiling. In colls. Dr. Staudinger and F. Moore.

Genus KATHA, *Moore*.

Katha, Moore, P. Z. S. 1878, p. 16.

KATHA TERMINALIS.

Katha terminalis, Moore, P. Z. S. 1878, p. 17, pl. i. fig. 14.

Male. Ochreous: fore wing darkest, with a distinct curved purplish band across exterior margin. Antennæ, front of head, tip of palpi, middle of thorax, and streak on tegulæ purplish-black; legs above purplish-brown.

Expanse 1$\frac{1}{12}$ inch.

Hab. Darjiling. In coll. Dr. Staudinger.

Allied to *K. apicalis* (*Lithosia apicalis*, Walker, Journ. Linn. Soc. Zool. vi. p. 104), from Borneo, but differs on the fore wing in the apical band not extending upward on to the costa, and in the absence of the slight apical patch on the hind wing, which is present in Bornean examples.

Genus SYSTROPHA, *Hübner*.

Systropha, Hübner, Verz. bek. Schmett. p. 166.

SYSTROPHA DORSALIS.

Systropha dorsalis, Moore, P. Z. S. 1878, p. 18.

Female. Fore wing straw-yellow, slightly ochreous along posterior margin: hind wing yellowish white. Head, thorax, legs, and abdomen at sides and beneath ochreous-yellow; abdomen above lilac-grey.

Expanse 1$\frac{2}{12}$ inch.

Hab. Darjiling. In coll. Dr. Staudinger.

Genus CAPISSA, *Moore*.

Capissa, Moore, P. Z. S. 1878, p. 19.

CAPISSA PALLENS.

Capissa pallens, Moore, P. Z. S. 1878, p. 19, pl. ii. fig. 3.

Female. Pale whitish-ochreous: fore wing glossy, costal edge ochreous. Underside—

costal border of both wings brighter-coloured; middle of fore wing pale ochreous-brown. Thorax, palpi, and abdomen beneath ochreous; legs dusky-brown above; palpi brown at tip; antennæ brown.

Expanse $1\frac{5}{8}$ inch.

Hab. Darjiling.   In coll. Dr. Staudinger.

Somewhat allied to *C. insolita* (*Lithosia insolita*, Walk. Catal. Lep. Het. B. M. ii. p. 497).

Genus DOLGOMA, *Moore.*

Dolgoma, Moore, P. Z. S. 1878. p. 20.

DOLGOMA BRUNNEA.

Dolgoma brunnea, Moore, P. Z. S. 1878, p. 20, pl. ii. fig. 8.

Male. Cinereous-brown: fore wing uniformly dark-coloured; hind wing paler.   Underside paler than above; costal border of fore wing and legs beneath yellowish.

Expanse 1 inch.

Hab. Darjiling.   In colls. Dr. Staudinger and F. Moore.

Genus MITHUNA, *Moore.*

Mithuna, Moore, P. Z. S. 1878, p. 21.

Wings short, rather broad.   Fore wing arched, apex pointed; first branch of subcostal vein arising near end of the cell, short, oblique, anastomosing with costal, but free at its end; second branch quadrifid, the upper and the lower branch starting together at one-third beyond the cell; upper discocellular very short, lower curved inward; radial from lower end of upper discocellular; two upper median branches starting at half distance beyond the cell, lower branch straight from one-third before its end.   Hind wing—subcostal and median branches at nearly one-half length beyond the cell.   Body short; palpi small, pilose beneath; antennæ setulose; legs smooth.

MITHUNA QUADRIPLAGA.

Mithuna quadriplaga, Moore, P. Z. S. 1878, p. 21, pl. ii. fig. 9.

Fore wing luteous-brown, with a distinct dusky-brown medial transverse band, which is angled outward at end of the cell, a similar band also crossing the disk; outer margin with a series of dusky spots, which also indistinctly cross the cilia: hind wing pale luteous-brown. Underside paler.   Head and thorax brown.

Expanse 1 inch.

Hab. Darjiling.   In colls. Dr. Staudinger and F. Moore.

Genus COSSA, *Walker*.

Cossa, Walker, Cat. Lep. Het. B. M. Suppl. i. p. 232.

COSSA BRUNNEA.

Cossa brunnea, Moore, P. Z. S. 1878, p. 22, pl. ii. fig. 11.

Female. Fore wing dark purple-brown, with a small black spot at end of the cell, and a short black streak on the costa near the middle: hind wing paler. Underside with the margins paler. Thorax, head, and legs dark brown.

Expanse 1¼ inch.

Hab. Darjiling. In coll. Dr. Staudinger.

COSSA QUADRISIGNATA.

Cossa quadrisignata, Moore, P. Z. S. 1878, p. 21, pl. ii. fig. 10.

Male. Fore wing dark purplish-brown, with slightly paler streaks along the veins; costal border pale purplish-cinereous, with two prominent short black streaks; a small spot at end of the cell: hind wing and abdomen above pale brownish-cinereous; anal tuft slightly ochreous. Underside pale cinereous-brown; middle of fore wing brown. Thorax, head, abdomen beneath, and legs brown.

Expanse 1¼ inch.

Hab. Darjiling. In coll. Dr. Staudinger.

Genus RANGHANA, *Moore*.

Ranghana, Moore, P. Z. S. 1878, p. 22.

Female. Fore wing very long, narrow; costa arched, apex pointed; exterior margin very oblique, angle and hind margin rounded; subcostal vein five-branched, first branch short, arising immediately above end of the cell and joining the costal, second and third at equal distance from first, third trifid; radial from below subcostal beyond the cell at equa distance between first and second branches; cell short; discocellulars straight; median vein straight, three-branched, the two upper at one-third from the exterior margin, lower recurving from below end of the cell; submedian vein extending to posterior angle. Hind wing long, narrow; subcostal vein forked at half length beyond the cell; cell short; discocellulars deeply curved; median straight, two-branched; submedian straight. Body short; palpi very short, stout; legs long, slender, middle and hind tibiæ spurred; antennæ setose.

RANGHANA PUNCTATA.

Ranghana punctata, Moore, P. Z. S. 1878, p. 22, pl. ii. fig. 12.

Female. Pale ochreous buff colour: fore wing with a marginal and apical series of nine small black spots. Underside paler, without marks.

Expanse 1½ inch.

Hab. Calcutta (July). In coll. Dr. Staudinger.

Genus TEGULATA, *Walker*.

Tegulata, Walker, Journ. Linn. Soc. Zool. vi. p. 110.

TEGULATA BASISTRIGA.

Tegulata basistriga, Moore, P. Z. S. 1878, p. 22, pl. ii. fig. 5.

Female. Upperside pale luteous-brown; fore wing brightest at the apex, slightly brown-speckled; a black streak along base of costal edge and along base of posterior margin: hind wing pale brownish fawn-colour externally. Underside darker; legs brown above.

Expanse 1 inch.

Hab. Ceylon. In coll. Dr. Staudinger.

TEGULATA PROTUBERANS.

Tegulata protuberans, Moore, P. Z. S. 1878, p. 23, pl. ii. fig. 6.

Female. Upperside pale luteous-brown: fore wing with numerous brown speckles and a dark brown prominent costal spot: hind wing and abdomen pale luteous-yellow. Underside paler; middle of fore wing brownish.

Expanse $1\frac{2}{12}$ inch.

Hab. Darjiling. In coll. Dr. Staudinger.

Genus NISHADA, *Moore*.

Nishada, Moore, P. Z. S. 1878, p. 23.

Male and female. Fore wing somewhat short, broad; costa considerably arched; apex very acute; exterior margin oblique and slightly convex, posterior margin short, in the male convex and fringed to near the base; subcostal vein five-branched, first branch arising at half length of the cell and slightly touching the costal vein, second at one-third before end of the cell, third at some distance beyond the cell, fourth bifid at half its length; cell narrow; discocellulars very slender, upper short, lower obliquely curved inward; radial starting from lower end of upper discocellular; median vein three-branched, the two upper at nearly half distance beyond end of cell, lower before its end; submedian vein recurved upwards from the base. Hind wing in the male very short and broad, somewhat quadrate; anterior margin produced upward from the base and folded over on to the underside, and there provided with a long recumbent fan-like plumose covering or appendage; the hind wing in the female is longer, being of the same length as posterior margin of fore wing, its anterior margin is nearly straight and is without the folded plumose appendage; costal vein following the anterior margin beneath the fold in the male; subcostal vein two-branched, the upper arising near base of the wing; cell broad; discocellulars recurved; median vein straight, two-branched, the branches very close together, lower branch from before end of the cell; submedian straight. Antennæ minutely and finely pectinate in male, setose in female; body slender, extending beyond hind wing; legs stout, smooth, fore femora thickened, middle and hind tibiæ spurred; palpi small, porrect, projected slightly beyond the head, pilose beneath, apex small, pointed.

NISHADA FLABRIFERA.

Nishada flabrifera, Moore, P. Z. S. 1878, p. 23.

Male and female. Yellowish-ochreous, palest in female: fore wing in male brighter ochreous, and slightly ferruginous at the base. Thorax, head, and antennæ brownish ochreous.

Expanse 1 inch.

Hab. Calcutta District. In colls. Dr. Staudinger and F. Moore.

Allied to the Bornean *N. rotundipennis* (*Lithosia rotundipennis*, Walk. Journ. Linn. Soc. Zool. vi. p. 104), which species is also probably identical with *Lith. chilomorpha*, Snell. Tijd. voor Ent. 1877, p. 67, pl. v. f. 1, from Sumatra.

Genus PRABHASA, *Moore.*

Prabhasa, Moore, P. Z. S. 1878, p. 25.

Wings long. Fore wing narrow, very slightly arched before the apex; exterior margin oblique, slightly convex; posterior margin long, nearly straight; subcostal vein at some distance from costal: first branch short, curving upward to costa before end of the cell, second straight, from end of the cell, third contiguous at base to second, trifurcate, looped to second near base; cell long, in the male folded and tufted with recumbent plumes above to beyond half its length, the median and submedian contiguous; discocellulars long, convex, radial from lower part; median vein three-branched, two upper curving hindward from lower end of cell, lower curving from below the cell at nearly half its length; submedian recurved. Hind wing produced at the apex; exterior margin recurved, abdominal margin short; subcostal two-branched beyond the cell; discocellulars concave; cell short, broad; median three-branched, two upper branches at half distance beyond the cell, lower from before half its length; submedian nearly straight. Body slender, longer than hind wing, tufted in male; antennæ slender, with very fine delicate pectinations; legs slender, long, smooth, spurred; palpi slender, long, slightly ascending beyond front of head.

PRABHASA VENOSA.

Prabhasa venosa, Moore, P. Z. S. 1878, p. 26, pl. ii. fig. 16.

Male and female. Upperside pale luteous-brown: fore wing with the veins darker brown; an indistinct brown transverse discal band, which is bent outward beyond the cell;

male with a greyish-brown plumose tuft overlapping and extending half the length of the cell; anal tuft in male ochreous. Underside paler; band across fore wing not visible.

Expanse, ♂ 1₁₀̄, ♀ 1₁₀̄ inch.

Hab. Darjiling. In coll. Dr. Staudinger and F. Moore.

PRABHASA FLAVICOSTA.

Prabhasa flavicosta, Moore, P. Z. S. 1878, p. 26, pl. ii. fig. 17.

Female. Upperside uniformly dark luteous-brown: hind wing and abdomen purplish brown; fore wing with the costal border pale yellow; discal band most prominent at costal end, broad, but not bent outward. Front of head, thorax at sides, palpi, and abdomen beneath yellow; legs above cinereous-brown, yellowish beneath. Underside uniform brown, costal border on both wings yellowish.

Expanse 1₁₀̄ inch.

Hab. Cherra Punji, Assam (October). In coll. Dr. Staudinger and F. Moore.

Genus LYCLENE, *Moore.*

Lyclene, Moore, Cat. Lep. E.I. C. ii. p. 300.

LYCLENE ARTOCARPI.

Lyclene artocarpi, Moore, P. Z. S. 1878, p. 30.

Male and female. Fore wing ochreous, with a dusky-grey subbasal series of short longitudinal streaks; an upright medial band, and outward discal oblique irregular band, both confluent on middle of the hind margin; a small blackish spot at base of wing, and an indistinct spot at end of the cell: hind wing and abdomen pale ochreous. Thorax with black spots. Underside paler; markings on fore wing less distinct; an indistinct medial dusky fascia across hind wing in the female.

Expanse ⅞ inch.

Hab. Darjiling. In coll. Dr. Staudinger and F. Moore.

Nearest allied to *L. humilis*, Walk. Cat. Lep. Het. B. M. ii. p. 554. " Larva feeds on the Jack tree (*Artocarpus incisa*), January."—*A. Grote.*

LYCLENE OBSOLETA.

Lyclene obsoleta, Moore, P. Z. S. 1878, p. 32, pl. iii. fig. 7.

Female. Yellow: fore wing slightly ochreous-yellow externally, with several purplish-brown subbasal spots, a deeply sinuous discal band, and outer contiguous series of spots: hind wing and abdomen yellow. Underside paler, markings on fore wing indistinct: hind wing with a short costal streak before the apex. Fore tibiæ with a blackish terminal streak.

F

Expanse $\frac{8}{10}$ inch.

Hab. Darjiling. In coll. Dr. Staudinger.

May be distinguished by the absence of the medial transverse band between the sublasal spots and sinuous band.

LYCLENE RADIANS.

Lyclene radians, Moore, P. Z. S. 1878, p. 30, pl. iii, fig. 2.

Male. Fore wing yellowish-white, with a black basal spot, three transverse series of spots, each series composed of three, and the middle series curved; a prominent spot at end of cell, and a marginal series of broad black lines, one on each vein, their inner ends confluent and forming a bordered line which is bent outward at its middle: hind wing with a less prominent series of short black marginal lines, which do not extend to the anal angle. Underside similarly marked, the basal spots on fore wing less distinct: hind wing with two indistinct spots from middle of costa. Body ochreous-yellow: legs black-streaked.

Expanse 1 inch.

Hab. Darjiling. In coll. Dr. Staudinger.

LYCLENE INDISTINCTA.

Lyclene indistincta, Moore, P. Z. S. 1878, p. 33, pl. iii. fig. 9.

Female. Fore wing dull yellow, with two very indistinct subbasal curved series of pale brown spots, a bent discal series of spots, and an outer series of short longitudinal streaks: hind wing and underside much paler. Fore tibiæ cinereous-brown above.

Expanse $1\frac{1}{4}$ inch.

Hab. Darjiling. In coll. Dr. Staudinger.

LYCLENE ASSAMICA.

Lyclene assamica, Moore, P. Z. S. 1878, p. 33, pl. iii. fig. 8.

Male. Pale yellow. Upperside: fore wing with a purple-brown basal spot, two subbasal transverse series of spots, a spot at end of the cell, and a discal dentate band, each point having a terminal spot, these spots forming a submarginal series. Thorax black-spotted.

Expanse $\frac{9}{10}$ inch.

Hab. Dibrughur, Assam. In coll. Dr. Staudinger.

Allied both to *L. undulosa* and to the Bornean *L. cuneigera,* Walker.

LYCLENE INTERSERTA.

Lyclene interserta, Moore, P. Z. S. 1878, p. 32, pl. iii. fig. 6.

Female. Upperside: fore wing ochrey-yellow, with a short black streak at the base, narrow subbasal transverse bent line, a discal line which is convex anteriorly and bent posteriorly; between these is an oblique angled line, the point being opposite to the angle of the subbasal line; an upper and a lower longitudinal line, the former along the cell, the latter below it and furcate, the forks bent backward; also an outer marginal confluent looped line; cilia black, with white border: hind wing pale yellow; cilia at apex black. Underside uniform yellow, markings indistinct. Thorax and head yellow, with black streaks; palpi and legs above blackish; abdomen paler yellow.

Expanse $1\frac{3}{10}$ inch.

Hab. Darjiling. In coll. Dr. Staudinger and F. Moore.

Near to *L. euprepioides*, Walk.

Genus BARSINE, *Walker*.

Barsine, Walker, Cat. Lep. Het. B. M. iii. p. 546.

BARSINE GLORIOSA.

Barsine gloriosa, Moore, P. Z. S. 1878, p. 29, pl. iii. fig. 16.

Female. Upperside: fore wing ochreous-red, veins broadly lined with yellow; a broad blackish well-defined subbasal cross band, a narrow discal recurved band, and an apical series of short black longitudinal streaks: two small black dots at base of wing; costal edge and cilia black: hind wing and abdomen pale pink. Underside: fore wing red, with short black costal and apical streaks; hind wing paler, apex dusky black. Thorax red, with a narrow black middle streak and two anterior spots: palpi and legs red: middle tibiæ with a black terminal streak.

Expanse $1\frac{4}{10}$ inch.

Hab. Khasia Hills (October). In coll. Dr. Staudinger.

Nearest allied to *B. cruciata*, Walker, from Borneo.

BARSINE INFLEXA.

Barsine inflexa, Moore, P. Z. S. 1878, p. 29, pl. iii. fig. 17.

Male. Upperside: fore wing ochreous-red, veins lined with yellow; costal edge, posterior margin, and cilia black; a black subbasal cross band, somewhat thickened within the cell, the outer lower arm of which is bent inward; a narrow discal transverse maculated band, and an outer series of short streaks, one on each vein: hind wing and abdomen pale pink. Underside: fore wing red, with black apex and margins: hind wing yellowish, with dusky black apex. Thorax and head red, streaked with yellow; legs red.

Expanse $\frac{7}{8}$ inch.

Hab. Darjiling. In coll. Dr. Staudinger and F. Moore.

BARSINE PUNICEA.

Barsine punicea, Moore, P. Z. S. 1878, p. 29.

Female. Upperside: fore wing ochreous-yellow, slightly reddish along the costal border, with a blackish subbasal cross band, a waved curved discal band, and an outer series of alternate long and short longitudinal streaks; cilia and apical edge of costa black: hind wing pale ochreous-red, with black marginal band. Underside dull ochreous-red; apex of fore wing and outer margin of hind wing blackish. Abdomen blackish, tip yellow; middle tibiæ with a terminal black streak.

Expanse ¾ inch.

Hab. Darjiling. In coll. Dr. Staudinger and F. Moore.

This species is nearest allied to *B. lineata*, Walk. Cat. Lep. Het. B. M. iii. p. 760, from Borneo.

BARSINE FLAVIVENOSA.

Barsine flavivenosa, Moore, P. Z. S. 1878, p. 30, pl. iii. fig. 18.

Male. Upperside: fore wing red, with all the veins very broadly lined with yellow; a blackish subbasal imperfect cross band with the lower portion of the outer arm obsolete: a curved discal band and an incurved apical series of short black streaks: hind wing, abdomen, and legs very pale pink. Underside pink; hind wing palest.

Expanse ⁶⁄₈ inch.

Hab. Darjiling. In coll. Dr. Staudinger.

Genus ÆMENE, *Walker.*

Æmene, Walker, Catal. Lep. Het. B. M. iii. p. 541.

ÆMENE SINUATA.

Æmene sinuata, Moore, P. Z. S. 1878, p. 34, pl. iii. fig. 11.

Male. Upperside ochreous-white: fore wing with five narrow black sinuous bands, a spot within and a lunule at end of the cell, and an outer marginal row of small spots; cilia with a black line and three widely separated spots: hind wing pale brownish grey; cilia brown-lined. Underside brownish grey, markings on fore wing indistinct: hind wing with a pale brown spot at end of the cell, and a submarginal fascia. A black terminal spot on tegulæ; palpi black; fore legs black-streaked.

Expanse 1 inch.

Hab. Cherra Punji, Assam (October). In coll. Dr. Staudinger.

ÆMENE MACULIFASCIA.

Æmene maculifascia, Moore, P. Z. S. 1878, p. 33, pl. iii. fig. 10.

Male. Fore wing white, crossed by six black maculated bands, the first basal, second curved, third and fourth discal, irregular, and with dusky-brown suffused interspace, the other two marginal; two spots within the cell, the one at the end large. Cilia spotted near apex: hind wing ochreous-white. Underside: fore wing brown; spots slightly visible: hind wing with a brown spot at end of the cell, an indistinct submarginal fascia, and spots on cilia near apex. Thorax black-spotted; palpi black; fore legs black-streaked.

Expanse 1 ₁⁷ inch.

Hab. Darjiling. In coll. Dr. Staudinger.

Genus SETINA, *Schrank.*

SETINA NEBULOSA.

Setina nebulosa, Moore, P. Z. S. 1878, p. 35.

Female. Upperside: fore wing yellow, with a black basal spot, a broad irregular-bordered discal ochreous-brown band and confluent subbasal irregular-shaped band: hind wing cinereous-white, with broad dusky indistinct submarginal fascia. Underside and legs paler. Thorax yellow; abdomen cinereous-white.

Expanse 1₁²₄ inch.

Hab. Darjiling. In coll. Dr. Staudinger and F. Moore.

SETINA DISCISIGNA.

Setina discisigna, Moore, P. Z. S. 1878, p. 35.

Female. Upperside pale yellow: fore wing tinged with ochreous along hind margin; a black spot near base, a smaller spot at base of costa, and a more prominent spot at end of the cell: hind wing suffused with purplish-brown near the apex and along abdominal margin. Underside: middle of fore wing dusky brown. Thorax and head yellow, both with black spots; abdomen and legs purplish-brown, tip yellow.

Expanse 1 inch.

Hab. Cherra, Assam (October). In coll. Dr. Staudinger and F. Moore.

ADREPSA, nov. gen.

Female. Fore wing long, narrow; costa arched at the base, apex acute, exterior margin oblique, posterior margin slightly convex at the base, angle convex; subcostal vein five-branched, first branch arising before end of cell, second beyond the cell at nearly half distance between it and the apex, third bifid, fifth at equal distance between the cell and third; discocellulars angled at upper and lower end, curved in the middle; radial from angle of upper discocellular; cell broad; median vein four-branched, the upper from angle of lower discocellular, second from end of the cell, third at equal distance before end of the cell, fourth at some distance off; submedian vein curving hindward, and contiguous to the margin. Hind wing broad, anterior margin straight, apex acute, exterior margin obliquely recurved; cell short, broad; subcostal vein two-branched beyond end of the cell; discocellulars long, upper bent inward, lower outward; a radial from near upper end of lower discocellular ; median vein three-branched, two upper beyond the cell ; a straight submedian and an internal vein. Body moderate : abdomen not extending beyond hind wing ; palpi small, slender, porrect, pilose beneath, apex pointed ; legs squamous ; fore femora and tibiæ slightly pilose beneath ; antennæ filiform.

ADREPSA STILBIOIDES, n. sp.   (Plate II. fig. 20.)

Female. Fore wing dark purplish-grey, with a curved transverse subbasal and recurved discal series of black-speckled spots, one on each vein ; a large spot within the cell, an oval spot at its end, and a linear series of half-oval spots on exterior margin, all slightly bordered on one side with a few white speckles : hind wing pale glossy yellow, with an indistinct, slender, black-speckled, maculated marginal line. Underside : fore wing and anterior border of hind wing pale purplish-brown ; a marginal series of small slender black-speckled streaks on both wings. Thorax, head, palpi, antennæ, and legs dark purplish-grey ; thorax white-speckled : abdomen ochreous-yellow.

Expanse 1⅝ inch.

Hab. Assam. In coll. Dr. Staudinger.

Family ARCTIIDÆ.

Genus GLANYCUS, *Walker.*

Glanycus, Walker, Catal. Lep. Het. B. M. iii. p. 634.

GLANYCUS TRICOLOR, n. sp.

Upperside bluish-black : fore wing with a small recurved discocellular hyaline white spot: hind wing with a large medial hyaline white spot ; a narrow crimson streak on abdominal margin. Underside as above. Body steel-blue, a broad band across fore part of thorax, another band at base of abdomen, anal tuft, a streak on side of thorax beneath, and abdomen beneath bright crimson : antennæ brown : palpi and legs steel-blue.

Expanse 1$\frac{2}{4}$ inch.

Hab. Darjiling. In coll. Dr. Staudinger and F. Moore.

Genus HYPERCOMPA, *Steph.*

Hypercompa, Steph., Brit. Ent. ii. 67.

HYPERCOMPA NYCTEMERATA, n. sp.

Similar to *H. equitalis,* Kollar, Hügel's Kasch. iv. pl. 20. f. 3, from the N.W. Himalayas. Differs on fore wing in the transverse discal series of spots being of more equal length and regular in position, with a series curving upward before the spot below the apex ; there is also an additional lengthened spot contiguous to and below the basal spot within the cell: hind wing white; veins and narrow outer marginal line slightly fuliginous; a subanal and slight subapical fuliginous streak, which is less defined in female. Cilia fuliginous in male, white near anal angle in male and at apical angle in female. Body ochrey-yellow, with black dorsal narrow bands, and three rows of small spots beneath; legs blackish above, ochrey-yellow beneath.

Expanse, ♂ 2⅛ inches, ♀ 2⅜ inches.

Hab. Darjiling. In coll. Dr. Staudinger and F. Moore.

Genus EUCHÆTES, *Clemens.*

Euchætes, Clemens, Proc. Acad. Nat. Sci. Phil. xii. p. 532.

EUCHÆTES SIKKIMENSIS, n. sp. (Plate II. fig. 12.)

Upperside: fore wing ochreous-brown, palest along the veins: hind wing fuliginous-brown with an indistinct paler submarginal band. Underside ochreous-brown, both wings with darker basal and marginal streaks between the veins, the disk thus showing a pale transverse fascia. Abdomen ochreous-red, with a dorsal and lateral row of black spots; anal tuft yellow; legs blackish.

Expanse 1⅝ inch.

Hab. Darjiling. In coll. Dr. Staudinger.

It is extremely interesting to find a species of the genus *Euchætes* occurring in Asia. The species hitherto described are American.

Genus ALPENUS, *Walker.*

Alpenus, Walker, Catal. Lep. Het. B. M. iii. p. 686.

ALPENUS FLAVENS, n. sp.

Uniform ochreous-yellow: fore wing with two medial transverse slightly recurved series of quadrate brown spots, and two less distinct spots below the apex: hind wing with a series of three small brown spots from anal angle, and a very small indistinct spot at end of cell. Underside of wings marked as above. Abdomen reddish, with indistinct blackish dorsal spots; antennæ, tip of palpi, and legs above brown.

Expanse 1⁴⁄₁₆ inch.

Hab. Cherra Punji, Assam. In coll. Dr. Staudinger and F. Moore.

Genus SPILARCTIA, *Butler.*

Spilarctia, Butler, Cist. Ent. ii. p. 39 (1875).

SPILARCTIA UNIFORMIS, n. sp.

Female. Upperside pale ochreous-buff: fore wing with a small black spot at base, a spot near end of submedian vein, a recurved discal, and a less distinct submarginal, vein-divided series of spots: hind wing with a large black cell spot, two larger spots above anal angle and smaller spot near apex. Underside uniformly coloured as above: fore wing with a black lunule at end of the cell: hind wing marked as above. Abdomen with dorsal and lateral black spots; antennæ black; palpi brown-streaked; legs yellow; terminal spot on femora, tibia, and tarsi above ochreous-brown.

Expanse $1\frac{9}{16}$ inch.

Hab. Calcutta (August). In coll. Dr. Staudinger.

Nearest allied to *Sp. punctata*, Moore, from Java.

SPILARCTIA HOWRA, n. sp.

Male and female. Pale yellowish-ochreous: fore wing with a short recurved discal series of minute black speckles: hind wing with a small black spot at end of the cell, and a spot near middle of hind margin. Underside with a small black spot at end of cell in both wings. Markings less distinct in the female.

Expanse, ♂ $1\frac{1}{4}$ inch, ♀ $1\frac{7}{8}$ inch.

Hab. Calcutta. In coll. Dr. Staudinger and F. Moore.

Nearest allied to *S. suffusa*, Walker, but differs in the wings being pale yellowish-ochreous, and in the absence of the ochreous-red on the hind wing above and on both wings beneath, as in that species.

SPILARCTIA OBLIQUIVITTA, n. sp. (Plate II. fig. 26.)

Male. Upperside: fore wing pale yellowish-ochreous, with a black medial oblique straight narrow band from middle of hind margin to beyond the cell, and terminating in a series of short longitudinal streaks before the apex: hind wing white. Underside white: fore wing slightly pale ochreous along costal border; oblique band of upperside visible; apical streaks black: hind wing with a small black-speckled spot below the apex, and a similar spot near the anal angle. Thorax ochreous-yellow; abdomen above brighter ochreous, with a lateral row of black spots, whitish beneath; antennæ black, shaft whitish; palpi black at the tip and sides; fore legs and tarsus of all the legs streaked with black above.

Expanse $1\frac{4}{8}$ inch.

Hab. Darjiling. In coll. Dr. Staudinger.

Genus ICAMBOSIDA, *Walker.*

Icambosida, Walker, Catal. Lep. Het. B. M. Suppl. 2, p. 401 (1865).

ICAMBOSIDA PUNCTILINEA, n. sp.

Male. Fore wing very pale ochreous-yellow; a recurved black maculated oblique band from middle of hind margin to lower end of the cell, composed of broad contiguous linear spots; a few narrow spots also before the apex: hind wing pure white, with small black cell-spot, two above anal angle, and one near apex. Underside as above. Thorax pale ochreous ; abdomen bright crimson above, white beneath ; sides of thorax beneath and pectus crimson ; antennæ, palpi, and side of head black ; legs white, fore legs above with a terminal spot on femora and tibiæ, and tarsi black.

Expanse $1\frac{6}{10}$ inch.

Hab. Darjiling. In coll. Dr. Staudinger.

Nearest allied to *I. rubitincta* (*Spilosoma rubitincta*, Moore, P. Z. S. 1865, p. 809).

CARBISA, nov. gen.

Fore wing somewhat short, trigonal, apex convex, exterior margin very oblique, hind margin short ; veins similar to those in *Icambosida*, but with the first and third median branches much curved : hind wing short and trigonal, apex convex, exterior margin slightly rounded. Body robust ; thorax broad, laxly pilose ; abdomen less pilose ; antennæ bipec-tinate ; palpi short, not extending beyond the head, decumbent, pilose ; legs pilose beneath.

CARBISA VENOSA, n. sp. (Plate II. fig. 10.)

Male. Upperside—fore wing pale brownish-ochreous ; veins pale ochreous-white. Cilia yellowish : hind wing pale yellowish-ochreous, with a very indistinct brown submarginal fascia. Underside pale ochreous-yellow ; a slight indistinct subapical brown streak on fore wing, and a similar-coloured streak near anal angle of hind wing. Thorax, palpi, and head ochreous-yellow ; abdomen above paler, with narrow black segmental bands : antennæ brown, shaft white ; fore legs black-streaked above, middle and hind femora with a black terminal spot.

Expanse $1\frac{3}{8}$ inch.

Hab. Darjiling. In coll. Dr. Staudinger.

PIMPRANA, nov. gen.

Male. Fore wing long, narrow ; costa arched at the base and before the apex, angle acute, exterior margin slightly oblique, posterior angle rounded ; subcostal vein five-branched, first and second arising before end of the cell, fourth and fifth from below the third beyond the cell ; discocellulars bent inward ; radial from upper end of the cell ; median vein four-branched, three upper branches from lower end of the cell : hind wing broad, trigonal, extending beyond fore wing, anterior margin slightly arched, apex very convex, abdominal margin short ; subcostal vein forked at a short distance from end of the cell, the radial from middle of discocellulars, median vein three-branched. Body moderate, pilose ; abdomen extending beyond hind wing, sides and apex tufted ; palpi stout, densely pilose, porrect, extending slightly beyond the head, third joint short ; antennæ long, setose ; femora and tibiæ of all the legs densely pilose, middle and hind legs spurred.

G

PIMPRANA ATKINSONI, n. sp. (Plate II. fig. 11.)

Upperside—fore wing ochreous-green, varied with ochreous-brown on hind border and exterior border, the base and lower veins white-speckled; a marginal apical series of blackish spots speckled with white: posterior angle white-streaked. Cilia blackish alternated at the veins with white; hind wing ochreous-yellow along costal and outer borders, the interior of the wing black; cilia and abdominal fringe yellow. Underside ochreous-yellow, the lower part of both wings black; fore wing with two, and hind wing with one black cell-spot. Thorax ochreous-green; abdomen black above; the tip and beneath yellow; antennæ brown; palpi black at tip and sides; legs yellow, tarsi grey.

Expanse $1\frac{3}{4}$ inch.

Hab. Darjiling. In coll. Dr. Staudinger and F. Moore.

PANGORA, nov. gen.

Fore wing long, apex rather acute, exterior margin short and oblique: hind wing broad, apex rather acute in male, rounded in female, exterior margin convex, abdominal margin short, anal angle rounded. Neuration similar to *Creatonotus*, but with the upper median branches on both wings closer together at their base. Body stout; palpi prominent, apical joint pointed, projecting beyond the head, covered with adpressed scales; legs squamose; antennæ minutely serrate in male, simple in female.

PANGORA DISTORTA, n. sp. (Plate II. fig. 14.)

Male and female. Upperside—fore wing dark dull olive-brown, with ochreous-white irregular-margined basal patch, a medial transverse oblique sinuous-margined irregular-shaped band, two small patches near the outer margin, one being near the apex, the other beneath it, and a very small spot above posterior angle; the basal patch extending into base of the cell and spotted with brown; the medial band with a short bent and angled inner extension at both ends: hind wing ochreous-red, brightest in female; cilia paler; a black spot at upper end of cell extending to the costal margin; a submarginal row of four large black spots, a narrow marginal streak at apex and small spot below it. Underside marked as above, the pale interspaces on fore wing ochreous-yellow; spots on hind wing in male indistinct. Head and thorax ochreous-white, black-spotted; abdomen ochreous-red, with dorsal and lateral row of black spots, and black bands beneath; legs black, femora bright red above; front of head and tip of palpi black, base of palpi pale ochreous.

Expanse, ♂ $2\frac{1}{2}$ inches, ♀ $2\frac{4}{8}$ inches.

Hab. Masuri (*Hutton*); Simla (*Atkinson*); N. W. Himalayas. In coll. Dr. Staudinger and F. Moore.

RAJENDRA, nov. gen. *

Form and neuration similar to *Creatonotus* (*C. interruptus*), but differs in the hind wing being shorter, less produced at the apex, and the greater convexity of the hind margin. Palpi prominent; antennæ minutely bipectinate in male, simple in female.

Type. *R. lativitta.*

RAJENDRA LATIVITTA, n. sp.

Male. Allied to *R. biguttata* (*Aloa biguttata*, Walker, Catal. Lep. Het. B. M. iii. p. 707) from the western Ghants. Differs in the fore wing having a much broader longitudinal band, which is also less bent below the end of the cell; the hind wing differs also in the black costal border being narrower and the marginal spot smaller.

Expanse $1\frac{3}{4}$ inch.

Hab. Parisnath Hill, Behar (September). In coll. Dr. Staudinger.

NAYACA, nov. gen.

Similar in form and neuration to *Alphea*, Walker. Body stout; thorax and abdomen thickly pilose; head prominent, pilose; abdomen extending beyond hind wings; antennæ long, rather broadly bipectinate in male, less so in female; palpi porrect, stout, first and second joints pilose beneath, third joint squamous; legs stout, squamous, spurred, femora pilose beneath.

Type *Nayaca imbuta* (*Arctia imbuta*, Walker, Catal. Lep. Het. B. M. iii. p. 614).

Hab. Himalayas.

NAYACA FLORESCENS, n. sp. (Plate II. fig. 13.)

Male. Fore wing pale ochreous-brown, veins ochreous-yellow; a series of white variously-shaped spots between the veins—disposed similar to those in *Alphea fulvohirta*, but of different shape; the spots between the lower median and submedian veins connected by a short peduncle: hind wing white, with three olive-brown spots on anterior margin, one at end of the cell, another below the apex, and a larger irregular-shaped spot from anal angle. Wings beneath paler, marked as above. Thorax ochreous-brown with pale ochreous edges; collar fringed with red: top of head ochreous-yellow; front of head and palpi olive-brown; abdomen bright ochreous-red above, beneath white with an abdominal streak and lateral spot olive-brown; legs olive-brown, paler beneath, femora red.

Expanse $1\frac{7}{8}$ inch.

Hab. Darjiling. In coll. Dr. Staudinger.

* This genus will embrace the species described by Mr. Walker as *Aloa integra*, *A. dentata*, *A. biguttata*, as well as those I described as *Aloa sipahi* and *A. nigricans*.

Family LIPARIDÆ.

Genus ORYGIA, *Ochsenheimer.*

Orygia, Ochs., Schmett. Eur. iii. 218.

ORYGIA OCULARIS, n. sp.

Male. Upperside umber-brown: fore wing with a large suffused-bordered black-lined spot enclosing a pale lunule at end of the cell; two black suffused lunular lines below the spot, and some black dentate marks before the apex. Underside paler; spot on fore wing indistinct.

Expanse $\frac{7}{8}$ inch.

Hab. Calcutta. In coll. Dr. Staudinger and F. Moore.

Allied to *O. postica* (*Lacida postica*, Walker, Catal. Lep. Het. Brit. Mus. pt. iv. p. 803), but differs in its smaller size, paler colour, and less prominent markings.

Genus CHARNIDAS, *Walker.*

Charnidas, Walker, Catal. Lep. Het. B. M. iv. p. 797.

CHARNIDAS OCHRACEA, n. sp.

Female. Upperside ochreous: fore wing with a few scattered black scales indistinctly visible, some of which are disposed in a maculated series across the disk; a more prominent black spot at end of the cell. Underside slightly paler, both wings with an indistinct black spot near upper end of the cell. Pectinations of antennæ dark brown.

Expanse $1\frac{3}{8}$ inch.

Hab. Calcutta. In coll. Dr. Staudinger and F. Moore.

"Larva with tawny brushes on the back. Feeds on Bamboo. Pupa October 19th; imago October 28th."—*MS. note,* A. Grote.

CHARNIDAS CINNAMOMEA, n. sp.

Male. Upperside—fore wing dark cinnamon-brown: hind wing paler, and with brighter brown borders. Fore wing with an indistinct dusky brown streak at end of the cell. Underside uniform cinnamon-brown. Thorax dark brown; abdomen cinnamon-brown; antennæ blackish, shaft brown.

Expanse $1\frac{2}{16}$ to $1\frac{4}{16}$ inch.

Hab. Pangi, Upper Kunawur, N.W. Himalayas. In coll. Capt. Lang, Dr. Staudinger, and F. Moore.

Nearest allied to *C. litura*, Walker, *l. c.* p. 797.

BARYAZA, nov. gen.

Male. Fore wing broad; costa almost straight, apex scarcely acute, exterior margin slightly convex, hind margin long, nearly of the same length as the costa; subcostal vein five-branched, first and second branches arising before end of the cell, third from the end and emitting the fourth from below half its length, fifth from end of the cell; discocellulars bent at upper and lower ends and concave in the middle, the radial emitted from the upper angle; median four-branched, upper branch starting from lower angle of discocellular, the two next from end of the cell; a submedian and an internal vein: hind wing broad, trigonal, apex and anal angle rounded; subcostal vein three-branched, third branch from below the second at one third its length; discocellulars angled at the middle; other veins as in fore wing. Body slender, abdomen extending beyond hind wing; antennæ bipectinate; palpi porrect, pilose, apical joint exserted, slender; legs rather slender, fore femora and tibiæ pilose, middle and hind tibiæ spurred.

BARYAZA CERVINA, n. sp. (Plate III. fig. 1.)

Male. Upperside—fore wing brownish fawn-colour, crossed by a straight subbasal dark brown band, and a narrow pale-bordered discal line; a prominent black lunule at end of the cell: hind wing and abdomen brown. Underside dull ferruginous-brown; both wings crossed by an indistinct darker discal narrow band; an indistinct streak at end of each cell. Thorax dark brown; palpi beneath and fore legs above blackish.

Expanse 1⅞ inch.

Hab. Darjiling. In coll. Dr. Staudinger.

Genus STILPNOTIA, *Westwood.*

Stilpnotia, Westw. & H., Brit. Moths, i. p. 90.

STILPNOTIA SERICEA, n. sp.

Male and female. Wings silky-white. Body white; abdomen in male yellowish-white. Palpi black at the side. Fore legs black above; middle and hind tibiæ and tarsi in female black-banded; antennæ pale ochreous, shaft white. Near to the European *S. salicis.*

Expanse, ♂ 1¾ inch, ♀ 1⅝ inch.

Hab. Masuri (*Capt. Lang*); Darjiling (*Atkinson*). In coll. Capt. Lang, Dr. Staudinger, and F. Moore.

STILPNOTIA OCHRIPES, n. sp.

Female. Pure white, silky; costal edge slightly pale ochreous. Body white; palpi at sides bright ochreous-yellow; fore legs, middle and hind tarsi bright ochreous-yellow; antennæ pale ochreous, shaft white.

Expanse 1⁶⁄₁₀ inch.

Hab. Darjiling. In coll. Dr. Staudinger.

In this species the wings are shorter and slightly broader than in *S. sericea.*

CARAGOLA, nov. gen.

Male. Fore wing triangular, apex produced, pointed, exterior margin oblique, straight, posterior angle acute ; subcostal vein five-branched, third and fourth branches arising near the apex, fifth at one third from end of the cell ; cell short ; discocellular veins sharply angled ; radial starting from first discocellular near its upper end ; median vein four-branched, angled before end of the cell ; second branch starting from the angle, third and fourth on a foot-stalk one third beyond the cell ; submedian vein extending to posterior angle : hind wing short ; convex at the apex ; abdominal margin rather long ; cell short ; subcostal vein two-branched, arising at one third their length from the end of the cell ; discocellulars sharply angled ; median vein four-branched, the three upper branches contiguous at their base from end of the cell ; submedian vein nearly straight. Body slender, pilose : antennæ broadly bipectinate, slender at tip ; legs thickly pilose beneath ; palpi moderate, pilose.

Allied to *Cariria* and to *Redoa.*

CARAGOLA COSTALIS, n. sp.   (Plate II. fig. 21.)

Male. Upperside pure silky-white : fore wing with black costal edge, the margin slightly folded over near the apex ; base of wing slightly tinged with ochreous ; veins raised above the surface. Body pale ochreous-white ; antennæ pale ochreous, shaft white ; palpi at side, and side of head beneath black ; fore legs black above.

Expanse 1⅝ inch.

Hab. Darjiling. In coll. Dr. Staudinger.

Genus REDOA, *Walker.*

Redoa, Walker, Catal. Lep. Het. B. M. iv. p. 826.

REDOA LACTEA, n. sp.

Male and female. Wings dull silky-white, sparsely silvery-scaled. Body white ; front of head, tip of palpi, and legs black-spotted : antennæ and costal border of fore wing tinged with pale ochreous.

Expanse, ♂ 1⅞ inch, ♀ 2⅝ inches.

Hab. Darjiling. In coll. Dr. Staudinger and F. Moore.

Differs from *R. submarginata,* Walker, in its larger size, the wings being less covered with silvery scales and not raised in slight corrugated folds as in that species.

REDOA DIAPHANA, n. sp.

Male and female. Wings dull white, diaphanous. Body white above, yellowish beneath; tibiæ of fore legs beneath slightly black-fringed; antennæ pale ochreous, shaft white.

Expanse, ♂ 1¾ inch, ♀ 2 inches.

Hab. Darjiling.   In coll. Dr. Staudinger and F. Moore.

Distinguished from *Redoa lactea* by the total absence of silvery scales on the wings, and the spots on the head, palpi, and legs.

Genus GAZALINA, *Walker.*

Gazalina, Walker, Catal. Lep. Het. B. M. Suppl. p. 398.

GAZALINA TRANSVERSA, n. sp.   (Plate 11. fig. 22.)

Female. White: fore wing with two transverse medial narrow black bands, the first short and inwardly oblique and ascending only to near lower part of the cell, the outer band slightly bent at lower end of the cell and at its lower end.   Underside white, bands of fore wing indistinctly visible.   Thorax slightly ochreous; abdomen with narrow blackish segmental bands; anal tuft large, golden brown; palpi and front of head fringed with black; fore legs above black, beneath white; middle and hind legs white; tarsi above black.

Expanse 1⅝ inch.

Hab. Darjiling.   In coll. Dr. Staudinger.

HARAPA, nov. gen.

Female. Fore wing long; costa arched; apex acute; exterior margin oblique, slightly convex hindward; posterior margin nearly straight; first and second subcostal branches arising before end of the cell, third from below second and coalescing with fourth beyond end of the cell, third forked near its end; discocellulars bent inward; radial starting from upper end of the cell; two upper branches of median from lower end of the cell: hind wing broad, apex and abdominal margin short; subcostal forked beyond end of the cell, two upper median branches from end of the cell, third before its end.   Body stout; legs pilose beneath; antennæ bipectinate; palpi long, porrect, projecting beyond the head, pilose beneath, apical joint slender.

HARAPA TESTACEA, n. sp.   (Plate 11. fig. 15.)

Female. Fore wing reddish-testaceous: hind wing pale cinereous-testaceous.   Underside uniform pale reddish-testaceous.   Body and antennæ reddish-testaceous; palpi ochreous.

Expanse 2⅝ inches.

Hab. Darjiling.   In coll. Dr. Staudinger.

48   LIPARIDÆ.

Genus EUPROCTIS, *Hübner*.

Euproctis, Hübn., Verz. bek. Schmett. p. 159.

EUPROCTIS SUBNIGRA, n. sp.

Male. Upperside—fore wing white: hind wing dusky-black, abdominal border whitish.
Cilia white. Underside of both wings uniform dusky-black. Cilia white. Antennæ dusky,
shaft white; anal tuft yellow.

Expanse 1¼ inch.

Hab. Cherra Punji (*Atkinson*): Jawai Hills (*Austen*). In coll. Dr. Staudinger and F.
Moore.

EUPROCTIS PYGMÆA, n. sp.

Male. Yellow: fore wing with two medial transverse scarcely perceptible paler
bands: hind wing pale yellow. Underside—fore wing tinged with brownish fawn-
colour.

Expanse ⁷⁄₁₆ inch.

Hab. Calcutta. In coll. Dr. Staudinger and F. Moore.

EUPROCTIS SEMIVITTA, n. sp.   (Plate II. fig. 25.)

Male. White: fore wing with a medial transverse oblique short black-speckled band
terminating within end of the cell: a few indistinct black speckles on hind margin at each
side of the band. Underside white. Thorax pale ochreous; abdomen ochreous-white,
slightly banded with black, anal tuft bright ochreous; second joint of palpi black; tip white:
fore legs ochreous in front; antennæ pale brown.

Expanse 1²⁄₁₀ inch.

Hab. Khasia Hills. In coll. Dr. Staudinger.

EUPROCTIS VARIEGATA, n. sp.   (Plate II. fig. 24.)

Male. Upperside white; cilia of fore wing maculated with brown: fore wing with six
transverse, partly ochreous and brown-speckled bands which are slightly confluent, the
outer band being marginal, broad, and occupying nearly a third of the wing: hind wing
with a brown broad marginal band and a speckled spot at end of the cell. Underside—fore
wing and marginal band on hind wing dusky-brown. Front of thorax ochreous; antennæ
brown; palpi black; fore tibiæ black-streaked.

Expanse 1¹⁄₁₀ inch.

Hab. Darjiling. In coll. Dr. Staudinger.

Genus PIDA, *Walker.*

Pida, Walker, Catal. Lep. Het. B. M. Suppl. p. 399.

PIDA LATIVITTA, n. sp. (Plate II. fig. 18.)

Female. Upperside pale ochreous, exterior borders whitish: fore wing with a broad medial recurved transverse darker ochreous-brown band, sparsely covered with black speckles and crossed by pale veins, and enclosing a black spot at upper end of the cell: outer border also sparsely black-speckled: hind wing with the middle densely brown-speckled, prominently so at end of the cell. Underside pale greyish-ochreous: both wings sparsely speckled at end of the cell. Abdomen dark brown, tuft ochreous above, white beneath; legs brown-speckled: antennæ brown, shaft whitish.

Expanse 2 inches.

Hab. Darjiling. In coll. Dr. Staudinger.

Genus CHÆROTRICHA, *Felder.*

Chærotricha, Felder, Nov. Voy. pl. 98 (1874).

CHÆROTRICHA BIPARTITA, n. sp. (Plate II. fig. 4.)

Female. Upperside pale ochreous-yellow: fore wing greyish, numerously speckled with minute brown scales: crossed by two medial recurved wavy pale bands; a darker speckled spot below the apex. Underside uniform pale ochreous. Abdomen bright ochreous.

Expanse 1¾ inch.

Hab. Darjiling. In coll. Dr. Staudinger.

CHÆROTRICHA MARGINATA, n. sp.

Male and female. Deep yellow: fore wing from the base to one third from exterior border ochreous-brown and minutely black-speckled; a yellow angular space at lower end of the cell, the division across the disk wavy and pointed below the apex; a submarginal row of distinct black spots. Underside uniform yellow. Thorax ochreous-brown; abdomen dark brown, apical tuft golden yellow.

Expanse 2¾ inches.

Hab. Darjiling. In coll. Dr. Staudinger and F. Moore.

Near to *C. plagiata* (*Euproctis plagiata*, Walker), but may be distinguished by the black spots along outer margin, and by the dark-coloured abdomen.

CHÆROTRICHA UNIFORMIS, n. sp.

Upperside—fore wing uniform ochreous-brown, minutely black-speckled. Cilia yellow:

H

hind wing yellow ; basal half brown and speckled.   Underside paler yellow ; basal portion of both wings brown.

Expanse 2 inches.

Hab. Darjiling.   In coll. Dr. Staudinger.

Allied to *C. plagiata* (*Euproctis plagiata*, Walker), distinguished from it by the absence of the yellow discal space on the fore wing, and by the dark colour of the hind wing.

CHÆROTRICHA QUADRANGULARIS, n. sp.   (Plate II. fig. 23.)

Female. Upperside pale yellow : fore wing from the base to near outer margin fawn-colour, minutely black-speckled ; an oblique quadrate yellow space at end of the cell ; four black marginal spots, two being below the apex, and two near posterior angle.   Underside pale yellow : fore wing with the basal portion pale fawn-colour, margin without spots. Thorax pale fawn-colour ; abdomen blackish fawn-colour ; anal tuft yellow, tip blackish.

Expanse 2 inches.

Hab. Munipur, Eastern Bengal.   In coll. Dr. Staudinger.

Allied to *C. marginalis*, but is of a different colour, and may be distinguished by the four spots on outer margin of fore wing.

Genus ARTAXA, *Walker.*

Artaxa, Walker, Cat. Lep. Het. B. M. iv. p. 794.

ARTAXA DISPERSA, n. sp.   (Plate II. fig. 6.)

Female. Fore wing dark yellow, with a few minute brown speckles forming an indistinct medial very oblique transverse band : hind wing very pale yellow.   Thorax yellow ; abdomen brownish black, apex grey-fringed.

Expanse 1¾ inch.

Hab. Darjiling.   In coll. Dr. Staudinger.

Allied to *A. subfasciata*, Walker, but of a much darker and brighter colour, the speckled band more oblique, and not extending to the hind margin.

ARTAXA VENOSA, n. sp.   (Plate II. fig. 5.)

Male and female. White : fore wing with subbasal, discal, and marginal yellow bands, crossed by white veins ; a short medial brown-speckled vein-divided band crossing from end of the cell ; the interspace along the disk also sparsely brown-speckled.   Antennæ, front of head, palpi, fore legs, and anal tuft pale ochreous.

Expanse, ♂ 1⅜, ♀ 1¾ inch.

Hab. Darjiling.   In coll. Dr. Staudinger.

ARTAXA TRIFASCIATA, n. sp.

Male and female. Upperside—fore wing pale primrose-yellow, with three indistinct transverse equidistant ochreous brown-speckled abbreviated bands ascending from the hind margin, the middle band most prominent: hind wing white. Underside white: costal border of fore wing brownish at the base. Thorax pale yellow; abdomen ochreous-brown above, tuft ochreous; palpi, fore femora, and tibiæ ochreous in front.

Expanse, ♂ 1⅜, ♀ 1⅝ inch.

Hab. Cherra Punji, Assam. In colls. Dr. Staudinger and F. Moore.

Allied to *A. subfasciata,* Walker.

ARTAXA BASALIS, n. sp. (Plate II. fig. 16.)

Male. Upperside—fore wing yellow, with a broad ochreous-brown black-speckled basal band extending to middle of wing, its outer border being angled beyond end of the cell; a small similar-coloured spot opposite to angle of band: hind wing whitish. Underside cinereous-white; basal band on fore wing dusky. Head, thorax, and palpi yellow; abdomen and legs cinereous-white.

Expanse 1⅔ inch.

Hab. Darjiling. In coll. Dr. Staudinger.

This species is allied to *A. atomaria* and *A. justiciæ.*

ARTAXA HOWRA, n. sp.

Male. Upperside ochrey-yellow, palest on hind wing: fore wing with a black-speckled short streak ascending from middle of hind margin: a small black-speckled spot at end of the cell. Palpi ochreous-brown above.

Expanse 1 inch.

Hab. Calcutta. In colls. Dr. Staudinger and F. Moore.

Allied to *Artaxa citrina,* from Ceylon. Differs in being darker-coloured, especially on the hind wing, and the fore wing having the black spot at end of the cell. It is also allied to *A. breviritta,* Moore, from Bengal.

DAPLASA, nov. gen.

Fore wing elongated, somewhat narrow; costa much arched at the base; apex slightly acute; posterior margin long; subcostal vein five-branched, first and second arising from before end of the cell, third trifurcate, the branches short, middle branch from below half length of upper; discocellulars bent in the middle; radial from anterior end of upper discocellular; median vein four-branched, two upper branches starting from lower end of the cell; submedian widely separated from median. Hind wing extending to angle of fore wing; costa nearly straight; apex somewhat pointed; costal vein extending to apex; subcostal two-branched, starting beyond end of the cell; other veins as in fore wing. Body slender; antennæ bipectinate; palpi minute, slender, ascending; legs slender, fore femora and tibiæ slightly pilose, other legs spurred.

DAPLASA IRRORATA, n. sp. (Plate II. fig. 17.)

Male. Fore wing white, irrorated with minute brown scales, composing apparently four oblique bands, which are shortest and most distinct on the hind margin, the outer band extending across the disk from apex. Hind wing and abdomen pale yellow. Underside pale yellow; costa of fore wing bright yellow. Thorax white, brown-speckled; palpi, and legs above, bright yellow.

Expanse $1\frac{1}{8}$ inch.

Hab. Darjiling. In coll. Dr. Staudinger.

MAHOBA, nov. gen.

Male. Fore wing short, broad; costa slightly arched; apex pointed; exterior margin long, oblique. Hind wing broad, extending beyond posterior angle, apex rounded; abdominal margin long; veins similar to those in *Lymantria*. Body narrow; abdomen shorter than hind wing; antennæ broadly bipectinate, less so in female; palpi porrect, extending beyond the head, flat, pilose, tip exposed; legs slender, pilose, spurred.

Female. Wings broader; outer margin of fore wing less oblique: hind wing broader and more rounded at apex.

MAHOBA PLAGIODOTATA. (Plate III. fig. 6, ♂.)

Cyclidia plagiodotata, Walker, Cat. Lep. Het. B. M. xxv. Geom. p. 1183 (1862), ♀.

Male and female. Whitish: fore wing numerously irrorated with ferruginous speckles, which form three confluent patches on the costa and indistinctly extend across the wing; also a ferruginous patch along exterior margin, a transverse discal pale-bordered sinuous streak at end of the cell: hind wing with a large brown medial, a subanal spot, and a patch at and below the apex. Underside creamy-white: fore wing with a large ferruginous-brown spot at end of the cell and patch at apex: hind wing with upper and lower spots as above; apical patch indistinct. Head and thorax hoary; abdomen ochreous, with black bands beneath; antennæ blackish; palpi and legs ochreous and black-speckled.

Expanse $2\frac{1}{2}$ inches.

Hab. Darjiling. In coll. British Museum and Dr. Staudinger.

MAHOBA IRRORATA, n. sp.

Female. Upperside ochreous-white: fore wing very sparsely irrorated with ferruginous brown scales; a pale spot at end of the cell enclosing an angled discocellular brown streak, which is slightly yellow-speckled: hind wing sparsely irrorated with ferruginous-brown scales along outer border. Underside as above, but speckled only along outer borders; an indistinct angled streak at end of each cell. Thorax pale ochreous; abdomen brighter; palpi and legs ochreous, brown-speckled; antennæ brown.

Expanse 2¼ inches.

Hab. Darjiling. In coll. Dr. Staudinger

NAGUNDA, nov. gen.

Female. Fore wing moderately long, narrow; costa arched near the end; apex rather acute; exterior margin oblique; lowest branch of subcostal vein emitted from upper angle of end of the cell; discocellulars long, bent acutely inward; three upper branches of median vein arising close together from lower angle of cell. Hind wing elongated, extending beyond posterior angle of fore wing; apex and exterior margin rounded. Body robust; abdomen extending beyond hind wing; ovipositor exserted; antennæ minutely bipectinate to tip, pectinations broad; palpi porrect, hairy beneath, tip minute; legs pilose.

Type *Nagunda semicincta* (*Alope semicincta*, Walker, Cat. Lep. Het. B. M. iii. p. 620).
Hab. N.W. Himalayas.

LOCHARNA, nov. gen.

Wings short. Fore wing arched; apex somewhat acute; exterior margin oblique, nearly straight; hind margin long; cell narrow; costal vein short; first branch starting at one half length of the cell, third subcostal much curved upward, lowest branch from upper angle of cell; discocellulars curved; other veins of both wings similar to those in *Lymantria*. Hind wing short, rounded at the apex and exterior margin; abdominal margin somewhat long. Body robust; abdomen extending to anal angle; thorax with a radiating tuft of loose hairs on each tegula, and an erect posterior crest of short spatulated hairs; antennæ thinly bipectinate, rather short; palpi thickly pilose, apical joint naked; fore femora and tibiæ densely pilose, tarsi naked; middle and hind legs sparsely hairy,

LOCHARNA STRIGIPENNIS, n. sp. (Plate III. fig. 11.)

Upperside—fore wing whitish-ochreous; veins ochreous, with numerous delicate, short, narrow transverse black strigæ between the veins, except at the lower part of base and upper part of the disk along the costa; cilia black-edged: hind wing uniform ochreous. Underside uniform ochreous. Thorax white; collar ochreous, the hairs black-tipped, with a posterior glossy-black curly spatulated crest; front of head, palpi, and fore legs ochreous, fringed with black; antennæ black, shaft white.

Expanse 1¼ inch.

Hab. Khasia Hills. In coll. Dr. Staudinger.

CADRUSIA, nov. gen.

Fore wing long; costa arched in the middle; apex acute; exterior margin oblique, straight; posterior margin long, and the angle somewhat abrupt. Hind wing arched in middle of anterior margin; apex rather acute; exterior margin slightly produced at the middle; veins similarly arranged as in *Lymantria monacha*. Body moderately short; antennæ minutely bipectinate; palpi ascending beyond front of head, densely pilose beneath, tip exserted; legs slender; fore legs thickly pilose to end of tarsi; middle and hind legs less pilose.

CADRUSIA VIRESCENS, n. sp. (Plate III. fig. 16.)

Female. Fore wing ochrey-grey, densely covered with dark sap-green and brown scales, forming broad basal indistinctly-confluent transverse bands, and a narrow sinuous pale-bordered discal band, beyond which is an outer blackish dentate pale-bordered line, and a wavy marginal line; a prominent angled blackish lunule with white-speckled interior at end of the cell; veins black-lined; hind wing ochreous, with a large triangular black spot at end of the cell, and two near outer border towards the angles; a wavy marginal black line. Underside pale ochreous; both wings with a prominent black triangular spot at end of the cell, and two indistinct spots near outer margin. Thorax greenish; abdomen, palpi, and legs ochreous; fore legs green in front.

Expanse 2¾ inches.

Hab. Darjiling. In colls. Dr. Staudinger and F. Moore.

IMAUS, nov. gen.

Male and female. Alike. Wings large, broad, short. Fore wing slightly arched at base of costa; apex somewhat rounded; exterior margin oblique and slightly convex. Hind wing somewhat quadrate; apex rounded, exterior margin produced in the middle. Body robust; abdomen not extending beyond hind wing; no anal tuft; ovipositor not exserted; antennæ in both sexes with broad bipectinations; palpi porrect, slender, clothed with closely adpressed hairy scales; femora and tibiæ slightly fringed with hair.

Type. *Imaus mundus* (*Lymantria munda*, Walker, Cat. Lep. Het. B. M. iv. p. 875).

Hab. Darjiling.

Genus PORTHETRIA, *Hübner.*

Porthetria, Hübner, Verz. bek. Schmett. p. 160.

PORTHETRIA LEPCHA, n. sp.

Male. Upperside grey, slightly suffused with fawn-colour; fore wing with four transverse equidistant sinuous black lines, the first three lines being medial, the fourth submarginal ; two indistinct basal series of spots ; discocellular angular streak and a small spot before it black ; a marginal row of black spots : hind wing reddish at base and along abdominal border, with an ill-defined pale brownish submarginal band. Underside pale greyish-ochreous, markings very indistinct. Antennæ dark brown ; spots on thorax above and short dorsal bands blackish ; palpi black at the side ; fore femora black and red-streaked above ; tibiæ and tarsi black-speckled.

Expanse $2\frac{9}{16}$ inches.

Hab. Darjiling. In colls. Dr. Staudinger and F. Moore.

PORTHETRIA UMBRINA, n. sp. (Plate III. fig. 4.)

Male. Upperside umber-brown : fore wing crossed by three narrow greyish basal wavy bands and two outer marginal sinuous bands ; veins greyish : a marginal row of black spots ; a blackish spot at end of the cell ; cilia alternated with black ; a slight ochreous streak at base of wing : hind wing with an indistinct marginal blackish pale-bordered maculated band, and a spot at end of the cell. Underside ochreous-brown, margins ochreous ; both wings with a dentate blackish spot at end of the cell ; fore wing with a row of black spots on exterior margin and cilia. Thorax longitudinally streaked with grey ; abdomen ochreous, with indistinct dorsal and lateral blackish spots ; palpi ochreous, with a lateral black spot ; body beneath and legs ochreous ; legs black-streaked ; antennæ dark brown.

Expanse $1\frac{7}{8}$ inch.

Hab. Darjiling (Selim Terai). In coll. Dr. Staudinger.

Similar in pattern of markings to *P. mathura*, Moore, also from Darjiling, but of entirely a different colour.

Genus LYMANTRIA, *Hübner.*

Lymantria, Hübner, Verz. bek. Schmett. p. 160.

LYMANTRIA GRISEA, n. sp. (Plate III. fig. 5, ♀.)

Male and female. Pale dull brownish-white : fore wings minutely brown-speckled, crossed by several pale brown sinuous bands, the interspaces white-speckled ; a marginal row of brown spots : hind wing with indistinct brownish marginal band. Underside pale greyish-ochreous, margins brighter. Thorax brown ; abdomen beneath pale ochreous ; palpi brown at side ; legs black-streaked above.

Expanse, ♂ $1\frac{4}{8}$, ♀ $1\frac{7}{8}$ inch.

Hab. Darjiling. In colls. Dr. Staudinger and F. Moore.

BARIIONA, nov. gen.

Female. Fore wing long, somewhat narrow ; apex much produced, acutely angled ; costa arched ; exterior margin very oblique. Hind wing rather longer, extending some distance beyond posterior angle of fore wing; costal margin somewhat straight, apex pointed; exterior margin very convex; abdominal margin short; veins similar to those in *Lymantria* (*L. dispar*), except that the two upper median branches are close together at their base. Body robust, short; antennæ minutely pectinate; palpi small, pilose; legs naked; fore femora slightly fringed beneath.

BARHONA CARNEOLA, n. sp.

Female. Upperside pale rosy flesh-colour; base of hind wing brighter red; fore wing with three short black streaks, one being at end of the cell, another obliquely on costa near the base, the third below it on the hind margin. Underside of wings more roseate than above; costal edge red. Antennæ black, basal tuft red; front of head and fore coxæ red; palpi black above; legs black in front; fore and middle femora red beneath; abdomen beneath with slight segmental red bands.

Expanse 3⅞ inches.

Hab. Darjiling. In coll. Dr. Staudinger and F. Moore.

DURA, nov. gen.

Fore wing short; costa slightly arched; apex somewhat rounded; exterior margin slightly oblique and convex, hind margin long. Hind wing short; apex rounded; exterior margin much produced and angled in the middle; abdominal margin short; veins similar to those in *Lymantria monacha*, excepting that in the fore wing the lowest subcostal branch is here emitted below the end of the cell (as a radial) from the angle of upper discocellular vein, the next subcostal branch being also emitted above it and at a short distance from the cell. Body moderate, not extending beyond anal angle; antennæ bipectinate; palpi porrect, pilose; legs slender, slightly pilose, spurred.

DURA ALBA, n. sp.

Male and female. White. Fore wing crossed by very indistinct pale brownish sinuous bands; a brown spot at end of the cell, and a marginal row of brown spots; hind wing with a brown spot at upper end of the cell, and a marginal row of spots. Underside white; cell-spot and marginal series as above. Palpi blackish at side; legs white.

Expanse, ♂ 1¾, ♀ 2⅔ inches.

Hab. Darjiling. In coll. Dr. Staudinger and F. Moore.

Allied to *D. albicans* (*Dasychira albicans*, Walk. Cat. Lep. Het. B. M. vii. p. 1739).

Genus PEGELLA, *Walker.*

Pegella, Walker, Cat. Lep. Het. B. M. Suppl. p. 1922.

PEGELLA BIVITTATA, n. sp.

Lymantria lineata ♀, Walk. Cat. Lep. Het. B. M. iv. p. 875.

Female. White. Fore wing silky, with a basal short curved sinuous brown streak and two transverse brown bands, the first band outwardly oblique, straight, and one fourth from the base ; second band inwardly oblique, broadest and lunular, its base touching the first on the hind margin ; an exterior marginal row of brown linear spots, which also cross the cilia. Thorax and a large silky anal tuft white ; abdomen reddish ; antennæ and palpi black ; collar and tuft at base of antennæ and front of head red ; legs black, fore legs red-streaked.

Expanse 3 inches.

Hab. Darjiling. In coll. British Museum and Dr. Staudinger.

HIMALA, nov. gen.

Male and female. Fore wing short ; costa slightly arched at the base ; apex somewhat pointed ; exterior margin oblique and slightly convex, posterior angle rounded : hind wing broad, apex and exterior border rounded. Veins similar to those in *Stilpnotia salicis.* Body robust, extending to length of hind wing ; antennæ bipectinate, broadly so in male ; palpi slender, porrect, pilose ; legs slender, femora and tibiæ pilose.

HIMALA ARGENTEA.

Redoa argentea, Walker, Cat. Lep. Het. B. M. iv. p. 827 (1855).
Dasychira ilita, Moore, Cat. Lep. Mus. E. I. C. ii. p. 341 (1858).

Male and female. Silvery-white. Fore wing with the veins black, except the costal and upper subcostal branches : hind wing with the veins externally and marginal line black. Palpi black at the side ; antennæ pale brown ; fore and middle legs black in front ; tarsi in female black beneath.

Expanse, ♂ 1⅜, ♀ 1⅞ inch.

Hab. Deyra Dhoon, N.W. Himalaya.

Genus DASYCHIRA, *Stephens.*

Dasychira, Steph. Brit. Ent. ii. p. 58.

DASYCHIRA BRUNNESCENS, n. sp.

Male. Upperside—fore wing greyish-white, irrorated with brown scales, crossed by a

I

brown irregular zigzag basal line and two subbasal lines, a sinuous discal line, and a sub-marginal pale-bordered lunular line; a brown dentate streak at end of the cell: hind wing greyish-brown, palest on outer margin; veins and a submarginal fascia darker brown. Underside grey, brownish-speckled; base of fore wing with ferruginous-brown hairs: both wings with a triangular brown spot at end of cell, and a pale brown submarginal fascia. Thorax, head, and legs grey, speckled with brown; abdomen pale brown; pectinations of antennæ ferruginous-brown, shaft grey.

Female. Marked as in male, excepting that the hind wings are much paler-coloured.

Expanse, ♂ 2⅜, ♀ 3¾ inches.

Hab. Darjiling. In coll. Dr. Staudinger.

This species has greater similarity of appearance to the European *D. pudibunda* than any other Indian member of the genus.

DASYCHIRA PERDIX, n. sp. (Plate III. fig. 3.)

Male. Grey; both wings and the veins irrorated with prominent blackish-brown speckles, forming on the disk of the fore wing indistinct transverse sinuous bands and patches, and a submarginal line; a suffused brown patch at end of the cell, and another ascending from middle of the hind margin: hind wing with a submarginal fascia; inner margin and abdomen above pale ferruginous-grey. Underside yellowish-white; veins lined with pale ochreous; costal base of fore wing broadly pale ferruginous. Thorax grey, speckled with brown; antennæ, side of palpi, and legs pale ferruginous.

Expanse 2¾ inches.

Hab. Darjiling. In coll. Dr. Staudinger.

DASYCHIRA FASCIATA, n. sp.

Male. Fore wing with two broad greyish-white bands, the first extending from middle of costa and narrowing to near posterior margin, the other along exterior margin, both intersected by brown veins; base of wing and space between the bands greenish-brown, crossed by grey lunules: hind wing creamy-white; anterior margin to below apex greyish-white: apex crossed by brown streaks. Underside creamy-white; costa and cilia of fore wing ferruginous. Head and front of thorax white; hind part of thorax greenish-brown; abdomen pale testaceous, tip dark brown; antennæ brown; palpi, pectus, and legs ferruginous.

Expanse 2₁⁸₄ inches.

Hab. Darjiling. In coll. Dr. Staudinger.

DASYCHIRA STRIGATA, n. sp.

Female. Upperside grey, suffused outwardly with fuliginous-brown; irrorated with black scales; crossed by four subbasal, two discal, and a marginal, black sinuous lines, the outer discal line being pale-bordered; a black lunule at end of cell; cilia black-spotted: hind wing greyish-brown, base pale ferruginous; an indistinct blackish submarginal fascia, and blackish narrow lunular marginal band and spots on cilia.   Underside pale ferruginous-brown; both wings with indistinct blackish narrow cell-streak, submarginal fascia, and cilial spots. Thorax, head, palpi, and legs grey, speckled with black; abdomen and antennæ brownish-ferruginous.

Expanse 2⅔ inches.

Hab. Gurhwal (*Atkinson*); Simla (*Hutton*).   In coll. Dr. Staudinger and F. Moore.

This species may ultimately prove to be the female of *D. kausalia.*

DASYCHIRA CINCTATA, n. sp.

Male. Upperside—fore wing glossy white, minutely irrorated with brown scales; a brown quadrate spot on costa one-third from the base; a transverse single medial and double discal wavy brown line terminating on costa in a small spot; a submarginal zigzag brown line: hind wing greyish-white, with a subdued brown spot in the cell, an irregular recurved submarginal band, and a marginal lunular line.   Underside greyish-white; both wings with a large dusky-brown spot at end of cell, and indistinct transverse discal brown streaks. Head, thorax, and legs grey; palpi black, tipped with grey; antennæ chestnut-brown, shaft white; legs black-speckled, and tarsi black-banded; abdomen ochrey-grey, with broad black bands above.

Expanse 2¼ inches.

Hab. Darjiling.   In coll. Dr. Staudinger.

DASYCHIRA ALBESCENS, n. sp.   (Plate III. fig. 2.)

Male. Upperside white. Fore wing minutely irrorated with brown scales, crossed by two subbasal and a discal irregular-zigzag indistinct brown lines: hind wing with indistinct brown submarginal fascia, darkest at anal angle, the brown colour also slightly extending inward between the veins; a brownish spot within the cell; cilia of both wings alternated with brown.   Underside white; middle of fore wing broadly, from base, brown; cell-spot on both wings distinct. Palpi at side and tarsi brown-streaked.

Expanse 2¼ inches.

Hab. Darjiling.   In coll. Dr. Staudinger.

Nearest allied to *D. cinctata.*

Family NOTODONTIDÆ.

Genus STAUROPUS, *Germar.*

Stauropus, Germ. Prod. p. 45.

STAUROPUS APICALIS, n. sp.

Male. Fore wing brownish-grey, crossed by a subbasal, discal, and submarginal zigzag grey lunular lines with brown borders; between the two former is a large indistinct greyish reniform mark; a series of blackish dots on exterior margin; an irregular longitudinal apical black streak curving below the costa: some irregular black marks also on middle and base of costa: hind wing dull purple-brown. Underside pale brown, middle of fore wing darker brown. Thorax grey; abdomen brownish-testaceous; front of thorax, palpi, and legs purple-brown.

Expanse 2⅜ inches.

Hab. Darjiling. In coll. Dr. Staudinger.

Genus HETEROCAMPA, *Doubleday.*

Heterocampa, Doubleday, Entom. p. 55.

HETEROCAMPA IRRORATA, n. sp.

Female. Fore wing grey, speckled with black and yellow, crossed by several zigzag rows of black dentate marks; a black streak obliquely from base, and outer marginal row of short transverse black streaks, one being between each grey-pointed vein: hind wing pale brown. Underside pale brown; costa darker brown. Thorax and head greenish-grey; abdomen yellow-brown.

Expanse 1⅝ inch.

Hab. Darjiling. In coll. Dr. Staudinger.

HETEROCAMPA BRUNNEA, n. sp.

Male. Fore wing pale umber-brown, crossed by a broad darker brown subbasal outwardly-oblique band with black-bordered lunular line, and an irregular submarginal indistinct band, outside which is a row of grey-speckled black spots: hind wing and abdomen brownish fawn-colour. Underside pale brown, palest on hind wing. Thorax, head, palpi, and legs chestnut-brown; antennæ pale brown.

Expanse 1⅘ inch.

Hab. Darjiling. In coll. Dr. Staudinger.

HETEROCAMPA MACULATA, n. sp.

Male. Fore wing with a broad basal band and posterior angle grey; apex and disk pale ferruginous, crossed by subbasal and submarginal black zigzag lunular spots; base interspersed with black spots; a row of submarginal black spots: hind wing pale ferruginous, clouded with brown at apex and anal angle. Underside pale testaceous, brownish and brown-blotched along costa. Thorax black; collar, head, palpi, and legs grey; abdomen pale testaceous, tipped with a black bar.

Expanse 1⅚ inch.

Hab. Darjiling. In coll. Dr. Staudinger.

HETEROCAMPA PLAGIVIRIDIS, n. sp.

Male. Fore wing purplish-brown, crossed by four green lunular bands with irregular dark brown borders—the first basal, second subbasal, both broad, third discal and formed by two long curves, the fourth submarginal, zigzag, and narrow : hind wing pale pinkish-brown, abdominal margin and abdomen yellowish-brown. Underside pale brown ; base of fore wing and abdominal margin yellowish. Antennæ brown ; thorax, palpi, and legs dark chestnut-brown.

Expanse 2 inches.

Hab. Darjiling. In coll. Dr. Staudinger.

HETEROCAMPA OBLIQUIPLAGA, n. sp.

Male. Fore wing crossed by two blackish-grey bands bordered by zigzag black margins, the first subbasal, outwardly oblique and broad, the other submarginal and indistinct at its middle, the broad interspace between them being greyish pink-brown ; reniform spot pale ; base and outer border of wing brown, with a marginal row of grey spots : hind wing pale pink-brown ; abdominal margin yellowish. Underside pale testaceous. Head, thorax, palpi, and legs blackish-grey ; abdomen pale testaceous, tipped with black-grey ; antennæ pale brown.

Expanse 2⅓ inch.

Hab. Darjiling. In coll. Dr. Staudinger.

HETEROCAMPA VARIEGATA, n. sp.

Male. Fore wing pale pinkish greyish-brown, darkest basally and at apex, palest obliquely before the apex ; crossed by narrow irregular lunular ochre-yellow black-bordered bands, clustered at the base and before the apex, and forming a border on exterior margin ; veins black-streaked : hind wing pinkish-brown, with an ochrey-yellow black-bordered anal spot ; cilia alternate black and grey. Underside pale brownish-white ; fore wing streaked with brown at base and apex. Thorax dark brown, tinged with ochre in front, and with a white posterior spot ; abdomen greyish brown ; antennæ and legs brown ; palpi black.

Expanse 2 inches.

Hab. Darjiling. In coll. Dr. Staudinger.

HETEROCAMPA BASALIS, n. sp.

Male. Fore wing greyish-white, crossed by a broad subbasal and discal dark umber-brown band with blackish lunular-bordered inner line ; two small black spots at end of cell ; outer margin crossed by an alternate black and white zigzag line : hind wing and abdomen greyish brown. Underside—fore wing brown ; hind wing very pale brown ; both wings

crossed by indistinct darker fasciæ. Thorax, head, palpi, and legs dark brown ; antennæ
pale brown.

Expanse 1⅝ inch.

Hab. Darjiling. In coll. Dr. Staudinger.

Genus ROSAMA, *Walker.*

Rosama, Walker, Cat. Lep. Het. B. M. v. p. 1066.

ROSAMA PLUSIOIDES, n. sp.

Male. Allied to the Javan *R. strigosa*, Walker; but differs in its much darker colour,
the costal and outer portion of the fore wing being very dark chestnut and washed with
grey, the hind part only being ochreous and darker-streaked ; silvery yellow spot before the
cell prominent ; the outer submarginal black line is also much less sinuous.

Expanse 1¼ inch.

Hab. Darjiling. In coll. Dr. Staudinger.

Genus SPATALIA, *Hübner.*

Spatalia, Hübner, Verz. bek. Schmett. p. 145.

SPATALIA GEMMIFERA, n. sp.   (Plate III. fig. 14.)

Male. Upperside—fore wing pale sienna-brown ; a series of silver-white spots bordered
with chestnut-brown below the cell at base, from which proceeds a dark brown streak to
below the apex ; a submarginal series of pale spots, and a few black dots before the apex ;
inner lobe of posterior margin black-tufted : hind wing tinted with pale sienna-brown.
Underside pale sienna-brown on fore wing, yellowish on hind wing.   Thorax, head, palpi,
and fore legs chestnut-brown ; crest of thorax whitish ; antennæ brown, shaft whitish ;
abdomen pale sienna-brown, tip chestnut-brown, anal tuft blackish.

Expanse 2 inches.

Hab. Darjiling. In coll. Dr. Staudinger.

Genus CELEIA, *Walker.*

Celeia, Walker, Cat. Lep. Het. B. M. Suppl. p. 463.

CELEIA DISRUPTA, n. sp.

Male. Upperside—fore wing dark purple-brown ; a narrow longitudinal silver streak
from base, followed by a white and black spot on a chrome-yellow patch, the silver streak
continuing and dilating broadly to outer margin below the apex ; a greenish tint below the

silver streak ; cilia alternate brown and white: hind wing brown; cilia edged with white. Underside dusky brown; hind wing and patch on fore wing below apex greyish-white. Thorax dark chestnut-brown, medial streak whitish ; front of head white ; palpi and legs dark chestnut ; abdomen brown, tip chestnut-brown ; antennæ brown, shaft whitish ; abdomen beneath whitish.

Expanse 2 inches.

Hab. Darjiling. In coll. Dr. Staudinger.

CELEIA SIKKIMA, n. sp.

Male. Upperside—fore wing pale chestnut-brown, streaked with black at base along the disk, at and below the apex ; a silver white interrupted streak below the cell : hind wing and abdomen brown ; anal tuft black-tipped. Thorax, antennæ, head, and palpi dark brown. Underside—fore wing dusky brown ; hind wing and abdomen beneath and legs pale brown.

Expanse 2 inches.

Hab. Darjiling. In coll. Dr. Staudinger.

NIGANDA, nov. gen.

Fore wing very long, narrow, exterior margin oblique : subcostal vein five-branched, first and second branches arising before end of the cell, second coalescing with third near its base ; fourth and fifth starting from below the third, each at one third its length from end of the cell ; first discocellular very short, oblique, second and third long, concave ; two discoidal veins ; median vein three-branched, lower branch arising at one third the length of the cell, second near to its end and upper branch from its angle with lower discocellular ; submedian vein parallel and close to hind margin : hind wing rather long, trigonal ; apex acute, outer margin slightly convex : costal vein long, extending to apex ; subcostal two-branched, each starting halfway beyond end of the cell ; two discocellulars, oblique ; median vein as in fore wing ; submedian and internal veins straight. Body long, narrow, attenuated to tip, extending half its length beyond hind wings, tufted : antennæ minutely biserrate in male, setaceous in female ; palpi thick, pilose, ascending : legs slender, pilose, spurred.

NIGANDA STRIGIFASCIA, n. sp. (Plate III. fig. 15, ♂.)

Male. Upperside pale ochreous-brown, greyish externally: fore wing with a narrow longitudinal striated white streak from middle of base to beyond end of the cell, the streak dentate at the end of the cell, a pale yellow fascia extending thence to apex ; a narrow ochreous streak below end of the cell ; an oblique discal series of indistinct black points and a less apparent series on outer margin ; front of thorax whitish-ochreous, thorax ochreous, abdomen paler.

Female. Upperside brownish-ochreous, brightest on fore wing : the longitudinal streak broad, occupying the cell, white at base and yellow to below the apex ; outer margin bordered with white, and brown-speckled. Underside paler in both sexes; fore wings with similar longitudinal streaks.

Expanse, ♂ 2 inches, ♀ 2⅝ inches.

Hab. Darjiling. In coll. Dr. Staudinger.

NIGANDA SIKKIMA, n. sp.

Male. Upperside—fore wing dull yellowish-ochreous, suffused with vinous-brown along hind margin; disk crossed by three recurved series of ochreous-brown dots, a contiguous inner row and two outer rows of fawn-coloured spots; a marginal row of blackish dots: hind wing vinous ochreous-brown. Cilia yellow. Underside pale dull ochreous-yellow, middle of both wings brownish. Head, thorax, and palpi ochreous-grey; abdomen ochreous, tip grey.

Expanse 2⅔ inches.

Hab. Darjiling. In coll. Dr. Staudinger.

NIGANDA AURATA, n. sp.

Male. Upperside—fore wing ochreous-yellow, the base and upper half glossy metallic golden-yellow; an ochreous narrow red bifid streak from base below the cell; three or more short oblique streaks on costa, two within the cell, and some speckles on hind margin; a black spot at end of the cell, a pure white sinuous marginal line, two submarginal indistinct dusky lunular lines, and a discal row of minute black points: hind wing pale vinous ochreous-brown. Cilia yellow. Underside—fore wing vinous-ochreous; costal border yellow: hind wing pale vinous ochreous-yellow. Thorax grey in front; head, palpi, legs, and thorax behind reddish-ochreous; abdomen greyish-ochreous, beneath and tip yellow.

Expanse 1⅞ to 2⅛ inches.

Hab. Darjiling (*Atkinson*); Khasia Hills (*Godwin-Austen*). In coll. Dr. Staudinger and F. Moore.

NIGANDA ALBISTRIGA, n. sp.

Male. Upperside ochreous-brown: fore wing with a medial longitudinal narrow ochreous line from base of wing, which widens on the disk and is streaked with white near the outer margin: a series of two subbasal, a transverse medial row, three discal rows, and two marginal rows of black points, the two latter with pale borders; a black spot within end of the cell. Cilia alternate brown and ochreous: hind wing and abdomen dark brown. Underside pale greyish-brown: fore wing clouded with brown ; a marginal row of blackish spots: hind wing with subbasal and discal transverse sinuous brown lines, and a marginal row of brown lunules. Head, thorax, and tip of abdomen ochreous-brown.

Expanse 1⅞ inch.

Hab. Darjiling. In coll. Dr. Staudinger.

NIGANDA DIVISA, n. sp.

Male. Upperside—fore wing brownish-ochreous; the veins on upper half of wing and longitudinal intermediate narrow streaks dark ochreous-brown; a narrow black longitudinal streak on hind margin near base, from the end of which an indistinct fascia ascends to costa; a curved subbasal transverse series of blackish-speckled spots, a discal series of black points, and a similar series on outer margin; a pale-bordered mark at end of the cell: hind wing dark umber-brown. Head, thorax, and tip of abdomen brownish-ochreous, crest of thorax, front of head, and palpi brighter ochreous; abdomen above dark brown Wings beneath dark umber-brown; apex of both wings, legs, and abdomen beneath pale brownish-ochreous.

Expanse 2⅛ inches.

Hab. Darjiling. In coll. Dr. Staudinger.

Genus CEIRA, *Walker.*

Ceira, Walker, Catal. Lep. Het. B. M. Suppl. p. 462.

CEIRA OCHRACEA, n. sp.

Female. Upperside—fore wing clear uniform ochreous-yellow, with an indistinct discal row of brown points: hind wing pale yellow, veins and outer margin darkest. Head and thorax ochreous; abdomen paler. Underside uniform paler ochreous-yellow.

Expanse 2⅞ inches.

Hab. Darjiling. In coll. Dr. Staudinger.

CEIRA JUNCTURA, n. sp.

Male. Upperside pale straw-yellow: fore wing with an ochreous-brown longitudinal band from base to apex, the band suffused broadly and darker-spotted below the cell and before reaching the apex, some dark speckles joining the band from hind margin; base of costa spotted and obliquely streaked, and the costal edge lined with ochrey-brown; two oblique submarginal speckled lines of the same colour. Underside dull pale ochrey-yellow, the veins of both wings and costal border of fore wing pale ochreous-brown. Body ochreous-yellow; thorax, head, palpi, and legs ochreous-brown.

Expanse 1⅞ inch.

Hab. Darjiling. In coll. Dr. Staudinger and F. Moore.

CEIRA BASISTRIGA, n. sp.

Male. Upperside—fore wing dark ochrey-yellow, with numerous ochrey-brown speckles disposed across the disk in oblique linear series; two short straight brown streaks speckled with greyish-white from base of wing below the cell; base of costa dotted with black; a marginal row of blackish dots, and a prominent upper discal black-speckled spot:

K

hind wing pale yellowish-cinereous.   Underside pale ochreous-yellow : fore wing suffused on
the disk with brown.   Thorax, head, palpi, and legs ochrey-brown ; abdomen pale ochreous.
    Expanse 2⅗ inches.
    Hab. Darjiling.   In coll. Dr. Staudinger.

CEIRA POSTICA, n. sp.

Male. Upperside pale straw-yellow : fore wing speckled with ochrey-brown ; a narrow
ochrey-brown ill-defined medial longitudinal band from base curving upward to apex ; a
triangular patch on hind margin near angle ; a series of darker points along outer margin.
Underside uniform whitish-ochreous.   Head, palpi, thorax, and legs ochrey-brown.
    Expanse 1⅜ inch.
    Hab. Darjiling.   In coll. Dr. Staudinger.

CEIRA DECURRENS, n. sp.

Female. Upperside dull straw-yellow : fore wing minutely brown-speckled ; two medial
and a submarginal transverse oblique brown-speckled bands, the two medial bands broadest
anteriorly, bent inward to costa and partly joined at the angle by a confluent fascia
extending to the apex ; a black dot at end of the cell.   Hind wing, body, and underside
paler whitish-ochreous.
    Expanse 2 inches.
    Hab. Cherra Punji, Assam.   In coll. Dr. Staudinger.

Genus PYDNA, *Walker.*
Pydna, Walker, Catal. Lep. Het. B. M. vii. p. 1753.

PYDNA FASCIATA, n. sp.

Male. Upperside pale ochreous-brown : fore wing with two transverse oblique recurved
darker fasciæ, which are broadly suffused on hind margin and anteriorly across to below
apex, the interspaces whitish-streaked ; a distinct transverse discal row of dark brown dots,
and a similar but less distinct series on outer margin : hind wing uniform ochreous-brown.
Underside paler : fore wing ochreous-brown at base : hind wing with an indistinct brown
sinuous discal band and a spot at end of the cell.
    Expanse 2⅗ inches.
    Hab. Darjiling.   In coll. Dr. Staudinger.
    Allied to *P. testacea,* Walker, but differs from it by the oblique suffused fascia and the
discal series of brown dots, and in the absence of the row of black dentate marks near the
outer margin.

PYDNA INDICA, n. sp.

Female. Upperside greyish-ochreous: fore wing brown-speckled; a discal transverse curved oblique series of brown points, one on each vein, and a transverse subbasal indistinct series: hind wing grey with indistinct broad ochreous streaks along the veins. Underside paler, whitish on hind wing: fore wing suffused with pale ochreous-brown at base.

Expanse 2½ inches.

Hab. Calcutta. In coll. Dr. Staudinger.

Genus LOPHOPTERYX, *Stephens.*

Lophopteryx, Steph. Brit. Ent. ii. p. 26.

LOPHOPTERYX ARGENTATA, n. sp.

Male. Fore wing silvery-grey, suffused with pale brown at base and chestnut brown on lobe of hind margin, at angle, and on cilia; crossed by two medial black zigzag lines, between which are two black streaks within the cell; some slight black streaks at the base and before the apex, and a row on exterior margin: hind wing and abdomen grey-brown. Underside grey-brown. Head, front of thorax, and antennæ chestnut-brown; sides of thorax silvery-grey.

Expanse $1\frac{5}{12}$ inch.

Hab. Darjiling. In coll. Dr. Staudinger.

LOPHOPTERYX FLAVISTIGMA, n. sp.

Male. Fore wing dark umber-brown, interspersed with hoary scales; a basal, subbasal, discal, and submarginal narrow zigzag yellowish transverse bands, each with black-bordered lines; a narrow yellowish reniform mark centred with a black streak: hind wing pale umber-brown, with a short yellow black-bordered hoary-margined streak above anal angle. Underside uniform umber-brown. Head and thorax hoary chestnut-brown; abdomen and antennæ pale umber-brown.

Expanse 1⅞ inch.

Hab. Darjiling. In coll. Dr. Staudinger.

LOPHOPTERYX FERRUGINOSA, n. sp.

Male. Fore wing light fuliginous-brown, interspersed with short yellowish longitudinal streaks; crossed by a subbasal and discal row of dark brown lunules, each row curving and nearly touching before reaching the hind margin, and terminating on the margin in a narrow yellow streak; two outer marginal rows of dark brown pale-centred dentate marks.

K 2

Hind wing and abdomen pale brown.   Underside pale brown ; hind wing with narrow darker brown transverse line.   Head and thorax light reddish-brown.

Expanse 1⅞ inch.

Hab. Darjiling.   In coll. Dr. Staudinger.

Genus RAMESA, *Walker.*

Ramesa, Walker, Catal. Lep. Het. B. M. v. p. 1016.

RAMESA APICALIS, n. sp.   (Plate III. fig. 12.)

Male. Upperside—fore wing pale chestnut-brown, with longitudinal black streaks from the base, and a straight transverse apical pale streak with white inner border and black outer margin ; streaks black-spotted between the veins : hind wing uniform brown.   Underside brown ; apex of fore wing and hind wing pale brown.   Thorax and head hoary blackish-brown : abdomen brown ; antennæ chestnut-brown ; legs brown, black-speckled.

Expanse 1⁹⁄₁₀ inch.

Hab. Darjiling.   In coll. Dr. Staudinger.

Genus HOPLITIS, *Hübner.*

Hoplitis, Hübn., Verz. bek. Schmett. p. 147.

HOPLITIS STRIGATA, n. sp.   (Plate III. fig. 13.)

Male. Upperside—fore wing longitudinally streaked with pale and dark chestnut-brown ; a basal patch and short oblique costal streaks before the apex grey ; from these a dark curved streak proceeds to the apex ; marginal lunular line brown : hind wing pinkish white, anterior border brownish ; a black grey-speckled spot at anal angle.   Underside—fore wing brown, whitish-streaked on hind margin : hind wing as above.   Thorax grey, front of head and streak down thorax blackish : abdomen and legs brown ; antennæ chestnut-brown, shaft grey.

Expanse 2½ inches.

Hab. Canara (*Ward*).   In coll. Dr. Staudinger and F. Moore.

Genus NOTODONTA, *Hübner.*

Notodonta, Hübn., Verz. bek. Schmett. p. 146.

NOTODONTA SIKKIMA, n. sp.

Male. Upperside—fore wing pale grey-brown ; a dark umber-brown broad basal band with wavy whitish border ; a patch on costa near apex ; a rounded spot midway between it and a geminated spot at posterior angle ; cilia white, alternated by a marginal row of brown spots ; reniform mark greyish white, brown-centred : hind wing greyish-white, with a brown submarginal curved band and marginal lunules ; abdominal margin broadly pale ochrey-brown. Underside greyish-white, base of wings pale ochrey-brown ; marginal spots on fore wing pale brown. Thorax blackish-brown ; head, collar, palpi, antennæ, abdomen, and legs pale ochrey-brown.

Expanse 2⅝ inches.

Hab. Darjiling. In coll. Dr. Staudinger.

Genus PHEOSIA, *Hübner.*

Pheosia, Hübn., Verz. bek. Schmett. p. 145.

PHEOSIA COSTALIS, n. sp.

Male and female. Upperside—fore wing with a broad longitudinal costal buff-white band, slightly brown-streaked along the veins ; hind part of wing and small patch at apex dark chestnut-brown ; a white trifid discocellular mark and a short silver-white streak on vein below : hind wing brownish fawn-colour ; cilia edged with white. Underside pale brownish fawn-colour ; base and apex streaked with pale buff. Head, palpi, thorax, and legs chestnut-brown ; a projecting buff-white crest on middle of thorax ; antennæ brown, shaft buff-white.

Expanse 2 inches.

Hab. Darjiling. In coll. Dr. Staudinger and F. Moore.

PHEOSIA ALBIFASCIA, n. sp.

Male. Upperside—fore wing chestnut-brown, with a broad longitudinal whitish fascia from base of costa to apex ; veins blackish ; a double brown-bordered white oblique discal and submarginal line ; exterior and posterior border dusky brown ; cilia and a pale-bordered marginal line brown : hind wing pale ochrey-yellow, darker on abdominal margin ; exterior margin and cilia chestnut-brown. Underside pale ochrey-yellow : fore wing washed with chestnut-brown ; costal border and outer margins of both wings dark chestnut-brown. Thorax, head, antennæ, and legs chestnut-brown ; tegulæ black-streaked ; abdomen pale ochrey-yellow.

Expanse 2⅓ inches.

Hab. Darjiling. In coll. Dr. Staudinger.

PHEOSIA SIKKIMA, n. sp.

Male and female. Upperside—fore wing fulvous-brown, darker fuliginous-brown from base to below apex, a paler longitudinal fascia traversed by darker veins on costa from its middle to the apex ; hind margin also paler and traversed by pale longitudinal streaks which are slightly whitish in male ; two indistinct medial zigzag darker lines, which are pale-bordered in the male : hind wing and abdomen pale brown ; thorax and crest chestnut-brown, a broad white band on front of thorax above. Underside and legs pale brown ; tarsi in female with pale whitish bands.

Expanse, ♂ 1¾ inch, ♀ 2¼ inches.

Hab. Darjiling. In coll. Dr. Staudinger.

RACHIA, nov. gen.

Male. Fore wing very long, narrow, subfusiform ; costa arched ; apex acute ; exterior margin very oblique, waved ; hind margin slightly convex at the base ; subcostal vein five-branched, first branch arising before end of the cell, third from the second at nearly half its length from end of the cell, forked, fourth from below the third at one third from its end, fifth from the second at half distance between third and end of the cell ; discocellulars slightly bent in the middle, whence springs the radial ; median vein three-branched, middle branch from before end of the cell ; submedian curved towards the hind margin. Hind wing long, extending beyond angle of fore wing, costa straight, apex slightly rounded, exterior margin waved, abdominal margin short, densely hairy ; costal vein extending to apex ; subcostal two-branched at half the length beyond the cell ; discocellulars, radial, and median as above ; a submedian and an internal vein. Body stout, abdomen extending beyond hind wing ; thorax thickly pilose ; antennæ long, plumose, very broadly bipectinate ; palpi small, short, not extending beyond the head, pilose and adpressed beneath, tip blunt : legs thickly pilose.

RACHIA PLUMOSA, n. sp.

Male. Upperside—fore wing dusky ferruginous-brown, with pale ochreous short longitudinal interspaces between the veins, the spaces streaked with dark chestnut-brown ; an indistinct transverse subbasal and a discal black zigzag line ; a marginal black denticulated line : hind wing duller brown, with two very indistinct transverse dusky pale-bordered submarginal sinuous lines. Antennæ black ; thorax dark brown, black in front ; abdomen paler. Underside paler, duller-coloured : fore wing unmarked : hind wing pale cinereous-brown along abdominal margin ; crossed by a discal brown band ; a brown lunule at end of the cell.

Expanse 2¾ inches.

Hab. Darjiling. In coll. Dr. Staudinger and F. Moore.

Genus PARAVETTA, *Moore.*

Paravetta, Moore, P. Z. S. 1865, p. 81 *l.*

PARAVETTA SIKKIMA, n. sp.

Male. Upperside dark brownish fawn-colour : fore wing with a wide transverse medial pale-margined outwardly bent ochreous-brown band, the upper part very broad, the lower part narrow, and inwardly oblique ; a narrow marginal line, the edge of cilia on both wings, and a short submarginal wavy line on hind wings ochreous. Underside slightly paler, with an ochreous bent submarginal and a pale marginal line.

Expanse $1\frac{2}{10}$ inch.

Hab. Darjiling. In coll. Dr. Staudinger and F. Moore.

Differs from *P. discinota* in its smaller size, darker colour on both wings, and in the absence of the prominent dark discal spots.

DANAKA, nov. gen.

Male. Fore wing long, narrow, costa nearly straight, apex acute, exterior margin recurved, posterior margin recurved, angle acute; first subcostal branch before end of the cell, second and third from the end, fourth from below the third at half its length, fifth from end of the cell; discocellulars obliquely curved ; median with the two upper branches confluent at their base and with the third from lower end of the cell ; a submedian and internal vein. Hind wing rather long, extending slightly beyond posterior angle of fore wing, apex rounded, exterior margin waved, abdominal margin short : subcostal three-branched : discocellulars very oblique, straight ; other veins as in fore wing. Body moderately stout ; abdomen slightly longer than hind wing, with lateral segmental and anal tufts ; antennæ serrate, teeth finely pectinate ; palpi pilose, conical ; femora and tibiæ thickly pilose.

DANAKA PYRALIFORMIS, n. sp. (Plate III. fig. 10.)

Male. Upperside—fore wing brownish-ochreous, crossed by a medial irregular-shaped violet-grey band, the lower part of base and a short streak within the cell also of the same colour: hind wing cinereous-brown, outer border ochreous-tinged, with an indistinct curved discal transverse darker pale outer-bordered band ; cilia edged with cinereous-white. Body ochreous ; antennæ and legs ochreous-brown. Underside paler, basal two thirds of both wings dusky black.

Expanse $1\frac{2}{8}$ inch.

Hab. Darjiling. In coll. Dr. Staudinger.

Genus BEARA, *Walker*.

Beara, Walker, Catal. Lep. Het. B. M. Suppl. p. 1703.

BEARA CASTANEA, n. sp. (Plate III. fig. 9.)

Male. Fore wing chestnut-brown, crossed by a subbasal, a discal, and a submarginal indistinct blackish sinuous band ; a blackish spot at lower end of wing, and large curved spot at end of the cell : both wings with a marginal row of dots : hind wing yellowish, tinted along outer border with very pale chestnut-brown. Thorax chestnut-brown, white beneath ; abdomen greyish-brown, beneath white ; legs chestnut-brown above, whitish beneath; palpi fringed with white at base; antennæ brown. Underside pale chestnut; hind borders of both wings yellow; a marginal dotted line indistinct.

Expanse 1¾ inch.

Hab. Darjiling. In coll. Dr. Staudinger.

This species has much the appearance of *B. dichromella*, Walker, but differs in being twice its size, and in the fore wing being devoid of the black speckles that are scattered throughout that wing; the hind wing also is of a different colour.

Family PSYCHIDÆ.

KÓPHENE, nov. gen.

Wings covered with minute compact scales: fore wing broad, trigonal, costa slightly concave in the middle, apex acute, exterior margin oblique, long, nearly straight, angle rather acute, posterior margin short; costal vein curved upward from the base; subcostal vein five-branched, first branch arising at some distance before and second at end of the cell, third and fifth from upper angle of the cell, fourth from below the third at one third its length; cell very broad at the end; discocellulars of equal length, lower slightly bent near its end; two discoidal veinlets emitted within the cell, upper from middle of the disco-cellulars parallel with the radial, the other from angle of lower discocellular, both joined together in the middle of the cell; and before their junction the lower throws out a short branch or spur to the median vein nearly in a line with its lower branch; median vein four-branched, the two upper from end of the cell, third near its end, fourth at some distance off; submedian vein angled and curved upward at half its length, having a lower curved branch extending to the base, the lower branch emitting a short spur at half its length which is directed outward. Hind wing broad, short, anterior margin arched, apex and exterior margin and posterior angle convex; subcostal slender, joined to the costal by an oblique upward cross branch; discocellulars oblique, curved; two discoidal veinlets emitted within the cell and joined together near the middle; median vein four-branched, angled at the lower branch, the two upper branches starting from end of the cell, third near its end; a submedian and two internal veins. Body pilose; thorax clothed with long hairs; abdomen slender; antennæ short, bipectinate, plumose; legs small, slender, femora and tibiæ clothed with long hairs.

KÓPHENE CUPREA, n. sp.

Wings covered with minute compact dark cupreous-brown scales; abdominal margin pale cinereous-brown. Body above cinereous-brown, beneath dusky-brown.

Expanse ⅞ inch.

Hab. Calcutta. In coll. Dr. Staudinger and F. Moore.

Wings covered with minute compact scales; pale cupreous-brown, palest beneath. Fore wing rounded at the apex. Legs sparsely pilose; body pilose.

Expanse ⅝ inch.

Hab. Calcutta (June). In coll. Dr. Staudinger.

Family LIMACODIDÆ.

Genus SCOPELODES, *Westwood*.

Scopelodes, Westwood, Nat. Libr. xxxviii. p. 222.

SCOPELODES VULPINA, n. sp. (Plate III. fig. 22.)

Upperside—fore wing bright golden velvety-brown; cilia hoary cinereous: hind wing fawn-colour, veins lined with pale yellow; base of wing yellowish; cilia cinereous. Underside brownish fawn-colour, veins on both wings lined with yellow; base of hind wing broadly pale golden-yellow. Head, thorax, antennæ, and legs hoary brown; abdomen golden-yellow, with an indistinct blackish dorsal tufted streak, tip black; palpi golden brown, tip black.

Expanse 2 inches.

Hab. Darjiling. In coll. Dr. Staudinger.

Differs from *S. renosa* in the wings being much more produced at the apex, as well as in colour and size.

Genus NAROSA, *Walker*.

Narosa, Walker, Catal. Lep. Het. B. M. v. p. 1151.

NAROSA RUFOTESSELLATA, n. sp. (Plate III. fig. 24.)

Upperside—fore wing silvery-yellow, ornamented with short transverse crimson streaks between the veins, some of which are brown-speckled; a prominent brown-speckled spot at end of the cell: hind wing bright crimson. Cilia yellow. Underside red, yellowish at apex and on hind margin of fore wing. Thorax silvery-yellow; abdomen reddish; palpi, legs, and body beneath yellow; collar, thorax beneath, and sides of palpi pale red.

Expanse 1¾ inch.

Hab. Darjiling. In coll. Dr. Staudinger.

L

Genus CONTHEYLA, *Walker.*

Contheyla, Walker, Catal. Lep. Het. B. M. Suppl. p. 384.

CONTHEYLA THORACICA, n. sp.   (Plate III. fig. 7.)

Upperside dark cinereous-brown; cilia paler: fore wing thickly black-speckled, interspersed with a few prominent lilac-white scales; a small pale brown spot on costa before the apex, and a similar-coloured spot above middle of hind margin. Thorax greyish-white; collar and tegulæ fringed with black; abdomen brown, segments slightly fringed with black, the fringe beneath long and tufted; palpi, thorax, body beneath, and legs brown; joints of legs tufted and fringed with white.

Expanse $1\frac{2}{10}$ inch.

Hab. Darjiling. In coll. Dr. Staudinger.

Genus MONEMA, *Walker.*

Monema, Walker, Catal. Lep. Het. B. M. v. p. 1112.

MONEMA OCELLATA, n. sp.

Female. Upperside pale golden-testaceous: both wings with a narrow brown marginal line and cilial border; base of hind wing fuliginous: fore wing with a clearly defined narrow black submarginal line, which is slightly tortuous anteriorly; a large medial ocellated spot composed of a silvery-grey centre, a bright cupreous-brown outer border on its discal half, and a black-speckled border on its inner half; a few silvery-grey and black speckles ascending obliquely towards the apex. Body above adorned with long thickened black-pointed scattered hairs. Underside slightly paler; fore wing suffused black in the middle: hind wing with slightly suffused black streaks from base.

Expanse 2 inches.

Hab. Darjiling. In coll. Dr. Staudinger and F. Moore.

Genus MIRESA, *Walker.*

Miresa, Walker, Catal. Lep. Het. B. M. v. p. 1123.

MIRESA CUPREA, n. sp.   (Plate III. fig. 8.)

Upperside—fore wing dark glossy cupreous-brown, with two short oblique streaks below the cell and a submarginal transverse fascia of bright cupreous-brown: hind wing brown, with slight cupreous tinge along outer margin. Underside dark brown. Body dark grey-brown; antennæ, front of head, palpi, fore legs, and tarsi ochreous-brown.

Expanse $1\frac{1}{8}$ inch.

Hab. Darjiling. In coll. Dr. Staudinger.

Genus SETORA, *Walker.*

Setora, Walker, Catal. Lep. Het. B. M. v. p. 1069.

SETORA DIVERGENS, n. sp. (Plate III. fig. 23.)

Male ochreous-red, *female* ochreous-brown, very sparsely black-speckled : fore wing crossed by two narrow blackish bands, the outer band from costa near apex and nearly erect, the other starting from its upper end and curving to near base of hind margin. Underside uniformly coloured, without bands: hind wing very sparsely black-speckled. Palpi fringed with black at the tip.

Expanse 1¾ inch.

Hab. Darjiling. In coll. Dr. Staudinger.

Genus BELIPPA, *Walker.*

Belippa, Walker, Catal. Lep. Het. B. M. Suppl. p. 508.

BELIPPA APICATA, n. sp.

Fore wing dark olive-brown, varied with purplish-brown along the costa, and pale ochreous along outer border; a very indistinct black-speckled spot at end of the cell, and a black spot with white speckles at apex : hind wing dark brownish fawn-colour, with an indistinct darker maculated margin, and a black-speckled spot at apex and at anal angle; cilia brownish-cinereous. Underside dark brown, veins paler, spot at apex and angle as above. Body dark rufous-brown ; abdomen with very indistinct black dorsal fringe ; anal tuft black at tip ; tibiæ with a black streak.

Expanse 1¾ inch.

Hab. Darjiling. In coll. Dr. Staudinger.

Family LASIOCAMPIDÆ.

Genus GASTROPACHA, *Curtis.*

Gastropacha, Curtis, Brit. Ent. i. p. 24.

GASTROPACHA SIKKIMA, n. sp.

Male. Ferruginous: fore wing crossed by indistinct darker narrow wavy bands, the exterior band zigzag and more prominent; a longitudinal indistinct yellowish fascia along hind margin, another at the apex, and also an intermediate vinous-grey fascia, the latter also extending along the exterior margin : hind wing with the anterior portion yellowish and crossed by wavy indistinct narrow ferruginous bands. Thorax yellowish-ferruginous.

L 2

Underside more uniform ferruginous, with indistinct darker zigzag discal band on fore wing, and less distinct anterior sinuous band and suffused patch at angle on hind wing.

Female. Yellowish-ferruginous; wavy bands less distinct; outer borders greyish-ochreous. Pectinations of antennæ and palpi blackish.

Expanse, ♂ 2⅜ inches, ♀ 3¼ inches.

Hab. Darjiling. In coll. Dr. Staudinger and F. Moore.

GASTROPACHA SINUATA, n. sp.

Male. Upperside greyish-ferruginous: fore wing minutely grey-speckled, with indistinct subbasal and discal wavy darker bands: hind wing darker ferruginous, with indistinct discal wavy grey bands; exterior margin very sinuous. Underside—fore wing uniform ferruginous: hind wing greyish, with anterior ferruginous wavy streaks; margin of both wings blackish-ferruginous. Cilia grey. Head and antennæ greyish.

Expanse 2¼ inches.

Hab. Darjiling. In coll. Dr. Staudinger.

GASTROPACHA TORRIDA, n. sp. (Plate III. fig. 19.)

Male. Upperside brownish-ochreous; fore wing with a whitish spot at end of the cell, and a medial transverse recurved indistinct dusky-black narrow band: hind wing darker-coloured at the base; with a narrow medial and a lunular submarginal indistinct dusky-black band. Underside as above; antennæ black.

Expanse 1⅗ inch.

Hab. Darjiling. In coll. Dr. Staudinger.

Genus ODONESTIS, *Germar.*

Odonestis, Germ. Prod. p. 49 (1811).

ODONESTIS SIGNATA, n. sp.

Male and female. Upperside reddish fawn-colour, hind wing and abdomen paler: fore wing with an indistinct black curved subbasal, and a recurved discal sinuous line; an outer discal zigzag row of dusky metallized spots; an oblique ochreous-white elongated spot at end of the cell and a small white dot above it: hind wing with a very indistinct dusky transverse medial fascia. Underside duller-coloured; both wings with a dusky-black transverse medial fascia; elongated spot at the end of cell yellowish-white. Antennæ pale ferruginous.

Expanse, ♂ 1⅗, ♀ 2⅜ inches.

Hab. Darjiling. In colls. Dr. Staudinger and F. Moore.

Genus EUPTEROTE, *Hübner*.

Eupterote, Hübner, Verz. bek. Schmett. p. 187.

(DREATA (part.), *Walker*.)

EUPTEROTE NILGIRICA, n. sp.

Female. Dull ochreous ; both wings with a broad discal transverse dark brown-speckled band and an outer submarginal narrow sinuous band, the interspace with row of triangular marks ; the discal space before the band sparsely and indistinctly brown-speckled. Under-side as above.

Expanse 2⅜ inches.

Hab. Nilgiri Hills. In coll. Dr. Staudinger.

Allied to *E. canaraensis*, Moore, but may be distinguished from that species by the en-tire absence of the exterior discal lunular band.

Genus SPHINGOGNATHA, *Felder*.

Sphingognatha, Felder, Nov. Voy. pl. 94.

SPHINGOGNATHA KHASIANA, n. sp.

Male. Upperside pale greyish-ochreous ; both wings apparently with five oblique trans-verse broad undulating brown-speckled bands, the first band being outside the cell on fore wing, and crossing the hind wing near the base, the fifth band grey-speckled ; a contiguous outer discal parallel narrow brown line, even on the fore wing, undulating on hind wing, curving from apex to one-third from posterior angle, and crossing the hind wing to one-third above anal angle ; beyond this is a submarginal greyish-white lunular band with black den-tate spots between the veins ; a small vitreous spot at end of the cell.

Female. Brighter-coloured, marked as in the male. Palpi and fore legs dusky above.

Expanse, ♂ 3⅜, ♀ 3⅞ inches.

Hab. Khasia Hills. In colls. Dr. Staudinger and F. Moore.

This is a much smaller species than *S. pallida* (*Jana pallida*, Walker, Cat. Lep. Het. B. M. iv. p. 912), which has very narrow and less oblique bands, the outer discal band being further from the margin, and its upper end on the fore wing some distance from the apex.

Genus EUTRICHA, *Hübner (Steph.)*.

Eutricha, Steph. Brit. Ent. ii. p. 36.

EUTRICHA FLAVOSIGNATA, n. sp.   (Plate III. fig. 17.)

Male. Upperside dark chestnut-red ; hind wing duller-coloured : fore wing with a pro-minent small yellow cell-spot, a slight yellow transverse subbasal band, and a very indistinct pale discal band, the interspaces between the outer veins and cilia blackish. Underside uniformly duller red ; both wings with a darker discal transverse fascia. Cilia slightly white-edged.

Expanse 1⅜ inch.

Hab. Darjiling. In coll. Dr. Staudinger.

MAHANTA, nov. gen.

Fore wing long, narrow ; costa arched near its end ; apex falcate, outer margin oblique and convex posteriorly ; hind margin short and deeply excavated at base ; subcostal vein five-branched, first arising before end of the cell, second beyond the cell, third and fourth near the apex, fifth from below the third at half its length between its base and fourth branch ; cell short ; discocellulars long, bent inwards ; a recurrent vein extending to base of cell ; upper radial vein anastomosing with subcostal to beyond end of cell ; lower radial starting from middle of lower discocellular ; three median veins ; submedian and internal veins straight. Hind wing short, ovate ; subcostal three-branched ; cell short ; upper discocellular short, lower long and outwardly oblique ; recurrent veins oblique ; one radial and three median veins at equal distances apart at base ; a submedian and two internal veins. Body stout ; abdomen long, extending beyond hind wing ; thorax broad ; head small ; antennæ slender, serrated to tip ; palpi adpressed, stout, ascending ; legs short, densely pilose to tip of tarsi.

MAHANTA QUADRILINEA, n. sp. (Plate III. fig. 20.)

Male. Upperside bright pale cinnamon-brown ; fore wing with two medial transverse obliquely recurved narrow dark-brown lines, each bordered outwardly from the hind margin with a short silky cinnamon-grey fascia ; a similar fascia extends upward from posterior angle, and another near base of wing ; cilia edged with silky white : hind wing golden yellowish broadly along anterior margin ; cilia whitish at anal angle. Underside paler cinnamon-brown ; veins of both wings and hind margin of fore wing golden yellow. Middle of thorax and antennæ cinnamon-grey ; a short straight white streak across each tegula.

Expanse 2 inches.

Hab. Darjiling. In coll. Dr. Staudinger.

Genus METANASTRIA, *Hübner.*

Metanastria, Hübner, Verz. bek. Schmett. p. 186.

METANASTRIA MINOR, n. sp.

Male and female. Dull ferruginous : fore wing indistinctly grey-speckled, and with two medial obliquely transverse straight narrow pale-bordered darker bands. Thorax with a slightly raised medial crest. Underside uniform ferruginous; apex of fore wing slightly grey-speckled.

Expanse, ♂ $1\frac{5}{10}$, ♀ $1\frac{7}{10}$ inch.

Hab. Darjiling. In colls. Dr. Staudinger and F. Moore.

ARGUDA, nov. gen.

Male and female. Fore wing short; costa arched at the middle; apex slightly falcate and pointed; exterior margin short, even; posterior margin long, straight, extending one third beyond hind wing, angle rounded. Hind wing abbreviated anteriorly, elongated posteriorly; costa short, exterior margin long, even, and very convex; abdominal margin long; veins similar to those in *Metanastria hyrtaca.* Body moderately stout; abdomen not extending beyond hind wing; palpi stout, conical, pilose, porrect; antennæ broadly bipectinate; legs stout; femora and tibiæ pilose.

ARGUDA DECURTATA, n. sp.

Male and female. Upperside ferruginous: fore wing speckled with ochreous-grey; two medial transverse oblique narrow black pale-bordered bands, the inner band straight, the outer slightly recurved before the costa; a small black cell-spot; an indistinct discal series of dusky zigzag lunules; hind margin black-fringed: hind wing speckled with ochreous grey, and crossed by a short zigzag dusky streak along anterior border. Underside bright dark ferruginous; exterior borders and two indistinct discal fasciæ across fore wing, and three basal dusky-bordered greyish wavy bands across hind wing, the latter wing also with a discal dusky zigzag band. Tip of palpi and a streak on vertex black.

Expanse, ♂ $1\frac{7}{8}$, ♀ $2\frac{1}{8}$ inches.

Hab. Darjiling. In coll. Dr. Staudinger and F. Moore.

RADIICA, nov. gen.

Male. Fore wing short, costa slightly arched, apex truncated; exterior margin waved, angled below the apex; posterior margin long, straight. Hind wing short, produced hindward; costa abbreviated; exterior margin very convex, waved; abdominal margin long. Veins, body, antennæ, palpi, and legs as in *Arguda.*

RADIICA FLAVOVITTATA, n. sp.

Male. Upperside yellowish-ferruginous: fore wing with wavy transverse subbasal and straight oblique medial pale-bordered narrow dusky bands; a prominent black cell-spot, and discal zigzag series of yellow-bordered dusky lunules: hind wing with a somewhat broad yellow dusky-inner-bordered subbasal band, which is most prominent anteriorly. A dark ferruginous streak on middle of thorax terminating in a lower greyish-black spot; palpi dusky black. Underside—fore wing paler, with a dusky cell-spot, medial transverse band, and marginal streaks: hind wing grey-speckled, crossed by a dusky black subbasal sinuous band, indistinct discal wavy yellow-bordered streaks, and a marginal row of black cilial streaks.

Expanse 2¼ inches.

Hab. Nynee Tal and Dhurmsala, N.W. Himalaya. In coll. Dr. Staudinger and F. Moore.

CHATRA, nov. gen.

Male and female. Wings ample: fore wing elongated, oval; costa considerably arched; apex acutely convex; exterior margin oblique; hind margin short, convex at the angle. Hind wing broad, short, apex convex; exterior margin produced and very convex, abdominal margin short. Veins similar to those in *Oeona,* Walker (Cat. Lep. Het. B.M. vi. p. 1418). Body stout; palpi short, compact and adpressed, broad, porrect; legs pilose; antennæ bipectinate.

CHATRA GRISEA, n. sp.

Male. Dark dull greyish-ferruginous: fore wing densely covered with grey scales, crossed by two subbasal and two medial indistinctly darker wavy bands and an outer discal zigzag series of prominent black grey-bordered lunules; a white subbasal cell-spot: hind wing with a very indistinct darker medial transverse fascia. Underside dusky ferruginous; fore wing densely grey-speckled on outer border, and the hind wing along anterior and outer borders; both wings crossed by two indistinct darker medial bands, and fore wing with a discal maculated band.

Female brighter-coloured above and beneath; markings as in male.

Expanse, ♂ 3⅛, ♀ 4⅝ inches.

Hab. Cherra Punji (*Atkinson*). Khasia Hills (*Godwin-Austen*). In coll. Dr. Staudinger and F. Moore.

Genus LEBEDA, *Walker.*

Lebeda, Walker, Catal. Lep. Het. B. M. vi. p. 1453.

LEBEDA PLACIDA, n. sp.

Male and female. Upperside uniform pale ferruginous; fore wing with four indistinct darker ferruginous-brown medial transverse narrow bands, the two inner waved, the two outer sinuous; an outer discal series of small black-speckled spots; a prominent white cell-spot; outer border and interspace between the middle bands washed with grey. Underside paler: fore wing with three indistinct darker transverse discal fasciæ; hind wing with two similar fasciæ.

Expanse, ♂ 3½, ♀ 5 inches.

Hab. Darjiling. In coll. Dr. Staudinger and F. Moore.

May be distinguished from *L. ferruginea* and *L. ampla* by its longer and narrow wings, and uniform colour.

LEBEDA VULPINA, n. sp.

Male. Fore wing very dark bright ferruginous, crossed by two indistinct rather broad medial darker fawn-coloured bands, and a similar oblique basal band; an outer discal zigzag series of grey-speckled dusky spots; a small prominent white cell-spot: hind wing and abdomen dull paler ferruginous. Underside pale dull ferruginous; both wings crossed by very indistinct dusky medial bands. Thorax, head, palpi, and legs dark ferruginous; pectinations of antennæ dusky ferruginous.

Expanse 2⅞ inches.

Hab. Darjiling. In coll. Dr. Staudinger.

LEBEDA LINEATA, n. sp.

Male. Ferruginous: fore wing with a longitudinal black narrow straight line from base to near outer margin; crossed by basal and discal narrow blackish sinuous lines; an outer discal row of black spots, and fawn-coloured speckled streaks between the veins on outer margin. Underside slightly paler; fore wing with dusky-black subbasal medial bands, discal and marginal spots, and hind wing with medial curved dusky band. Antennæ dusky-ferruginous; palpi at tip and side, and legs above, black.

Expanse 2½ inches.

Hab. Darjiling. In coll. Dr. Staudinger and F. Moore.

LEBEDA FULGENS, n. sp.

Male. Fore wing dark bright ferruginous, somewhat yellowish at the base of hind margin and at the posterior angle, and washed with greyish-ferruginous on exterior border; crossed by apparently six darker oblique bands, which are wavy anteriorly and sinuous posteriorly, the outer discal one formed of somewhat zigzag dusky spots; a prominent small white triangular cell-spot: hind wing and body dull ferruginous; anal tuft dusky. Underside uniform ferruginous; veins paler. Head, palpi, antennæ, and legs dark dull ferruginous.

Expanse 2¾ inches.

Hab. Darjiling. In coll. Dr. Staudinger and F. Moore.

The fore wings in this species are narrower than in the other Indian species of the genus.

M

Genus TRICHIURA, *Stephens.*

Trichiura, Steph. Brit. Ent. ii. p. 42.

TRICHIURA KHASIANA, n. sp. (Plate III. fig. 21.)

Upperside dark vinous-brown: fore wing with ferruginous-brown veins and black longitudinal interspaces speckled with grey; crossed by three black pale-bordered lines—the first subbasal, second discal, and third marginal, the middle band wavy, the outer zigzag and ochreous-speckled: hind wing with very indistinct darker subbasal band. Underside uniform brown: fore wing with a few grey speckles at the apex; hind wing with indistinct subbasal darker band. Antennæ, head, palpi, thorax, and legs dark brown; abdomen paler. Expanse 1⅗ inch.

Hab. Khasia Hills. In coll. Dr. Staudinger.

Genus MUSTILIA, *Walker.*

Mustilia, Walker, Catal. Lep. Het. B. M. Suppl. p. 580.

MUSTILIA CASTANEA, n. sp.

Male. Fore wing dark chestnut-red; an oblique blackish line curving from apex to hind margin near the angle; three broad transverse bands formed of greyish chalybeate scales, retracting inward to the costa; a small black spot at end of the cell: hind wing dark ferruginous, palest at base; three short discal indistinct narrow dusky lines curved from above the anal angle. Underside ferruginous, palest at base: fore wing with short blackish oblique line from apex; a spot at end of cell: hind wing grey at anal angle; two transverse discal brown lines; a black spot at end of the cell. Body dark chestnut-red, paler beneath; thorax washed with chalybeate; antennæ pale ferruginous.

Expanse 2⅛ inches.

Hab. Darjiling. In coll. Dr. Staudinger.

Differs from *M. falcipennis*, Walk. (which is also from Darjiling), in being of a darker colour both above and beneath. The hind wing has the outer margin convex, and is not produced before the anal angle; and there are three transverse lines which are curved, two widely separated lines only being present in *M. falcipennis.* On the underside the markings in *M. castanea* are very indistinct.

MUSTILIA HEPATICA, n. sp. (Plate III. fig. 18.)

Male. Upperside ochreous-brown, greyish-ochreous at base of wings and on thorax: fore wing with an oblique narrow wavy blackish line from apex to hind margin one third from the angle, and a transverse subbasal zigzag line; a large oblique blackish spot at end of the cell; two brown streaks on costa before the apex; three confluent pale ochreous spots above posterior angle: hind wing with a transverse medial curved wavy blackish line from middle of costa to above anal angle ; three confluent pale ochreous spots above anal angle. Underside pale chestnut-brown, palest on hind wing ; a blackish transverse discal pale outer-bordered line from apex of fore wing to above anal angle ; a black spot at end of each cell. Shaft of antennæ whitish ; fore legs brown above, ochreous beneath ; mid and hind tarsi with brown bands.

Expanse 2⅘ inches.

Hab. Darjiling.   In coll. Dr. Staudinger.

Family BOMBYCIDÆ.

Genus OCINARA, *Walker*.

Ocinara, Walker, Cat. Lep. Het. B. M. vii. p. 1768.

OCINARA DIAPHANA, n. sp.

Male. White ; diaphanous ; both wings crossed by three paler diaphanous bands. Abdomen greyish-brown above ; antennæ brown, shaft white : sides of head, palpi, pectus, and fore legs above blackish.

Expanse 1⅖ inch.

Hab. Khasia Hills.   In coll. Dr. Staudinger.

Differs from *O. lactea*, Hutton, Trans. Ent. Soc. 1865, p. 326, from the N.W. Himalayas, in the absence of the black markings on the fore wing and on the abdominal margin.

Family DREPANULIDÆ.

Genus DREPANA, *Schrank*.

DREPANA ALBONOTATA, n. sp.

Male. Upperside yellow, washed with silvery yellow at the base ; female pale silvery yellow : fore wing with two transverse discal very indistinct confluent lunular suffused brown lines, enclosing a small pure white spot at end of the cell ; a submarginal series of black and silvery-speckled spots, blotched with brown below the apex : hind wing with a small brown-bordered white spot at end of the cell, a very indistinct brown lunular discal line, and marginal row of small brown silvery-speckled spots.   Underside paler, without markings.

Expanse,   1⅕, ♀ 1⅔ inch.

Hab. Mount Parisnath, Behar.   In colls. Dr. Staudinger and F. Moore.

Drepana flava, n. sp.   (Plate II. fig. 7.)

Female. Upperside yellow: fore wing with an indistinct subbasal transverse lunular brown line with the points inward, a submarginal double lunular line with the points outward, the interspace below the apex showing a silvery-white spot; an indistinct small brown spot at end of the cell: hind wing with three very indistinct brown outer lunular lines; a prominent silvery-speckled spot at lower end of the cell, and a smaller spot at its upper end. Underside yellow: fore wing with prominent oblique brown band from apex to near middle of hind margin; cell-spots indistinct. Antennæ and legs ochreous-brown.

Expanse $2\frac{1}{8}$ inches.

Hab. Darjiling. In coll. Dr. Staudinger.

Drepana pallida, n. sp.

Male and female. Upperside pale ochreous-grey, almost diaphanous. Both wings with an upper and lower blackish-speckled spot at end of the cell, most prominent on the fore wing; a pale ochreous-brown pale-bordered discal narrow duplex band, straight and bent before the apex on fore wing, and lunular on hind wing; both wings with subbasal transverse indistinct brown lunular lines, a submarginal lunular line with darker dentate points, and a narrow brown marginal line. Underside paler, oblique bands and cell-spots on fore wing indistinct. Antennæ brown.

Expanse $1\frac{5}{8}$ inch.

Hab. Darjiling. In colls. Dr. Staudinger and F. Moore.

Drepana bioccularis, n. sp.   (Plate II. fig. 9.)

Male. Upperside pale ochreous-brown, palest on upper part of hind wing, numerously speckled with silvery scales; an indistinct darker double line crossing from apical angle to middle of abdominal margin; a submarginal series of small indistinct black-speckled spots: fore wing with a large and prominent oval black pale-centred spot at end of the cell: cilia blackish and silver-edged. Underside paler, without markings.

Expanse $1\frac{7}{16}$ inch.

Hab. Darjiling. In coll. Dr. Staudinger.

Drepana postica, n. sp.   (Plate II. fig. 8.)

Male. Upperside—fore wing pale ochreous-brown, suffused with purple, the base and posterior angle yellowish; an oblique transverse nearly straight narrow brown discal line, and indistinct subbasal line; a submarginal series of small black spots, the spots below the apex grey-speckled: hind wing pale yellow, with narrow brown subbasal and discal transverse line; both wings with a pure white small spot at end of the cell. Underside brighter-coloured; oblique discal band and apical spots on fore wing distinct. Body and antennæ yellowish.

Expanse $1\frac{3}{8}$ inch.

Hab. Darjiling. In coll. Dr. Staudinger.

DREPANA VINACEA, n. sp.

Male. Upperside ochreous-brown, suffused with purple; a duplex narrow blackish band indistinctly bent to the costa before the apex and crossing to middle of abdominal margin; a black streak from apex to below angle of the band; both wings with indistinct transverse subbasal and submarginal wavy darker lines, and two black-speckled grey-bordered spots at end of the cell: cilia brown. Underside ochreous-yellow, with indistinct spots at end of the cell, and fore wing with dusky-brown fascia from apex. Antennæ and fore legs above blackish.

Expanse $1\frac{1}{4}$ inch.

Hab. Darjiling. In colls. Dr. Staudinger and F. Moore.

Genus ORETA, *Walker.*

Oreta, Walker, Catal. Lep. Het. B. M. v. p. 1166.

ORETA SANGUINEA, n. sp.

Upperside—fore wing with the basal two thirds and outer border deep red, the former traversed by short transverse yellow and brown strigæ, the latter partly covered with grey streaks; the outer discal space yellow: hind wing with the basal portion and a patch at apex red; both wings with two submarginal indistinct series of black speckles. Underside of both wings as above. Body and legs deep red; abdomen yellow at sides; antennæ pale red.

Expanse $1\frac{7}{8}$ inch.

Hab. Darjiling. In colls. Dr. Staudinger and F. Moore.

Family COSSIDÆ.

Genus COSSUS, *Fabr.*

Cossus, Fabr. Ent. Syst. iii. 2. 1.

COSSUS TESSELLATUS, n. sp.

Male. Upperside—fore wing greyish-ochreous, crossed by several ochrey-brown bands, darkest below the cell and on a conspicuous spot at end of the cell, the outer bands alternately lunular and maculated, those at the base outwardly oblique: hind wing pale greyish-ochreous, sparsely crossed by brown strigæ. Cilia alternately pale and dark cinereous.

Underside paler; markings as above but less apparent. Body greyish-cinereous; antennæ, legs, and anal tuft pale ochrey-brown.

Expanse 1⅜ inch.

Hab. Calcutta. In coll. Dr. Staudinger and F. Moore.

Near *C. quadrinotatus*, Walk.

COSSUS CASHMIRENSIS, n. sp.

Male. Fore wings short, very broad, truncate, outer margin short and nearly straight, hind margin deeply angled near the base. Upperside—fore wing cinereous-grey, darkest at the base, crossed by several narrow outwardly-oblique pale greyish-black irregular wavy lines, and two blacker short oblique lines across the middle to some oblique wavy lines before the apex : hind wing fuliginous-brown. Underside dark-brown, reticulations and streaks on costa black. Body cinereous-grey, palest beneath : antennæ brown ; palpi, frontal tuft, and legs blackish-brown.

Expanse 1¾ inch.

Hab. Tawi, Kashmir. In coll. Dr. Staudinger.

Genus ZENZERA, *Latreille.*

Zenzera, Latr. Hist. Nat. xiv. p. 175.

ZENZERA PARDICOLOR, n. sp.

Male. Pure white, veins margined with pale brown : fore wing with a prominent series of black spots between the veins, the spots largest at base and end of the cell: hind wing with black confluent spots between the veins from middle of the base to apex; the anterior margin white, anal margin broadly pale brown. Thorax spotted with black; antennæ and frontal tuft light brown ; legs white banded with black. Underside as above.

Expanse 2 inches.

Hab. Darjiling. In coll. Dr. Staudinger.

ZENZERA STIGMATICA, n. sp.

Male. Upperside pale golden-yellow : fore wing with a large prominent curved geminate black and red spot below the cell ; some confluent brown reticulations along the costa, a clouded series obliquely from the apex to the spot, and a paler series on the outer margin : hind wing glossy at the base ; without markings. Underside duller-coloured ; reticulations on fore wing and upper portion of spot fuliginous-brown ; lower or red portion of the spot obsolete. Palpi and frontal tuft black ; antennæ brown ; legs at the joint and tarsi with black bands.

Expanse 1¹⁰⁄₁₂ inch.

Hab. Darjiling. In coll. Dr. Staudinger.

ZENZERA ALBOFASCIATA, n. sp.

Male. Upperside purplish cinereous-white: fore wing with a broad unmarked fascia from base to apex, costal border crossed with short black confluent strigæ, lower part of wing from base below the cell to outer margin below the apex cinereous fawn-colour, and crossed by numerous narrow black strigæ which are reticulated on outer margin: hind wing with the lower half pale cinereous, crossed with delicate cinereous-brown strigæ. Cilia pale cinereous. Underside cinereous-brown, strigæ less prominent, both wings with a pale fascia. Front of head brown; vertex and antennæ blackish; collar on thorax above pale ochreous; hind part of thorax and abdomen dark cinereous-brown; legs black.

Expanse 2¾ inches.

Hab. Darjiling. In coll. Dr. Staudinger.

Allied to *Z. asylas,* Cram., pl. 137. f. c, from the Cape.

Genus PHRAGMATŒCIA, *Newman.*

Phragmatœcia, Newman. (Walker, Catal. vii. 1542.)

PHRAGMATŒCIA SACCHARUM, n. sp.

Male. Greyish-ochreous: fore wing brownish-ochreous on the upper half; a small indistinct brown spot above submedian vein, and a row of spots on the cilia. Underside paler. Thorax, palpi, and fore legs brownish-ochreous; antennæ brown, shaft ochreous.

Expanse 1⅜ inch.

Hab. Darjiling. In coll. Dr. Staudinger.

PHRAGMATŒCIA MINOR, n. sp.

Male. Fore wing pale ochreous-brown, darkest along costa, indistinctly reticulated with narrow brown lines: hind wing pale greyish-ochreous, veins brown externally. Underside— fore wing uniform brown, not reticulated: hind wing as above. Body, palpi, and legs ochreous-brown; antennæ brown, shaft ochreous.

Expanse 1 inch.

Hab. Silhet. In coll. Dr. Staudinger.

" Inhabits the Jheels: the moth appearing in September."—*Atkinson,* MS. note.

Family HEPIALIDÆ.

Genus HEPIALUS, *Fabr.*

Hepialus, Fabr. Syst. Ent. p. 589.

HEPIALUS UMBRINUS, n. sp.

Semidiaphanous, pale umber-brown with a vinous tint externally, fringe at base of wings fulvous-yellow; both wings sparsely irrorated externally with dark brown scales; costal margins and cilia dark purple brown, with a darker row of lunular spots on outer margin of fore wing. Underside as above. Body, legs, palpi, and antennæ dark umber-brown; base of thorax with fulvous-yellow fringe.

Expanse 2 inches.

Hab. Darjiling. In coll. Dr. Staudinger.

DESCRIPTIONS

OF

INDIAN LEPIDOPTERA HETEROCERA

FROM THE

COLLECTION OF THE LATE MR. W. S. ATKINSON.

BY

FREDERIC MOORE, F.Z.S. ETC.,
LATE ASSISTANT CURATOR, INDIA MUSEUM, LONDON.

—   —

Family CYMATOPHORIDÆ.

Genus THYATIRA, *Ochs.*

THYATIRA DECORATA.

Thyatira decorata, Moore, P. Z. S. 1881, p. 328, pl. xxxvii. fig. 1.

Male and female. Fore wing dark brown; with a peach-coloured basal trilobed spot, a large oval spot within the cell, a smaller spot at its end, an oblique irregular-shaped costal spot before, and a similar spot at the apex, a large deep-coloured spot at the posterior angle, and two small marginal spots above it, followed by lunules to the apical spot; an elongated spot on middle of the hind margin; some indistinct pale-bordered black sinuous lines crossing the wing between the spots: hind wing pale brown, with a small marginal pale yellowish patch near anal angle. Body pale ochreous-brown; palpi and legs with dark-brown bands.

Expanse, ♂ $1\frac{5}{8}$, ♀ $1\frac{6}{8}$ inch.

Hab. Darjiling. In coll. Dr. Staudinger.

PART II.—*Nov.* 28, 1881.   B

Genus HABROSYNE.

Habrosyne, Hübner, Verz. bek. Schmett. p. 236 (1818–25).
Gonophora, Bruand (1849)*.

HABROSYNE PLAGIOSA, n. sp.

Male and female. Pale ferruginous indistinctly marked with a few pale strigæ; crossed by a pinkish-grey, narrow, curved antemedial and an outer marginal band, a similar-coloured line curving from apex to end of the cell, and a short white tufted streak extending upward from the base; a double sinuous transverse discal brown line, and a discocellular lunule: hind wing and abdomen pale ochreous-brown. Thorax, head, and legs pale ferruginous.

Expanse 2 to 2⅛ inches.

Hab. Darjiling. In coll. Dr. Staudinger and F. Moore.

HABROSYNE ARMATA, n. sp.

Habrosyne armata, Moore, ' Aid,' pt. . pl. .

Male. Fore wing dark ferruginous-brown ; a narrow pink band curving from near base to posterior end of a discal blackish double zigzag line ; a pink streak curving inward from before the apex and a submarginal line ; a discocellular lunule, and two narrow outer marginal lines ; a short white curved tufted streak at base, above which are two straight streaks at base of the cell : hind wing and abdomen brown. Thorax, head, and palpi ferruginous-brown ; legs paler.

Expanse 1⁷⁄₁₀ inch.

Hab. Khasia Hills (October). In coll. Dr. Staudinger.

HABROSYNE SANGUINEA, n. sp.

Male and female. Fore wing very dark purple-red ; the basal third, a spot along posterior margin to near the angle, and some curved streaks on costa pale ferruginous, irregularly bordered with white ; two contiguous discal double zigzag dark-brown pale-bordered lines ; a discocellular lunule, a straight submarginal and a sinuous marginal pale line ; a basal white-tufted streak : hind wing and abdomen dark brown. Thorax and head dark ferruginous ; palpi and legs paler.

Expanse 1⁷⁄₁₀ inch.

Hab. Darjiling. In coll. Dr. Staudinger and F. Moore.

Genus RISOBA, *Moore.*

Risoba, Moore, P. Z. S. 1881, p. 328.

Allied to *Thyatira.* Fore wing somewhat shorter ; costal margin arched at apex, angle acute, exterior margin less oblique ; second subcostal branch trifurcate, its lowest fork (or

---

* Used in 1834 for Coleoptera.

fourth branch) near the apex, fifth branch curved upward from end of the cell and touching the third near its base; upper discocellular obliquely concave, slightly bent at its lower end; upper radial straight from end of the cell, lower radial and upper median branch from angles close above end of the cell: hind wing very convex exteriorly; costal and subcostal veins joined together at their base, two subcostal branches from end of the cell, radial and upper median branch from end of the cell, middle median branch from angle before end of the cell. Body slender; thorax broad, crested in front; antennæ long, minutely pectinated in male.

Type *R. repugnans* (*Thyatira repugnans*, Walk.).

RISOBA OBSTRUCTA.

Risoba obstructa, Moore, P. Z. S. 1881, p. 328.
Bolina obstructa, Walker, MS. Brit. Mus. Cabinet.

Male and female. Fore wing greyish-brown, indistinctly black-speckled; a white basal oblique slender band, which is slightly brownish hindward and has a black sinuous border; a discal greyish-white fascia bordered externally by an irregular black line; a black dot within the cell, and a circle enclosing another dot at its end; exterior border traversed by a broad blackish whitish-bordered indistinct band terminating at the apex in blacker streaks; a marginal row of black lunules bordered inwardly with white: hind wing pale cinereous-brown, with pale-brown marginal band.

Expanse, ♂ 1, ♀ 1⅔ inch.

Hab. Ceylon (*Mackwood*), Calcutta (*Atkinson*). In coll. F. Moore and Dr. Staudinger.

RISOBA BASALIS, n. sp.

Male and female. Fore wing pale brownish-ochreous, greyish along the costa and below the apex; a whitish basal oblique band, bordered outwardly by dark brown; disk crossed by an irregular indistinct black sinuous pale-bordered line terminating hindward in white; interior of the wing crossed by an indistinct black sinuous fascia; apical patch dark brown and black, streaked with whitish tips at costal end and below angle; a slight marginal pale-bordered black lunular line: hind wing cinereous-white, with broad brown marginal band, and a discal spot. Body cinereous-brown, white-speckled; legs brown above.

Expanse, ♂ 1¼₀, ♀ 1¼₀ inch.

Hab. Cherra (*Austen*); Darjiling (*Atkinson*). In coll. F. Moore and Dr. Staudinger.

RISOBA VIALIS, n. sp. (Plate IV. fig. 1.)

Male. Fore wing ochreous-brown, greyish at the apex, indistinctly crossed by sinuous black lines; with a white outwardly oblique band from end of the costa to one third of posterior margin, and a postmedial oblique narrow band, these forming a broad letter **W** across both the wings, the former brown-speckled, the latter bordered by a sinuous black line; two confluent and a lower narrow black streak at the apex, bordered inwardly by a curved streak; a submarginal series of narrow sinuous white lunules; a black circular mark

B 2

enclosing a dot at end of the cell: hind wing greyish-white, with broad greyish-brown marginal band, and a slight spot at end of the cell. Body greyish-brown; tegulæ white-speckled; front of thorax, palpi, tibiæ and tarsi above, brown.

Expanse 1⅔ inch.

Hab. Darjiling. In coll. Dr. Staudinger.

RISODA CONFLUENS, n. sp.

Female. Fore wing very dark brown; a very large basal pale brownish-ochreous oblique-bordered patch occupying nearly half the wing; a contiguous smaller irregular spot extending to posterior angle; a marginal series of four very small spots, and a large sinuous-bordered apical spot: hind wing and body brownish-ochreous. Front of head, palpi and legs above dark brown.

Expanse 1$\frac{7}{10}$ inch.

Hab. Darjiling. In coll. Dr. Staudinger and F. Moore.

Genus KERALA, *Moore.*

Kerala, Moore, P. Z. S. 1881, p. 329.

Fore wing long, narrow; costa slightly arched near the base, apex acute, exterior margin slightly oblique and excavated near posterior angle; posterior margin somewhat recurved; second branch of subcostal trifurcate near its end, fifth bifurcate, curving upward and touching the second at one fourth from its base, the fork (or upper radial) emitted from beyond end of the cell; discocellular angled at its lower end, lower radial and two upper median branches from the angles at end of the cell. Hind wing moderately broad, triangular; costal and subcostal veins joined together at their base, subcostal bifurcate at half its length beyond the cell; discocellular angled inward at its middle, and outward near its lower end, the radial from lower angle; two upper median branches on a foot-stalk beyond the cell. Body slender, abdomen smooth; antennæ slender, setose; palpi slender, squamose, apex short and conical; legs squamose.

KERALA PUNCTILINEATA.

Kerala punctilineata, Moore, P. Z. S. 1881, p. 330.

Male. Yellowish or greyish-ochreous, costal area reddish-ochreous; all the veins to beyond the middle with alternate black and yellow or grey spots, those on the subcostal branches forming streaks; a transverse black dentated band with white inner border; a pale-yellow orbicular and reniform spot, the latter with black outer border: hind wing and abdomen pale cinereous-brown, discal area ochreous. Thorax, head, palpi, and legs yellowish-ochreous; legs black-banded.

Female reddish-ochreous, marked as in male: hind wing uniformly pale cinereous-brown.

Expanse, ♂ 1⅝, ♀ 1⅝ inch.

Hab. Darjiling. In coll. Dr. Staudinger and British Museum.

KERALA MULTIPUNCTATA, n. sp. (Plate IV. fig. 4.)

Male. Fore wing dull greyish-ochreous; veins with black ringlet spots, terminating in a transverse discal line and two marginal series of larger dentate white-bordered spots, a subbasal black sinuous line, and a reniform spot at end of cell: hind wing pale brownish cinereous, discal area ochreous, with an indistinct brown marginal border. Thorax and legs greyish ochreous; collar, head, and palpi reddish-ochreous; tarsi black-banded.

Female differs only in the spots along the costal area being more confluent, and in the hind wing being uniformly cinereous.

Expanse 1 7/16 inch.

Hab. Darjiling. In coll. Dr. Staudinger.

Genus PALIMPSESTIS, *Hübner.*

PALIMPSESTIS ALTERNATA.

Palimpsestis alternata, Moore, P. Z. S. 1881, p. 331, pl. xxxvii. fig. 2.

Allied to *P. ocularis* of Europe.

Male. Fore wing pale metallic cupreous-brown, crossed by a broad basal, a medial, and two narrow submarginal greenish-grey indistinct bands; some black basal spots, an ante- and postmedial transverse black sinuous line, and black and white streaks externally along the veins: hind wing and abdomen pale cupreous-brown. Thorax greenish grey; collar, front of head, palpi, and legs rufous-brown; legs with black bands.

Expanse 1 7/10 inch.

Hab. Darjiling. In coll. Dr. Staudinger.

PALIMPSESTIS CUPRINA.

Palimpsestis cuprina, Moore, P. Z. S. 1881, p. 331, pl. xxxvii. fig. 3.

Male and female. Fore wing pale metallic-brown, slightly cupreous anteriorly, and the area below the cell greenish; two or three black transverse antemedial lines angled at median vein; some basal spots, a black discocellular recurved mark and a spot within the cell; five or six transverse discal indistinct sinuous lines with black and white dentate marks on the veins; a submarginal pale lunular line, and a marginal black line: hind wing pale cupreous-white, with pale cupreous-brown marginal band. Thorax, head, and palpi brown, abdomen paler; collar black; fore and middle legs with black bands.

Expanse 1 3/10 inch.

Hab. Darjiling. In coll. Dr. Staudinger.

PITRASA, nov. gen.

Fore wing elongated, narrow; costa slightly arched at base, apex acuminate, exterior margin obliquely convex, posterior margin convex towards the base; costal vein very long, extending to three fourths the margin; first subcostal emitted at one half before end of the cell, second at one sixth, trifurcate, third at one fifth, and fourth at one half from base of second, fifth curving upward from end of the cell and touching the third near its base; discocellulars deeply concave in the middle, radials from the upper and lower ends; cell extending more than half the wing; middle median from immediately before end of the cell, lower at one fourth before the end; submedian curved at the base. Hind wing oval, short, apex very convex: costal vein recurved; two subcostal branches emitted from end of the cell; discocellulars bent obliquely outward from the middle; radial near its lower end; middle median from immediately before end of the cell, lower at one third before the end; submedian and internal straight. Body stout, long; collar erect; palpi small, porrect, second joint broad, laxly squamose, third joint short, cylindrical; antennæ very long, minutely bipectinate to near tip in the male; legs long; fore and middle femora and tibiæ laxly squamose above, femora finely pilose beneath.

PITRASA VARIEGATA, n. sp. (Plate IV. fig. 2.)

Male. Fore wing dark purplish-ochreous, fasciated with purplish-grey on the disk and posterior border; an oblique waved black line from the apex across the disk, bordered within and before the apex by a white fascia; some indistinct white streaks along the outer veins, and an outer marginal black line bordered by a white line; some indistinct black basal transverse streaks, and a linear white-bordered streak from base of median vein: hind wing whitish-cinereous, with broad brown outer band. Body, palpi, and legs brown.

Expanse 1 ⅗ inch.

Hab. Darjiling. In coll. Dr. Staudinger.

PITRASA VITELLINA, n. sp. (Plate IV. fig. 3.)

Female. Fore wing ochreous-yellow; costal border slightly speckled towards the apex with purplish-ochreous and white, the veins across the disk also speckled with purple and white, the ends of the three at lower end of the cell and their interspaces entirely white; an oblique transverse discal row of black dots, one on each vein, a purple streak extending for a short distance along the vein from the exterior margin, the margin itself with a narrow white and purple line; some pale-bordered purple streaks across the veins at the base, and a short linear streak from base of median vein: hind wing whitish-cinereous, with broad brownish-cinereous outer band. Body cinereous; thorax brownish; palpi and legs above brownish.

Expanse 1 ¼ inch.

Hab. Darjiling. In coll. Dr. Staudinger and British Museum.

TYCRACONA, nov. gen.

Allied to *Acronycta.*

Fore wing moderately long, broad anteriorly; costa nearly straight, apex slightly acute, exterior margin convexly oblique; second subcostal branch bifid, fourth anastomosed to third near the base, the fifth starting from the point of junction above end of the cell, sixth (or upper radial) from angle at upper end of the cell; lower radial and two median branches from broad angle at lower end of the cell. Hind wing broad, slightly angular below the apex, exterior margin convex; costal and subcostal joined at their base, two subcostal branches from end of the cell, radial from the middle of discocellulars; two median branches from end of the cell, lower contiguous. Body moderately stout, squamous; palpi slender, squamous, third joint conical, half the length of the second; legs squamous; antennæ slender, setose.

TYCRACONA OBLIQUA, n. sp.

Tycracona obliqua, Moore, Aid, pt. . pl. .

Male and female. Fore wing covered with dark-grey and dark vinous-brown scales; crossed by irregular marron and grey subbasal and discal zigzag narrow bands, the marron colour predominating apically; a prominent oblique subapical black streaky band crossing from middle of the costa to middle of exterior margin: hind wing and abdomen pale cupreous-brown, with a grey marginal lunular line. Collar, head, and palpi dark marron, thorax marron and grey-speckled; legs grey with marron bands.

Expanse, ♂ $1\frac{4}{10}$, ♀ $1\frac{7}{10}$ inch.

Hab. Cherra and Darjiling. In coll. Dr. Staudinger.

TYCRACONA TRANSVERSA, n. sp. (Plate IV. fig. 5.)

Upperside vinous-brown. Fore wing crossed by irregular grey-bordered short longitudinally-disposed black bands, two beyond the cell being most prominent, recurved and sinuous; a marginal row of black triangular spots; orbicular and reniform spots grey: hind wing pale cupreous-brown. Head, palpi, and thorax grey-speckled; legs brown-streaked above.

Expanse $1\frac{4}{10}$ inch.

Hab. Darjiling. In coll. Dr. Staudinger.

SYDIVA, nov. gen.

Near to *Acronycta.* Fore wing long, narrow; apex pointed, exterior margin oblique and slightly convex; costal vein extending two thirds the margin; first subcostal branch emitted before half length of the cell, second branch at one fourth from the end, trifurcate, third and fourth branch at equal distances apart from second beyond the cell, fifth ascending from end of the cell and touching the third near its base; cell long; upper radial from angle close to subcostal; lower discocellular concave, bent at its lower end, lower radial from its angle; upper median branch from angle above end of the cell, middle branch from its end, lower branch from opposite middle of first and second subcostals. Hind wing long, exterior margin oblique, recurved, apex convex; costal vein extending to apex; cell short, broad; two subcostal branches from end of the cell; discocellulars concave, radial from below the middle; two upper median branches from end of the cell, lower at one third before its end. Body stout, laxly pilose, head pilose; palpi projecting out in front of the head, second joint long, extending half its length beyond the head, third joint one fourth of an inch long, slender, cylindrical; legs pilose; antennæ minutely pectinated in male.

SYDIVA NIGROGRISEA, n. sp.

Fore wing black, speckled with grey; crossed by a basal, subbasal, medial, and discal more distinct grey-bordered black sinuous line, and a submarginal line with greyer dentate marks; six or seven grey-white spots on the costal border, and a row along the cilia; orbicular and reniform spots indistinct, the latter with two grey-white dots on its lower end: hind wing brown, palest basally; with an indistinct transverse discal darker line; cilia bordered with cinereous. Thorax, palpi, and legs above black speckled with grey; tarsi with pale bands; abdomen brown.

Expanse 2 inches.

Hab. Darjiling. In coll. F. Moore and Dr. Staudinger.

Family LEUCANIIDÆ.

Genus ALETIA*, *Hübner.*

Aletia, Hübn. Verz. bek. Schmett. p. 239 (1816).
Hyphilare, Hübn. ibid.

ALETIA ANGULIFERA, n. sp.

Aletia angulifera, Moore, P. Z. S. 1881, p. 333.

Male. Allied to *A. conigera* (*Noctua conigera*, Schiff.). Fore wing ochreous-yellow, indistinctly clouded with red-ochreous scales; crossed by a reddish-black antemedial outwardly bent line, and an oblique postmedial line which is bent inward at costal end; a whitish spot at lower end of the cell, a pale yellow reniform and orbicular space; marginal and cilial line black-speckled: hind wing pale dusky brown; cilia pale yellow. Thorax, head, palpi, and legs ochreous-yellow; abdomen paler.

Expanse $1\frac{3}{10}$ inch.

Hab. N.W. Himalaya; Cashmere (Sind valley). In coll. F. Moore and Dr. Staudinger.

* Type *A. vitellina*, Hübn.

Allied to *A. lithargyria.*

Male and female. Fore wing dusky reddish-ochreous, sparsely speckled with black scales ; veins greyish-white, most prominent on median vein to end of cell, beyond which is a cluster of black scales ; a slightly paler orbicular and reniform space within the cell : hind wing dusky-brown, darkest marginally. Cilia brown-bordered. Thorax, head, palpi, and legs dusky ochreous-brown, abdomen paler.

Female. Veins greyer, a black speckle on each vein, forming a transverse discal series. Expanse, ♂ 1$\frac{4}{10}$, ♀ 1$\frac{5}{10}$ inch.

Hab. Darjiling. In coll. Dr. Staudinger and F. Moore.

Allied to *A. lithargyria*, Esp., of Europe.

Male and female. Fore wing reddish-ochreous, sparsely speckled with black scales, in parts disposed in ill-defined strigæ, which are slightly clustered on the disk, forming two transverse recurved very indistinct bands, the speckles on the outer band disposed on the veins and those on the inner band between the veins ; a brighter ochreous reniform and orbicular spot; veins (except the costal and first and second subcostal branches) lined with white, most distinct along the median to end of the cell, where there is a whiter spot externally clustered with black scales ; costal edge white : hind wing and abdomen pale brownish-ochreous, darkest marginally, veins prominent. Thorax, head, palpi, legs, and anal tuft reddish-ochreous.

Expanse 1$\frac{6}{8}$ inch.

Hab. Darjiling ; Gurhwal ; Dharmsala.   In coll. Dr. Staudinger and Rev. J. H. Hocking.

Aletia distincta, Moore, P. Z. S. 1881, p. 333, pl. xxxvii. fig. 4.

Allied to *A. vitellina.*

Male. Fore wing deep ochreous-yellow, sparsely crossed by short irregularly disposed red strigæ, more or less black-speckled : a prominent antemedial and a postmedial transverse curved narrow band of the same colour but more blackly speckled, beyond which is a discal row of black spots followed by a white streak—one on each vein ; a marginal series of black dots, one between each vein ; orbicular and reniform space marked by red strigæ, the space beyond the cell to the postmedial band clouded with dusky red ; cilia from apex to near end of hind margin thickly black-speckled : hind wing pale ochreous-brown posteriorly, yellowish anteriorly. Cilia pale yellow. Thorax, head, palpi, and legs deep ochreous ; a slight black-speckled collar, a few speckles on tegulæ, and a cluster at base of pale ochreous abdomen.

Expanse 1-$\frac{3}{10}$ inch.

Hab. Darjiling. In coll. Dr. Staudinger.

Genus BOROLIA, *Moore.*

Borolia, Moore, P. Z. S. 1881, p. 334.

Fore wing narrow; costa slightly arched at the base, apex pointed, exterior margin slightly oblique, posterior angle convex; cell narrow; first subcostal branch emitted at half length of the cell and extending close alongside the costal, second emitted before end of the cell, trifid, the third starting from near its base above the cell and touching the fifth, fourth short, fifth emitted and curving upward from end of the cell; upper discocellular short, outwardly oblique, lower concave; radial from end of upper discocellular; lower radial and two upper median branches from angles at lower end of the cell, fourth at some distance before the end; submedian recurved. Hind wing with long costal margin, exterior margin very oblique, abdominal margin short; subcostal straight, both branches emitted from end of the cell; discocellulars concave; radial from their middle; median straight, two upper branches from end of the cell. Cilia broad. Antennæ setulose; palpi ascending, pilose beneath, second joint long, third short and slender; femora pilose beneath.

Type *B. furcifera.*

BOROLIA FASCIATA.

Borolia fasciata, Moore, P. Z. S. 1881, p. 334, pl. xxxvii. fig. 12.

Male and female. Upperside—fore wing pale brownish-ochreous, palest along the costa; an irregular-bordered chestnut-brown fascia from the base to below the apex, which is bordered below the base by a short darker streak; a black dot at base of the wing, and two transverse discal indistinct rows of dots, which in the female are apparently joined together by an intervening line; a marginal row of black dots: hind wing pale brown. Underside paler; fore wing with a slightly darker discal area: hind wing with an indistinct discocellular spot, transverse discal linear streaks, and marginal spots. Palpi brown at the side.

Expanse, ♂ 1$\frac{2}{8}$, ♀ 1$\frac{3}{8}$ inch.

Hab. Darjiling. In coll. Dr. Staudinger.

BOROLIA FURCIFERA, n. sp. (Plate IV. fig. 16.)

Upperside—fore wing brownish-ochreous, with a transverse subbasal black zigzag line and an oblique discal sinuous line, the interspace more or less clouded with chestnut-brown and crossed longitudinally by a white bifid streak (formed by the median vein and a portion of its two middle branches), the orbicular and reniform spots paler; base of wing and apex streaked with chestnut-brown ; a black spot at base of wing, and marginal row of black dots: hind wing pale brown. Cilia pale reddish-ochreous. Body brownish-ochreous. Underside pale dull reddish-ochreous; fore wing with the discal area dusky brown, crossed by a short darker streak on each vein; outer border slightly black-speckled ; a marginal row of black dots: hind wing black-speckled, crossed by a discal speckled and streaked band, a distinct discocellular spot, and a marginal row of dots.

Expanse 1⅝ inch.

Hab. Darjiling. In coll. Dr. Staudinger.

Genus LEUCANIA.

LEUCANIA BISTRIGATA.

Leucania bistrigata, Moore, P. Z. S. 1881, p. 334, pl. xxxvii. fig. 18.

Female. Fore wing pale ochreous; with a suffused brown fascia from the base, through the cell, to end of the costa, another fascia extending obliquely from below the apex and along the exterior margin : a very prominent white black-bordered terminally-bent streak at lower end of the cell ; the outer veins also whitish; a black basal streak below the cell, and another streak below the submedian vein ; three very indistinct black dots on the costal edge, some across the disk, a marginal series of minute black points bordered with pale outer cilial lines : hind wing pale ochreous-brown. Cilia ochreous-white, with a brown line. Thorax with a black-and-white-speckled crest, collar, and tegulæ; palpi and legs brown-speckled : abdomen pale ochreous.

Expanse 1⁷⁄₁₀ inch.

Hab. Darjiling. In coll. Dr. Staudinger and British Museum.

Allied to *L. decissima*, Walker. May be distinguished from it by having a linear discoidal streak much more clearly defined, the two basal black streaks, and by the cilia of fore wing being distinctly lined.

LEUCANIA MODESTA.

Leucania modesta, Moore, P. Z. S. 1881, p. 335, pl. xxxvii. fig. 11.

Male and female. Fore wing pale ochreous, external area below the apex slightly darker; with an antemedial transverse indistinct sinuous black line, a postmedial series of points, and marginal row of black dots : hind wing pale brownish-ochreous and yellow along the costal border, with an indistinct darker discocellular lunular spot and marginal spots. Tegulæ very slightly black-speckled.

Expanse 1½ inch.

Hab. Darjiling. In coll. Dr. Staudinger.

LEUCANIA ADUSTA.

Leucania adusta, Moore, P. Z. S. 1881, p. 335.

Male and female. Fore wing ochreous-yellow; median and submedian veins white;

c 2

borders of the vein and a line between each vein ochreous-brown; costal border and an oblique apical streak ochreous-brown; a few black speckles disposed on the costa, also within and below the cell; a discal series of black-speckled spots, one on each vein, and a marginal row of spots; cilia brown; hind wing ochreous-white, external area pale cupreous-brown; marginal spots black; cilia white. Underside ochreous-white, suffused with pink externally; both wings speckled with brown on anterior and exterior borders, and with a marginal row of black spots; fore wing with a small black costal spot before the apex.

Expanse, ♂ 1¾, ♀ 1⅞ inch.

Hab. Manpuri, N.W. India (*Horne*); Darjiling, Khasia Hills (*Atkinson*). In coll. F. Moore and Dr. Staudinger.

LEUCANIA CONSIMILIS.

Leucania consimilis, Moore, P. Z. S. 1881, p. 336, pl. xxxvii. fig. 19.

Male. Near to *L. decissima*, but differs in being larger, paler, and more uniformly coloured. Fore wing with paler linear markings between the veins, the white portion of the discoidal streak half its length and less distinct, the transverse discal spots and the subbasal series more loosely speckled; the marginal series less distinct, the basal two thirds of the wing also sparsely sprinkled with black speckles; hind wing paler brown, and whitish at the base. Underside of both wings ochreous-white, with an indistinct transverse discal brown line. Thorax, head, and palpi brown-speckled; legs with pale-brown streaks.

Expanse 1⁷⁄₁₀ inch.

Hab. Darjiling. In coll. Dr. Staudinger.

LEUCANIA COMPTA.

Leucania compta, Moore, P. Z. S. 1881, p. 336, pl. xxxvii. fig. 8.

Male. Fore wing pale brownish-ochreous; veins ochreous-white, with intervening pale white and brown lines; a transverse discal row of distinct minute black points; subcostal branches and median with its branches slightly black-bordered to the discal points; two indistinct black spots on the costa, one below the cell, and two on the submedian vein; marginal row of black dots indistinct; hind wing ochreous-white, veins and the apical border ochreous-brown. Thorax, palpi, and legs above brownish-ochreous, with darker speckles; abdomen ochreous-white.

Expanse 1¹⁄₈ inch.

Hab. Pudda River (*Atkinson*). In coll. Dr. Staudinger.

LEUCANIA NAINICA.

Leucania nainica, Moore, P. Z. S. 1881, p. 337, pl. xxxvii. fig. 15.

Male and female. Fore wing very pale reddish-ochreous, palest along the veins; a few minute black speckles along the costal and posterior borders; median vein and its two upper branches white, bordered with a brown streak above and below the cell; a black spot at end of cell; a transverse discal series of minute black points, and a marginal row of dots: hind wing slightly paler, outer margin suffused with brown; cilia whitish. Thorax in front, palpi, and legs above brown-speckled.

Expanse, ♂ $1\frac{2}{10}$, ♀ $1\frac{4}{10}$ inch.

Hab. Naini Tal, N.W. Himalayas. In coll. Dr. Staudinger.

LEUCANIA ALBISTIGMA.

Leucania albistigma, Moore, P. Z. S. 1881, p. 337, pl. xxxvii. fig. 9.

Female. Fore wing pale brownish-ochreous, greyish along the costal border and obliquely below the apex; veins speckled with grey and brown, most prominently along the median and its branches; an indistinct curved discal series of minute black points; a small brown spot in middle of the cell, and a black streak extending through and beyond its end, which is crossed by a white discocellular spot; a slender black streak below end of the cell: hind wing ochreous-white; veins lined with brownish-ochreous. Thorax, palpi, and legs greyish-ochreous, the palpi and legs brown-speckled; abdomen ochreous-white.

Expanse $1\frac{2}{10}$ inch.

Hab. Darjiling. In coll. Dr. Staudinger.

LEUCANIA HOWRA.

Leucania howra, Moore, P. Z. S. 1881, p. 337, pl. xxxvii. fig. 16.

Female. Fore wing pale brownish-ochreous, greyish along base of the costa, below the apex, and below the cell, brown-speckled; veins ochreous-white, the subcostal branches and the median with its branches lined with brown to the discal black points; a brown line between all the veins, and a prominent black basal streak below the cell; a marginal row of minute black dots: hind wing ochreous-white; veins brownish-ochreous; some marginal black dots. Thorax, palpi, and legs greyish-ochreous, speckled with brown; abdomen ochreous-white.

Expanse $1\frac{1}{4}$ inch.

Hab. Calcutta. In coll. Dr. Staudinger.

LEUCANIA DHARMA.

Leucania dharma, Moore, P. Z. S. 1881, p. 338, pl. xxxvii. fig. 17.

Female. Near to *L. album.* Fore wing pale ochreous-brown; costal and posterior borders and oblique fascia below the apex paler; a pale and brown line between the veins; veins whitish, the median and its branches to the transverse discal indistinct black points most prominent; a black spot within end of the cell; a few black speckles on costal and posterior borders: hind wing and abdomen paler brown, whitish at the base. Underside uniformly paler and brown-speckled: fore wing with a black costal spot and indistinct transverse brownish fascia: hind wing with a transverse discal series of brown points, one on each vein, and discocellular lunular spot. Thorax grey, with black speckles; front of head, palpi and legs above pale brown with darker speckles.

Expanse 1¼ inch.

Hab. Darjiling. In coll. Dr. Staudinger.

LEUCANIA ALBICOSTA.

Leucania albicosta, Moore, P. Z. S. 1881, p. 338, pl. xxxvii. fig. 10.

Female. Similar to same sex of *L. dharma*, but paler-coloured. Fore wing with the costal border, apical fascia, and posterior margin ochreous-white, the veins and lines between them not so prominent, the median vein being of the same uniform colour as the others; an indistinct blackish spot on the costa, and another spot below the cell. Underside—fore wing with discal area broadly brownish: hind wing brown-speckled, and with a prominent blackish discocellular spot and marginal row of dots. Thorax very pale, and not speckled; palpi and legs above pale ochreous-brown, and not speckled.

Expanse 1¼ inch.

Hab. Darjiling. In coll. Dr. Staudinger.

LEUCANIA SINUOSA, n. sp.

Male and female. Fore wing pale ochreous, with a longitudinal dark-brown irregular suffused fascia from the base below the cell to apex below the angle; an antemedial and a postmedial transverse series of black-pointed streaks, which form an indistinct almost continuous sinuous line; median vein to the discal point of the two middle branches distinctly lined with white and black; a small orbicular spot and prominent reniform spot dark brown; three dots on the costal edge and a marginal series of dots black; cilia waved, with black and brown speckles: hind wing uniformly ochreous-brown; cilia pale ochreous, lined with brown. Collar, palpi, and tegulæ with a lateral brown streak and speckles; legs brown-streaked.

Expanse 1⅝ inch.

Hab. Darjiling. In coll. Dr. Staudinger and F. Moore.

LEUCANIA RUFESCENS, n. sp.

Nearest to *L. decissima*. Differs in the fore wing being of a more reddish-ochreous, with yellowish streaks between the veins, the discoidal linear white streak shorter, the basal end of the median vein black, and outside end of the cell is a prominent blackish spot; the transverse discal black points and the other black spots are more distinct. Underside of both wings with a transverse discal dusky band with black streaks, discocellular lunular spot, and outer marginal row of dots—the underside of *L. decissima*, in both sexes, being entirely covered with silver scales. The abdomen of male has no basal tuft.

Expanse 1⅝ inch.

Hab. Darjiling. In coll. Dr. Staudinger and F. Moore.

LEUCANIA NIGRILINEOSA, n. sp.

Male and female. Fore wing pale purplish greyish-ochreous, palest along costal border and obliquely from the apex; veins greyish-white, with a medial black and outer brown streaks between the veins; an indistinct oblique discal series of black spots, and marginal row of minute dots; a black spot near base of costa, and a few black dots disposed on posterior border: hind wing greyish-ochreous, with a brownish suffused streak along the veins. Body greyish-ochreous, darker on thorax. Underside duller-coloured; discal area of fore wing dusky-brown: hind wing with a short brown streak on each vein, forming a transverse discal series; a few brown speckles along anterior margin. Allied to *L. comma.*

Expanse 1$\frac{9}{12}$ inch.

Hab. Khasia Hills (October). In coll. Dr. Staudinger and F. Moore.

Genus SESAMIA, *Guén.*

SESAMIA FRATERNA, n. sp.

Very similar to *S. proscripta* and *S. inferens.* Fore wing with a slightly darker shade longitudinally through the median veins, and a small distinct black spot below the cell: hind wing white.

Expanse, ♂ 1$\frac{1}{10}$, ♀ 1$\frac{3}{10}$ inch.

Hab. Dharmsala. In coll. F. Moore and Dr. Staudinger.

Genus SIMYRA, *Ochs.*

SIMYRA CONSPERSA.

Simyra conspersa, Moore, P. Z. S. 1881, p. 340.

Female. Upperside—fore wing pale whitish-ochreous, numerously covered with minute brown speckles: hind wing white. Thorax ochreous, abdomen paler. Underside uniformly pale ochreous-white. Near *S. confusa.*

Expanse 1$\frac{7}{12}$ inch.

Hab. Manpuri, N.W. India (*Horne*); Calcutta (*Atkinson*). In coll. F. Moore and Dr. Staudinger.

Genus AXYLIA, *Hübner.*

AXYLIA RENALIS, n. sp.

Fore wing pale reddish-ochreous, with the costal border iron-grey; orbicular and reniform spot very prominent, black-lined and centred with iron-grey; veins speckled with grey and black; a black fascia extending through the cell to outer margin; a black subbasal transverse sinuous double line, some discal spots, and a marginal row of dentate spots: hind wing pale brownish white, darker along costal border. Thorax black above, ochreous laterally and in front; abdomen pale brown; palpi black, tipped with ochreous; legs black; fore and middle tibiæ ochreous.

Expanse $1\frac{1}{5}$ inch.

Hab. Sind valley, Kaschmir (*Atkinson*); Solun, Punjab (*Reid*). In coll. Dr. Staudinger and F. Moore.

Allied to the European *A. putris*.

Family GLOTTULIDÆ.

CALYMERA, nov. gen.

Female. Fore wing narrow; costa arched from the base, apex slightly pointed, exterior margin oblique and convex below the apex, posterior margin short, convex at base; costal and subcostal veins widely apart, second subcostal branch emitted near end of the cell, bifid, fourth from end of the cell, and anastomosed for some distance with third, fifth from below the point where anastomosed; upper radial from end of the cell; discocellular very slender, slightly concave; cell long, narrow; lower radial and upper median branch from angles above end of the cell, middle at an equal distance before its end, lower from opposite first subcostal; submedian running close to margin. Hind wing long, narrow, somewhat triangular; cell short; two subcostal branches from end of the cell; discocellular slender, concave, very oblique hindward; radial from near its lower end; two upper median branches from acute point at end of the cell, lower branch from some distance before the end. Body short, stout; palpi long, ascending above the head, second joint long, compactly clothed, third joint short, cylindrical; femora slightly pilose beneath; antennæ setose.

Near to genus *Eudryas*.

CALYMERA PICTA, n. sp. (Plate IV. fig. 7.)

Female. Fore wing pale brownish-ochreous, with a large chestnut-red patch occupying the disk from costal to posterior margin, the outer border of the patch is very convex from the costa, and thence sinuous to the posterior angle, the inner border is irregular and streaked with black below the cell, the interior of the patch being purple at the end and below the cell, and there also speckled with grey and ochreous scales, and marked by a distinct whitish discocellular lunule; base of wing with black streaks and speckles; outer margin lunulated with black; cilia brown: hind wing brown; cilia paler. Body brownish-ochreous; thorax, front of head, a broad collar in front, large spots on top of thorax, and speckles on tegulæ blue-black; fore and middle legs above brown-speckled; palpi tipped with black.

Underside of both wings pale reddish-ochreous, with dusky discal areas.

Expanse $1\frac{1}{2}$ inch.

Hab. Darjiling. In coll. Dr. Staudinger.

Family APAMIIDÆ.

Genus HYDRÆCIA.

HYDRÆCIA KHASIANA.

Hydræcia khasiana, Moore, P. Z. S. 1881, p. 312, pl. xxxvii. fig. 5.

Male. Fore wing dull reddish-brown, washed with purple-grey, with a brighter brown pale-bordered subbasal sinuous line, a large orbicular and reniform spot, and discal sinuous line; outer border also brighter, and traversed by a pale waved line; median and submedian veins dusky-brown: hind wing cinereous-white, with pale-brown veins and indistinct marginal fascia. Thorax, head, palpi, and legs reddish-brown, grey-speckled; abdomen paler. Underside much paler.

Expanse $1\frac{6}{10}$ inch.

Hab. Khasia Hills, E. Bengal. In coll. Dr. Staudinger.

Near to *H. petasitis.*

Genus DIPTERYGIA, *Guén.*

DIPTERYGIA SIKKIMA, n. sp.

Allied to *D. indica.* Differs in the fore wing being slightly narrower, with a pale brownish-ochreous transverse subbasal narrow sinuous line, a less distinct discal and marginal line terminating in a pale patch at the posterior angle, the patch being less prominent, and traversed with dark-brown streaks, but not extending along the margin towards the costa; orbicular and reniform spots more distinct: hind wing much paler, nearly brownish-white at the base, and showing an indistinct discal transverse line.

Expanse $1\frac{1}{2}$ inch.

Hab. Darjiling. In coll. Dr. Staudinger.

Genus SASUNAGA, *Moore.*

Sasunaga, Moore, P. Z. S. 1881, p. 312.

Fore wing very narrow; costa almost straight; exterior margin oblique, convex, waved; posterior margin convex at the base; costal vein extending two thirds the margin; first subcostal branch emitted at two thirds, and second at one fourth before end of the cell, third branch one third and fourth at one half from below second, fifth curving from end of the cell and free from the third; upper radial from end of the cell, discocellular obliquely concave, lower radial from near its middle; cell long, very narrow at the base; upper median branch from angle above end of the cell, middle branch from the end, lower at one fourth before the end; submedian curved downward near the base. Hind wing very broad, triangular; costa convex near the base, apex convex; exterior margin very oblique, waved; abdominal margin long; costal vein slightly arched near the base, extending to apex; two subcostal branches from end of the cell; discocellular slender, concave; radial very slender,

D

emitted from below middle of discocellular; cell short, broad; two upper median branches from end of the cell, lower at one fourth before the end; submedian and internal vein straight, the latter extending to anal angle. Thorax robust; abdomen long, somewhat slender and extending beyond the wing; palpi ascending, slender, second joint squamose, reaching to vertex; third joint half its length, cylindrical; femora pilose beneath, tibiæ tufted above; antennæ setose.

Near to *Dipterygia*. The American genus *Magusa* (Walker, Catal. Lep. Het. B. M. xi. p. 762) is a very closely-allied form.

SASUNAGA TENEBROSA.

Hadena tenebrosa, Moore, P. Z. S. 1867, p. 59.
Sasunaga tenebrosa, Moore, P. Z. S. 1881, p. 343.

Fore wing dark brown, with short ochreous-bordered blackish costal streaks, lengthened longitudinal upper discal and less distinct lower discal streaks, and oblique streaks below the cell, the latter bordered by an ascending lower discal curved duplex sinuous pale-pointed black line, and a submarginal less distinct pale-pointed sinuous line; the ochreous borders palest before the apex; orbicular spot small, ochreous with black border, reniform spot less distinct: hind wing glossy cupreous-brown, palest at the base; cilia cinnamon-brown. Thorax dark brown, black-speckled; abdomen pale brown, tuft ochreous; palpi and legs ochreous, brown-speckled; tarsi with blackish bands.

Expanse, ♂ 1⅜, ♀ 1⅝ inch.

Hab. Darjiling. In coll. F. Moore and Dr. Staudinger.

Remark. Some specimens of this insect are pale ochreous-brown, with less distinct darker brown and black streaks, and also show a darker subapical costal patch.

KARANA, nov. gen.

Fore wing—costa straight, apex pointed; exterior margin oblique, convex and sinuous; posterior margin convex at base; cell long and narrow; first subcostal emitted at half length of the cell and widely separated from the costal, second branch before end of the cell, trifurcate, the third being thrown off beyond and above the cell, fourth short, fifth from end of cell and curving up to the third and almost touching it at its base; discocellulars very slender; radial and upper median vein from angles above end of the cell, middle median from its end, lower some distance before the end. Hind wing triangular; anterior margin long, abdominal margin short, exterior margin convex and waved; subcostal straight, its two branches from end of the cell; discocellulars very slender, radial from their middle; the middle median from angle close to end of the cell. Thorax moderate, abdomen long; palpi ascending, slender, second joint clothed with short hair; femora and tibiæ slightly pilose; antennæ setose.

Type *K. decorata*.

KARANA DECORATA, n. sp.

Fore wing chestnut-brown; with a short white basal cross bar, a subbasal transverse band with an inner cross bar and two contiguous outer spots, a broad quadrate subapical costal spot with pointed angles, all bordered with black; some metallic-green speckles at base along the costa and posterior margin, within and below the cell, and a marginal series of greyish-white black-bordered lunular spots, and a less distinct transverse discal row: hind wing pale brown, with indistinct darker pale-bordered discal narrow band. Cilia white-streaked. Thorax chestnut-brown, green-and-grey speckled; abdomen with chestnut-brown dorsal tufts; head and palpi grey-speckled; legs dark brown banded with white.

Expanse $1\frac{1}{2}$ inch.

Hab. Darjiling. In coll. Dr. Staudinger and F. Moore.

Genus NEURIA, *Guén.*

NEURIA SIMULATA.

Neuria simulata, Moore, P. Z. S. 1881, p. 343, pl. xxxviii. fig. 1.

Female. Allied to *N. separata.* Differs in its larger size and paler colour, the fore wing having less distinct transverse sinuous markings; orbicular and reniform spots, and a shorter and broader dentate mark below the cell.

Expanse 2 inches.

Hab. Darjiling. In coll. Dr. Staudinger.

NEURIA SEPARATA, n. sp.

Fore wing dark brown; with a transverse subbasal and a discal sinuous double black line with yellowish centre; a less distinct basal line and a submarginal line, the interspace between the latter and the discal line black-streaked; a black dentate mark below the orbicular spot; a brown fascia curving across the middle; some white dots on the costa before the apex, and an outer marginal row of black lunules; a white-lined orbicular and reniform spot: hind wing and abdomen pale brown. Thorax, head, palpi, and legs brown; palpi and legs with pale ochreous bands.

Expanse, $\male$ $1\frac{7}{12}$, $\female$ $1\frac{8}{12}$ inch.

Hab. Sikkim. In coll. Dr. Staudinger and F. Moore.

Allied to *N. saponaria.*

Genus APAMEA, *Ochs.*

APAMEA CUPRINA.

Apamea cuprina, Moore, P. Z. S. 1881, p. 345, pl. xxxviii. fig. 2.

Allied to *A. leucostigma.* Fore wing dark cupreous-brown, with indistinct grey

D 2

transverse sinuous fasciæ; costal and median veins grey-speckled; orbicular and reniform marks grey, the latter also black-speckled: hind wing pale cupreous-brown. Thorax, palpi, and legs above dark brown; abdomen ochreous-brown.

Expanse 1$\frac{5}{10}$ inch.

Hab. Sikkim (*Blanford*, 1870). In coll. Dr. Staudinger.

APAMEA MUCRONATA.

Apamea mucronata, Moore, P. Z. S. 1881, p. 345, pl. xxxviii. fig. 8.

Male and female. Fore wing pale ferruginous, with a broad medial transverse darker ferruginous band bordered by an antemedial and postmedial double black sinuous line, the latter with very long outer discal points; orbicular and reniform spots black-lined and pale-centred; a ferruginous black-speckled spot and some contiguous streaks at base of the wing; some streaks on costal border, and a marginal irregular fascia bordered by the black lunular points with pale tips: hind wing and abdomen pale pinkish-brown. Thorax, palpi, and legs pale ferruginous; palpi laterally and legs above slightly brown-streaked.

Expanse 1$\frac{4}{10}$ inch.

Hab. Darjiling. In coll. Dr. Staudinger.

Near to *A. undicilia*, Walker.

APAMEA STRIGIDISCA.

Apamea strigidisca, Moore, P. Z. S. 1881, p. 346, pl. xxxviii. fig. 9.

Male. Fore wing dark ferruginous; costal border, some basal streaks, and discal area paler ferruginous; a transverse antemedial and postmedial sinuous pale-bordered black lines: orbicular and reniform spots black-lined, with a pale inner border and blackish centre; a black conical mark below the orbicular spot; a submarginal transverse pale line, the middle portion zigzag and longitudinally crossed by a black line between the median veins; base of cell and posterior margin black-streaked; submedian vein black; some black streaks on the costal border, and a marginal waved line with a row of black points: hind wing pale ferruginous-brown. Body ferruginous; collar and tegulæ, front of head, palpi laterally, and bands on legs black.

Expanse 1$\frac{1}{2}$ inch.

Hab. Darjiling. In coll. Dr. Staudinger.

The markings in this species are somewhat similar to those in *Mamestra adusta*, Esper.

APAMEA NUBILA.

Apamea nubila, Moore, P. Z. S. 1881, p. 346, pl. xxxviii. fig. 10.

Male and female. Fore wing dark purple-brown, washed with chalybeate-grey; crossed by a subbasal, antemedial, postmedial, and a submarginal sinuous black-bordered grey line; orbicular and reniform spots greyish, their interspace and a streak below base of the cell, a mark below the orbicular spot, and a streak above posterior angle black : hind wing and abdomen pale ochreous-brown. Thorax dark purple-brown ; palpi and legs ochreous-brown, the latter with black bands.

Expanse $1\frac{3}{16}$ inch.

Hab. Darjiling. In coll. Dr. Staudinger.

APAMEA SIKKIMA, n. sp.

Male and female. Fore wing brownish-ferruginous, brightest on discal area ; base of costa and a broad medial transverse band dusky greyish-ferruginous ; crossed by a subbasal, and the band bordered by a, double black sinuous pale line ; orbicular and reniform spots black-lined, with a pale border and dusky centre ; a black-pointed mark below the orbicular spot ; an indistinct submarginal transverse sinuous pale fascia, and a marginal row of black lunular points : hind wing and abdomen pale purplish-brown. Thorax, palpi, and legs reddish-ferruginous ; legs with blackish bands.

Expanse, ♂ $1\frac{1}{4}$, ♀ $1\frac{5}{8}$ inch.

Hab. Darjiling In coll. Dr. Staudinger and F. Moore.

APAMEA DENTICULOSA, n. sp. (Plate IV. fig. 13.)

Male and female. Fore wing ochreous-brown, crossed by a subbasal, an antemedial and postmedial sinuous yellow-centred double black line, and a submarginal waved yellow line broadest at the apex, with contiguous yellow dentate streaks bordering the marginal row of black dentate marks ; orbicular and reniform spots black-lined ; some black streaks on costal border, and a broad pointed mark below the orbicular spot : hind wing pale cinereous brown, darkest externally. Thorax, palpi, legs, and anal tuft ochreous ; tegulæ, palpi at side, abdomen, and bands on legs above brown.

Expanse $1\frac{7}{16}$ inch.

Hab. Darjiling. In coll. Dr. Staudinger.

APAMEA OBLIQUIORBIS, n. sp.

Male and female. Fore wing pale ferruginous, with darker ferruginous streaks between the veins ; a black subbasal sinuous double line, an outwardly oblique orbicular spot, and a paler reniform spot ; a pale yellowish transverse lunular discal line, and a blackish irregular dentate-bordered marginal band : hind wing pale brown, with indistinct darker discal and a marginal band.

Expanse $1\frac{1}{2}$ inch.

Hab. Darjiling. In coll. Dr. Staudinger and F. Moore.

Allied to *A. flavistigma* (*N. flavistigma*, Moore, P. Z. S. 1867, p. 50). Distinguished from it by the oblique position of the orbicular spot, and in the central streak below the cell being simply a longitudinal zigzag black line.

MOTAMA, nov. gen.

Fore wing short, costa slightly arched at base, apex pointed, exterior margin even, oblique hindward; first subcostal emitted at one half length of the cell, second close to the end, trifurcate, fifth from the end of cell and touching third near its base; upper discocellular very short, lower slightly concave, radial from end of upper; two upper medians from angles at end of cell, third from near its end, fourth from opposite first subcostal. Hind wing short; two subcostal branches from end of the cell: upper discocellular long, concave, lower very short, radial from lower end of upper and close to median vein; two upper median branches from end of the cell, lower from half distance before its end. Body short with an anal tuft; palpi ascending, squamose, second joint stout, third slender; legs squamose; antennæ setose.

MOTAMA CIDARIOIDES, n. sp. (Plate IV. fig. 9.)

Fore wing dark purple-brown, with a transverse subbasal sinuous and an irregular waved discal black-bordered golden-yellow line; discal area below the apex golden-yellow, with a marginal lunular black line and a patch in its middle; orbicular spot indistinct, reniform spot golden-yellow and brown-streaked: hind wing cinnamon-brown. Thorax, palpi, and legs in front purple-brown; abdomen and legs beneath paler.

Expanse 1,$\frac{1}{10}$ inch.

Hab. Darjiling. In coll. Dr. Staudinger.

MOTAMA AURATA, n. sp. (Plate IV. fig. 10.)

Male. Fore wing golden-yellow, with an interrupted subbasal and discal transverse sinuous black line, the latter slightly white-bordered towards the costa; within the band is a dark golden-brown patch between the orbicular and reniform spots, and a grey-speckled patch below the cell; some dark golden-brown streaks at base of wings, a short irregular interrupted marginal band below the apex, and an outer marginal lunular line: hind wing uniform pale brown Thorax golden-yellow, speckled with dark golden-brown; palpi and legs above blackish-brown with pale bands: abdomen pale brown, tuft brighter.

Expanse 1,$\frac{1}{12}$ inch.

Hab. Darjiling. In coll. Dr. Staudinger.

MOTAMA DECORATA, n. sp. (Plate IV. fig. 11.)

Male. Fore wing dark golden-brown; costal border and the veins white-speckled; orbicular and reniform spot and a sinuous bordered interrupted discal band golden-yellow; some streaks at base, and speckles below the cell, and a sinuous submarginal line also golden-yellow: hind wing uniform dark brown. Thorax dark golden-brown, grey-speckled; palpi and legs above blackish with pale bands; abdomen brown.

Expanse 1$\frac{1}{4}$ inch.

Hab. Darjiling. In coll. Dr. Staudinger.

Genus MAMESTRA, *Ochs.*

MAMESTRA RENALBA, n. sp.

Fore wing dark cupreous-brown; basal half thickly black-speckled, crossed by two sub-basal and two discal indistinct transverse sinuous grey bands, a more prominent black outer discal waved band, and a marginal row of spots with intervening white dots; reniform mark distinctly white-speckled: hind wing pale brown, whitish towards the base. Thorax, head, palpi, and legs dark brown, the latter with pale bands.

Expanse $1\frac{4}{10}$ inch.

Hab. Darjiling. In coll. Dr. Staudinger and F. Moore.

MAMESTRA DECORATA, n. sp. (Plate IV. fig. 8.)

Fore wing purple-brown; veins grey-speckled; with a broad medial pale-bordered dark golden band, the inner border of which is oblique and the outer irregularly waved; orbicular and reniform spots pale-lined and centred with ochreous-brown: a brown basal streak on posterior margin, and outer marginal transverse zigzag line with grey-speckled borders; posterior margin fringed with ochreous: hind wing ochreous-brown; cilia brighter, edged with grey; an indistinct transverse discal brown fascia. Thorax, head, palpi and legs above purple-brown; palpi and legs speckled with ochreous-brown; abdomen ochreous-brown.

Expanse $1\frac{7}{10}$ inch.

Hab. Darjiling. In coll. Dr. Staudinger.

Genus PROSPALTA, *Walker.*

PROSPALTA STELLATA, n. sp.

Fore wing dark brown; crossed by some medial pale-bordered black sinuous lines; three clusters of pure white spots within the cell and another cluster below its end, a lobate spot at base of the wing; three spots on the costal edge, some on base of posterior border, a row of smaller spots crossing the disk, and two interrupted marginal rows; two series of white spots also on the cilia: hind wing brown, with indistinct paler tranverse discal waved line; cilia bordered with white. Thorax dark brown, abdomen paler brown, with a white-speckled dorsal crest; palpi blackish with pale bands; legs above black, banded with white.

Expanse, ♂ $1\frac{4}{8}$ inch, ♀ $1\frac{3}{8}$ inch.

Hab. Darjiling. In coll. Dr. Staudinger and F. Moore.

Allied to *P. leucospila*, Walker, Cat. Lep. Het. B. M. xiii. p. 1114, and to *P. albomaculata* (*Mamestra albomaculata*, Moore, P. Z. S. 1867, p. 52).

Genus ILATTIA, *Walker.*

ILATTIA APICALIS, n. sp.

Male and female. Fore wing dark brown, crossed by an indistinct subbasal, an ante-medial, and a more distinct postmedial pale-bordered black sinuous lunular band, the latter bordered externally by white-tipt black points; a recurved medial black fascia and a sub-marginal waved fascia, the latter terminating at the apex in a patch of white speckles; some pale ochreous streaks on the costal edge, a constricted reniform spot with a white-speckled border; a marginal row of minute white-pointed black dots: hind wing pale brown. Thorax and palpi dark brown. legs brown-speckled, with darker tarsal bands.

Expanse 1$\frac{1}{12}$ inch.

Hab. Darjiling. In coll. F. Moore and Dr. Staudinger.

ILATTIA MONILIS.

Ilattia monilis, Moore, P. Z. S. 1881, p. 348, pl. xxxviii. fig. 11.

Fore wing yellowish-ochreous, crossed by a slender indistinct whitish basal, antemedial and postmedial sinuous bands, a more distinct white submarginal macular band, and a mar-ginal row of black white-bordered dots; some whitish streaks on the costal edge, a prominent white spot at base of the cell. and white-bead-bordered orbicular and reniform marks: hind wing pale brownish-ochreous, with indistinct transverse narrow discal band and discocellular streak. Thorax and palpi ochreous; abdomen and legs paler; tarsi with brownish bands.

Expanse 1 inch.

Hab. Darjiling. In coll. Dr. Staudinger.

ILATTIA CERVINA.

Ilattia cervina, Moore, P. Z. S. 1881, p. 348, pl. xxxviii. fig. 12.

Male and female. Fore wing greyish-brown, crossed by an indistinct darker brown zigzag antemedial and a recurved postmedial line, the latter bordered externally by indistinct brown points; a short subbasal line, and a submarginal pale-bordered fascia; orbicular spot small and circular, ochreous-brown; reniform spot partly ochreous-brown and yellow, bordered with white speckles above and below: hind wing pale greyish-brown.

Expanse 1$\frac{1}{2}$ inch.

Hab. Darjiling. In coll. Dr. Staudinger.

ILATTIA CUPREIPENNIS, n. sp.

Male and female. Fore wing dark cupreous; veins speckled with grey and black, the speckles more strongly marked antemedially and postmedially, thus forming a more distinct transverse linear series; orbicular and reniform marks white-and-black speckled; costal edge grey-speckled: hind wing pale brown. Thorax and palpi dark cupreous-brown; legs speckled with brown.

Expanse, ♂ $1\frac{2}{14}$, ♀ $1\frac{4}{14}$ inch.

Hab. Darjiling. In coll. Dr. Staudinger and F. Moore.

Nearest to *I. cephusalis*, Walker, from Ceylon.

ILATTIA RENALIS, n. sp.

Male and female. Fore wing dark ferruginous, crossed by a subbasal, an antemedial, and a postmedial distinctly white-speckled bordered black sinuous line, the latter with external white-tipt black points; an indistinct darker medial and a submarginal waved fascia, the latter terminating at the apex in a white-speckled patch; some white costal streaks; reniform spot constricted, the upper half with white-speckled border, the lower slightly ochreous: hind wing dark brown. Thorax dark ferruginous; palpi dark brown; legs brown-speckled, and with blackish tarsal bands.

Expanse $\frac{11}{14}$ inch.

Hab. Darjiling. In coll. Dr. Staudinger and F. Moore.

Closely allied to *I. apicalis.*

Genus CELÆNA, *Steph.*

CELÆNA SIKKIMENSIS.

Celæna sikkimensis, Moore, P. Z. S. 1881, p. 348, pl. xxxviii. fig. 16.

Male. Fore wing brown, with an indistinct black antemedial and postmedial transverse pale-bordered sinuous line, the interspace blackish-streaked; a submarginal series of longitudinal black streaks disposed between the veins, and crossed by a whitish-speckled lunular line; orbicular and reniform spots white, the latter most prominent: hind wing paler brown, with paler discal and submarginal fascia. Body brown; palpi ochreous-brown; legs with pale bands.

Expanse $1\frac{1}{8}$ inch.

Hab. Sikkim (*Blanford*). In coll. Dr. Staudinger.

CHANDATA, nov. gen.

Fore wing somewhat short, narrow; costa nearly straight, apex acute, exterior margin slightly oblique and convex in middle; cell long. Hind wing with the costal vein straight, extending to apex; two subcostals from end of cell; upper discocellular concave, lower short and oblique, radial from end of upper; cell short, two upper medians from end of the cell. Body moderate; thorax robust; palpi porrect, thickly pilose, third joint short, cylindrical; antennæ bipectinate to tip in male, setose in female.

E

CHANDATA PARTITA, n. sp. (Plate IV. fig. 16.)

Male and female. Fore wing bluish-white, with three blue-black small basal costal spots, followed by a larger central spot crossing the cell, and succeeded by a dentate spot, with an intervening lower oval spot beyond the cell; a lower basal triangular mark and an oblique streak at posterior angle; an indistinct indigo-blue postmedial transverse sinuous line, a lower postmedial double line, and an upper submarginal line; a marginal row of black lunules and cilial streaks: hind wing and abdomen uniformly brown; cilia edged with white. Thorax and head white, with black collar and border to tegulæ; palpi and legs black, second and third joint of palpi white-tipt; legs with white bands.

Expanse $1\frac{9}{14}$ inch.

Hab. Darjiling. In coll. Dr. Staudinger.

Genus LUPERINA, *Boisd.*

LUPERINA PARDARIA, n. sp. (Plate IV. fig. 12.)

Male and female. Brownish-ochreous: fore wing with a transverse erect antemedial and an oblique postmedial duplex sinuous black line; orbicular and reniform spots indistinctly black-lined, and with central streak; a black-lined conical mark below the orbicular spot; some black sinuous basal streaks, and short streaks on the costal border; veins across the disk slightly black-speckled; a submarginal pale zigzag line slightly bordered on each side with blackish dentate streaks; a marginal slender dentate black line: hind wing cupreous-brown; cilia ochreous. Thorax, palpi, and legs ochreous, speckled with black; abdomen brown; tarsi with black bands.

Expanse $1\frac{1}{2}$ inch.

Hab. Darjiling. In coll. Dr. Staudinger.

Allied to *L. luteago.*

LUPERINA OLIVASCENS, n. sp.

Fore wing olivaceous; markings interspersed with brown; with a black transverse basal, antemedial, and a postmedial sinuous duplex line, and costal streaks; a submarginal blackish-bordered sagittate fascia, and a prominent marginal dentated line; orbicular and reniform marks black-lined; a black conical mark below the orbicular: hind wing and abdomen cupreous-brown. Thorax, palpi, and legs olivaceous, slightly black-speckled.

Expanse $1\frac{3}{8}$ inch.

Hab. Darjiling. In coll. Dr. Staudinger and F. Moore.

LUPERINA LAGENIFERA, n. sp.

Male and female. Ochreous: fore wing with a transverse medial ochreous-red flagon-shaped band, bordered on each side by a waved double dark-brown pale line orbicular spot pale ochreous, reniform with brown centre; intervening space and streaks on the costa very dark brown; below the orbicular spot is a brown lunule, and a brown fascia from the reniform spot to inner end of the band; a subbasal sinuous brown double pale line; veins across the discal area black-streaked, and a black spot before the apex: hind wing with a dusky narrow brown transverse discal line. Thorax grey; collar, frontal tuft, palpi, and legs ochreous, the latter with brown bands.

Expanse, ♂ 1¼, ♀ 1½ inch.

Hab. Darjiling. In coll. Dr. Staudinger and F. Moore.

Genus PACHÆTRA, *Dup.*

PACHÆTRA HETEROCAMPA, n. sp. (Plate IV. fig. 15.)

Fore wing black, variegated with greyish-purple and ochreous; with pale ochreous black-streaked irregular sinuous transverse basal bands; a bilobed spot obliquely below the cell, and an outer marginal dentate band; orbicular and reniform spots, and a narrow sinuous white discal line with black borders: hind wing pale cinereous-brown, with indistinct whitish marginal dentate band, slightly black-freckled above from the anal angle. Thorax black; tegulæ, crest, collar, and tip of palpi ochreous; legs black, with ochreous-white bands; antennæ bipectinated and brown; abdomen brown, with slight dorsal black-and-white crest.

Expanse 1¾ inch.

Hab. Darjiling (May). In coll. Dr. Staudinger.

Family CARADRINIDÆ.

Genus CARADRINA, *Ochs.*

CARADRINA ARENACEA.

Caradrina arenacea, Moore, P. Z. S. 1881, p. 349.

Fore wing pale greyish-ochreous, with an indistinct transverse sinuous brown line and three discal lines, the outer line bordered below the apex with ochreous-yellow; reniform spot brownish, speckled with white and ochreous-yellow; a small indistinct brown orbicular spot; some black short streaks on the costa; a marginal row of dentate points: hind wing whitish-ochreous, with pale-brown border. Body whitish-ochreous; thorax with a few blackish speckles; palpi blackish laterally, white at tip; fore and middle legs with blackish speckles and tarsal bands.

Expanse 1₁³⁄₂ inch.

Hab. Masuri (*Lang*); Darjiling (*Atkinson*). In coll. Dr. Staudinger and F. Moore.

Caradrina delecta, Moore, P. Z. S. 1881, p. 349, pl. xxxviii. fig. 15.

Fore wing pale greyish-ochreous, with an indistinct black-speckled subbasal and three medial transverse sinuous lines, and an outer discal row of points; a submarginal transverse straight pale line; orbicular and reniform spots black-speckled: hind wing ochreous-white. Body ochreous-grey; palpi and legs above brown-speckled.

Expanse 1⅛ inch.

Hab. Darjiling. In coll. Dr. Staudinger.

Allied to *C. kadeni.*

Genus ACOSMETIA, *Stephens.*

Acosmetia nebulosa, Moore, P. Z. S. 1881, p. 350, pl. xxxviii. fig. 13.

Upperside pale ochreous-brown: fore wing darkest, with several transverse indistinctly darker waved narrow fasciæ; some pale spots on costal edge near the apex. Underside paler.

Expanse 1 inch.

Hab. Darjiling. In coll. Dr. Staudinger.

Genus AGROTIS, *Ochs.*

Near to *A. corticea. Male* differs in the fore wing having a darker costa, darker strigæ, a more prominent subbasal and antemedial transverse waved black line, and a shorter pointed mark below the cell, a less-defined transverse discal sinuous line, and no submarginal fascia, but a more distinctly formed marginal brown border; and the hind wing is generally paler. *Female* differs in having the medial area between the transverse lines and the outer margin either brown or dark slaty-grey, the basal and submarginal areas being much paler.

Expanse 1⅔ to 1¾ inch.

Hab. Darjiling; Punjab Hills. In coll. Dr. Staudinger and F. Moore.

Agrotis costigera, Moore, P. Z. S. 1881, p. 350.

Male and female. Fore wing dark purple-brown, the basal and discal areas suffused with grey; costal border pale purplish-ochreous; a transverse basal, antemedial, and a postmedial sinuous black pale-centred line; orbicular and reniform spots black-lined, dark-centred, and with a pale ochreous inner border, the reniform slightly angled at the upper end of its inner border; an oval black mark below the cell; a pale ochreous submarginal line, with irregular alternate purple-brown and reddish inner border; a marginal row of distinct black dentate marks; some pale spots on costa near apex: hind wing cinereous-brown, palest at the base. Thorax and palpi dark purple-brown, abdomen and legs greyish-brown, legs with brown bands.

Expanse, ♂ 1$\frac{4}{12}$, ♀ 1$\frac{6}{12}$ inch.

Hab. Solun, Punjab (*Reid*), Cherra Pungi, Assam (*Atkinson*). In coll. F. Moore and Dr. Staudinger.

Allied to *A. christophi*, Staudinger.

AGROTIS PLACIDA, n. sp. (Plate IV. fig. 19.)

Male and female. Dark grey: fore wing crossed by a black short subbasal double line, an antemedial and postmedial sinuous line, a submarginal yellowish lunular line, bordered inwardly by blackish dentate marks; a distinct black marginal lunular line; orbicular and reniform spots black-lined, with black interspaces; a black blunt dentate mark below the orbicular spot from the sinuous line: hind wing and body brownish grey. Thorax and legs dark grey, black-spotted; tarsi with black bands.

Expanse 1$\frac{5}{12}$ inch.

Hab. Kaschmir (Sind Valley); Calcutta. In coll. Dr. Staudinger.

Has somewhat the appearance of *A. multangula*, Hübn.

Genus TIRACOLA, *Moore.*

Tiracola, Moore, P. Z. S. 1881, p. 351.

Fore wing long, narrow, costa slightly arched towards the end, apex acute; exterior margin oblique, convex, and sinuous; cell long; first subcostal branch emitted at half length of the cell, second at three fourths its length, trifurcate, third emitted at one third beyond base of the second, and fourth at about one half from base of second, fifth from end of the cell, and touching the third near its base; discocellular deeply concave, upper radial from angle close to upper end of the cell, lower radial from angle near lower end of the cell; upper median from angle above and middle median from end of the cell, third at one third before its end; submedian concave near its base. Hind wing triangular, apex convex; exterior margin oblique, recurved, slightly sinuous; cell short; two subcostal branches from end of the cell; discocellular very slender, angled inward in the middle, a slender radial from the angle; two upper median branches from lower end of the cell, third branch at half before its end. Body stout, abdomen long; palpi short, ascending, compactly clothed, second joint thick, third joint short; legs compactly pilose above; antennæ setose.

Type *T. plagiata.*

TIRACOLA PLAGIATA.

Agrotis plagiata, Walker, Catal. Lep. Het. B. M. xi. p. 740 (1857).
Agrotis plagifera, Walker, *l. c.* p. 741.

Hab. Ceylon; S. India; Darjiling (*Atkinson*).

Genus GRAPHIPHORA, *Ochs.*

GRAPHIPHORA FLAVIRENA.

Graphiphora flavirena, Moore, P. Z. S. 1881, p. 352, pl. xxxviii. fig. 3.

Allied to *G. neglecta,* Hübner. Fore wing dark ferruginous; crossed by an indistinct brown-bordered pale-waved antemedial and a postmedial line, a medial brown fascia angled at lower end of the cell; orbicular spot obsolete; reniform spot narrow, yellowish, and dusky at its lower end; hind wing ferruginous-brown. Body and legs ferruginous.

Expanse 1½ inch.

Hab. Darjiling. In coll. Dr. Staudinger.

GRAPHIPHORA NIGROSIGNA.

Graphiphora nigrosigna, Moore, P. Z. S. 1881, p. 352, pl. xxxviii. fig. 4.

Fore wing brownish-ochreous, crossed by indistinct basal, antemedial, and postmedial zigzag brown duplex lines and a waved submarginal pale line; veins across the disk with indistinct black-pointed pale spots; orbicular and reniform marks paler, with brown border, lower lobe of reniform dusky; a prominent black triangular spot below the orbicular mark: hind wing and abdomen pale brownish-ochreous. Thorax ochreous; second joint of palpi at the side and legs above dark brown, third joint of palpi and tip of second joint ochreous.

Expanse 1¼ inch.

Hab. Tonglo, Sikkim. In coll. Dr. Staudinger.

GRAPHIPHORA VULPINA, n. sp.

Fore wing dull deep chestnut-brown, suffused with grey; costal margin and the veins grey-and-black speckled; a large more or less distinct yellowish quadrate reniform spot: hind wing pale chestnut-brown; cilia brighter. Thorax, head, and palpi deep chestnut-brown; thorax in front, palpi at sides, and legs above black-and-grey speckled.

Expanse 1⅝ inch.

Hab. Darjiling. In coll. Dr. Staudinger and F. Moore.

Allied to *G. sobrina* and *G. festiva* of Europe.

GRAPHIPHORA INTERSTINCTA, n. sp.

Male and female. Fore wing dark ochreous-brown, crossed by an antemedial and post-medial narrow black waved line, the latter bordered externally by a row of black-pointed dots, a medial transverse dusky waved fascia, a subbasal pale-bordered dark-brown sinuous line, and a submarginal red-brown pale-bordered waved fascia; reniform spot red-brown, its external edge pale ochreous and white-speckled, orbicular spot indistinct: hind wing paler brown. Thorax, palpi, and legs ochreous-brown; tarsi with pale bands; abdomen paler brown.

Expanse 1 7/16 inch.

Hab. Darjiling. In coll. Dr. Staudinger and F. Moore.

Allied to the European *G. bella,* Bork.

GRAPHIPHORA STELLATA, n. sp.

Male and female. Fore wing greyish-brown, crossed by a darker brown subbasal, an antemedial, and a postmedial waved line, the latter bordered by a contiguous row of dark-brown fine points, one on each vein ; a medial transverse and a submarginal waved fascia ; reniform spot in male pale ochreous-yellow, with upper and lower white dots, in female white and more distinct: hind wing pale greyish-brown. Body greyish ochreous-brown ; tarsi with pale bands.

Expanse 1¼ inch.

Hab. Khasia Hills ; Darjiling. In coll. Dr. Staudinger and F. Moore.

GRAPHIPHORA COGNATA, n. sp.

Male and female. Fore wing greyish-brown, crossed by seven darker brown waved lines; reniform spot indistinct, small, pale ochreous, with upper and lower white dots: hind wing brownish-white ; veins and marginal line brown. Body greyish-brown, first and second joints of palpi partly black ; legs blackish-speckled, tarsal bands pale.

Expanse ⅞ inch.

Hab. Darjiling. In coll. Dr. Staudinger and F. Moore.

Genus MEGASEMA, *Hübn.*

MEGASEMA CINNAMOMEA.

Megasema cinnamomea, Moore, P. Z. S. 1881, p. 352, pl. xxxviii. fig. 6.

Fore wing dull cinnamon-brown, with an indistinct darker basal, subbasal, and a discal, transverse sinuous line ; a more distinct outer discal pale-bordered irregular line ; a large pale-bordered orbicular and reniform spots, their interspace within the cell and a transverse medial fascia darker brown ; outer margin also darker brown, with a pale-bordered lunular line: hind wing paler. Underside pale brighter cinnamon-brown. Thorax, head, palpi, and legs above dark cinnamon-brown.

Expanse 1⅞ inch.

Hab. Darjiling. In coll. Dr. Staudinger.

Genus HERMONASSA, *Walker.*

HERMONASSA CHALYBEATA.

Hermonassa chalybeata, Moore, P. Z. S. 1881, p. 353, pl. xxxviii. fig. 17.

Smaller than *H. consignata.* Fore wing dark brown, with indistinctly darker markings, which are all bordered with chalybeous-grey speckles: hind wing pale cinereous-brown. Underside cinereous-brown.

Expanse 1 inch.

Hab. Darjiling. In coll. Dr. Staudinger.

HERMONASSA SINUATA.

Hermonassa sinuata, Moore, P. Z. S. 1881, p. 353, pl. xxxviii. fig. 17.

Near *H. cuprina*. Fore wing paler and of a brighter cupreous-brown, the interspace between the transverse basal and subbasal lines wider, the latter more acutely sinuous, and its lower end almost touching that of the discal line, which latter is also more sinuous: hind wing very pale brown. Underside also much paler.

Expanse 1⅝ inch.

Hab. Darjiling. In coll. Dr. Staudinger and British Museum.

HERMONASSA INCISA, n. sp.

Nearest *H. lanceola*. Somewhat larger. Fore wing greenish-brown or greyish ochreous-brown, the mark projecting below the cell from the transverse subbasal line less lanceolate in form, being shorter and much broader, the transverse discal double line more sinuous: hind wing paler, and of a cinereous-brown.

Expanse 1¾ inch.

Hab. Darjiling. In coll. Dr. Staudinger and F. Moore.

HERMONASSA CUPRINA, n. sp.

Fore wing dark cupreous-brown; the basal markings, orbicular and reniform spots darker brown and with grey outer lines; discal and marginal line also greyish: hind wing duller cupreous-brown. Underside uniformly bright cupreous-brown.

Expanse 1¼ inch.

Hab. Darjiling. In coll. Dr. Staudinger and F. Moore.

HERMONASSA LUNATA, n. sp.

Fore wing dark purple-brown; with black basal and costal streaks, a mark below the cell, a triangular-shaped orbicular and a lunate reniform spot, marginal dots, and an indistinctly darker outer discal sinuous fascia: hind wing pale cinereous-brown. Collar and palpi at sides black, the latter with pale bands.

Expanse 1½ inch.

Hab. Kaschmir (Sind Valley). In coll. Dr. Staudinger and F. Moore.

Family ORTHOSIIDÆ.

Genus ORTHOSIA, *Ochs.*

ORTHOSIA RECTIVITTA.

Orthosia rectivitta, Moore, P. Z. S. 1881, p. 353.

Fore wing mottled with greyish-ochreous and dark brown, posterior and exterior areas more uniformly brown ; a transverse sinuous antemedial and a straight postmedial pale line with black-speckled borders ; a submarginal zigzag pale-bordered black-speckled fascia ; orbicular and reniform spots pale, with black-speckled border ; a marginal row of black dots : hind wing brown ; cilia pale ochreous-brown. Thorax and head grey-speckled ; palpi with broad brown lateral bands ; legs dark brown with ochreous speckles and tarsal bands.

Expanse 1⅔ inch.

Hab. Darjiling (*Russell & Atkinson*). In coll. F. Moore and Dr. Staudinger.

RANAJA, nov. gen.

Wings somewhat short : fore wing slightly arched at the base, apex acute, exterior margin straight, lower angle convex, posterior margin convex at the base ; cell short ; first subcostal branch widely separated from costal, second emitted at one fourth before end of the cell, bifid, the third starting from one third its length, fourth from end of the cell and anastomosed to third for a short distance, fifth from below the fourth opposite the third ; radial straight from upper end of the cell ; discocellular concave, outwardly oblique, three upper median branches from angles at end of the cell, lower from opposite second subcostal. Hind wing somewhat quadrate, anterior margin straight, exterior margin angular in the middle and truncate at posterior angle, abdominal margin short ; cell short ; two subcostal branches from end of the cell ; discocellular very concave, radial from angle above lower end of the cell and immediately contiguous to the two upper median branches from end of the cell. Palpi compactly pilose beneath, second joint stout, third slender ; antennæ filiform ; tibiæ pilose beneath.

RANAJA FASCIATA, n. sp. (Plate IV. fig. 18.)

Fore wing greyish purple-brown, with an antemedial transverse outwardly-oblique straight sap-green line, and two postmedial similar wavy lines, the antemedial line with grey-speckled exterior margin ; basal area and within the first postmedial line greenish ; a slight pinkish patch near base of posterior margin : hind wing dull brown. Thorax, head, and palpi greenish-brown ; fore legs above dark-brown- and grey-speckled, white beneath ; middle and hind legs white with a brown knee-spot and speckles above, tarsi brown and banded with white.

Expanse 1½ inch.

Hab. Darjiling. In coll. Dr. Staudinger.

DIMYA, nov. gen.

Fore wing—costa nearly straight, apex acute ; exterior margin oblique, slightly sinuous, and almost angular hindward, posterior margin short ; first subcostal emitted at one half length of the cell, second at one fourth, trifurcate, third starting from immediately above end of the cell, the fourth at one half length from base of second and terminating at apex, fifth curving up from end of the cell and touching third near its base ; radial from upper end of

F

the cell, discocellular slightly concave; two upper medians from angles above lower end of the cell, third from its end, fourth from opposite second subcostal; submedian close to margin: hind wing short, exterior margin slightly sinuous and almost angular beyond the middle, abdominal margin short; costal vein recurved, extending to apex; two subcostals from a short distance beyond end of the cell; upper discocellular obliquely concave, lower shortest and obliquely straight; radial from end of upper; two upper medians from end of the cell, lower from some distance before its end. Body stout; palpi porrect, very laxly pilose, third joint long; antennæ setose; legs pilose beneath.

Near *Iliptelia*.

DIMYA SINUATA, n. sp. (Plate IV. fig. 17.)

Male. Greyish-ochreous, minutely brown-speckled: fore wing with a brown transverse waved subbasal line, a sinuous discal line, a large pale-centred orbicular and reniform spot, a medial transverse fascia, and a submarginal series of darker brown points; a darker marginal line, and a few blackish speckles at lower end of reniform spot: hind wing with the posterior half pale dusky brown. Underside—fore wing with the basal area and indistinct transverse discal line pale dusky brown: hind wing with a transverse discal waved brown line. *Female* darker-coloured; markings less distinct. Body ochreous-brown; palpi and legs darker brown.

Expanse $1\frac{2}{7}$ inch.

Hab. Jongei, 13,500 feet (October); Darjiling, 10,000 feet. In coll. Dr. Staudinger.

Genus TÆNIOCAMPA, Guén.

TÆNIOCAMPA CASTANEIPARS, n. sp.

Allied to *T. leucographa*, W. V. Fore wing dark purplish chestnut-brown, indistinctly washed with chalybeate-grey; an antemedial and a postmedial transverse sinuous indistinct black line: hind wing pale cinereous-brown. Underside much paler. Thorax, head, palpi, and legs above dark chestnut-brown; abdomen cinereous-brown.

Expanse $1\frac{1}{4}$ inch.

Hab. Darjiling. In coll. Dr. Staudinger and F. Moore.

Family COSMIIDÆ.

Genus COSMIA, Ochs.

COSMIA HYPENOÏDES.

Cosmia hypenoïdes, Moore, P. Z. S. 1881, p. 354, pl. xxxviii. fig. 19.

Male and female. Fore wing dark purple-brown, crossed by an outwardly-oblique antemedial, a medial, and an outwardly-angled postmedial transverse pale-bordered black line; a black dot at base of cell, a submarginal indistinct lunular line, and a marginal black line; orbicular and reniform spot very indistinct: hind wing and abdomen dull brown. Thorax purple-brown; palpi grey, with broad black band on second joint; fore and middle legs with blackish bands.

Expanse 1_1^1 inch.

Hab. Parisnath Hill (Bengal). In coll. Dr. Staudinger.

Genus IPIMORPHA, *Hübn.*

IPIMORPHA DIVISA, n. sp.

Near to *I. subtusa.* Fore wing ochreous-brown; crossed by a straight subbasal, an antemedial, a waved postmedial, and a straight submarginal ochreous line; orbicular and reniform spots lined with ochreous: hind wing greyish-brown. Body ochreous-brown; palpi, and bands on legs dark brown.

Expanse 1 inch.

Hab. Darjiling; Simla. In coll. Dr. Staudinger and F. Moore.

Family HADENIDÆ.

Genus DIANTHECIA, *Boisd.*

DIANTHECIA CONFLUENS.

Dianthecia confluens, Moore, P. Z. S. 1881, p. 354, pl. xxxviii. fig. 20.

Fore wing dark purple-brown, crossed by a subbasal black-bordered pale ochreous sinuous line, a similar antemedial and a postmedial narrow waved line, and a submarginal brown-bordered pale line; orbicular and reniform spots pale ochreous with brown centre; the two lower median branches pale ochreous to the postmedial line; a black dentate mark below the cell; a marginal row of dentate lunules; cilia brown: hind wing greyish dusky-brown, palest at base, and with a pale-bordered brown transverse discal line; cilia ochreous. Body and legs dark purple-brown, black-speckled.

Expanse $1\frac{1}{4}$ inch.

Hab. Darjiling. In coll. Dr. Staudinger.

DIANTHECIA STELLIFERA, n. sp.

Male. Fore wing ænescent-ochreous, crossed by a subbasal and antemedial zigzag paler yellow line and a postmedial sinuous line, the two former with broad clouded-black

P 2

outer border, and the latter with a similar inner border; a submarginal yellow zigzag line with interrupted black border; some yellow spots on costal edge; a marginal row of black dentate spots; orbicular and reniform quadrate marks ill-defined, the latter with a bright yellow contiguous spot at lower angle of the cell: hind wing and abdomen ænescent purplish brown. Thorax, head, palpi, and legs ænescent-ochreous.

Expanse $1\frac{6}{10}$ inch.

Hab. Darjiling. In coll. Dr. Staudinger.

DIANTHECIA LITERATA, n. sp.

Fore wing brownish-ochreous, crossed by a subbasal, antemedial, and a postmedial yellow zigzag black-bordered line and a less defined submarginal zigzag fascia; a broad obliquely-quadrate black-bordered yellow orbicular and reniform marks, which are confluent hindward and extend at an equal width below the cell to submedian vein, thus forming a large letter Y; costal edge with black and yellow spots; exterior border with a marginal row of yellow-bordered dentate marks: hind wing pale purplish-cinereous. Thorax, head, palpi, and legs brownish-ochreous, yellow-speckled; collar blackish.

Expanse $1\frac{3}{10}$ inch.

Hab. Darjiling. In coll. Dr. Staudinger and F. Moore.

DIANTHECIA VENOSA, n. sp.

Similar to *D. literata*. Fore wing paler-coloured, the veins speckled with lilac-grey; with bright olive-yellow transverse markings, but the space between orbicular and reniform marks more triangular, and the portion below the cell narrower; a slight zigzag blackish fascia below the apex only, and no black dentate marks on exterior margin, but replaced by olive-yellow marks: hind wing pale purplish-cinereous. Thorax yellow.

Expanse $1\frac{2}{10}$ inch.

Hab. Darjiling. In coll. Dr. Staudinger and F. Moore.

DIANTHECIA CALAMISTRATA, n. sp.　(Plate IV. fig. 23.)

Near to *D. perplexa*, Schiff. Fore wing pale ochreous-brown, with two subbasal and a discal transverse white-bordered black zigzag line, an outer discal white line with black dentate borders; a white inner-bordered black ringlet below the cell, and a white-bordered black orbicular and reniform mark; some short black streaks on the costa, and a marginal lunular spotted line; cilia alternate black and white: hind wing ochreous-brown; cilia paler. Thorax and collar black-speckled; palpi black at the side; abdomen beneath and legs above with black bands.

Expanse $1\frac{4}{10}$ inch.

Hab. Darjiling. In coll. Dr. Staudinger.

Genus HECATERA, *Guén.*

HECATERA TRANSVERSA, n. sp.

Allied to *H. serena. Male.* Fore wing with the basal and discal area greyish-white ; a medial transverse black-speckled band, which is broad on the costa and narrow on the hind margin, its inner border being outwardly oblique and nearly straight, the outer border sinuous and concave below end of the cell ; orbicular and reniform spots greyish-white, lined with black speckles ; a black transverse streak at base of wing, some streaks on the costa, and a narrow marginal dentate-bordered band : hind wing uniformly brown ; cilia greyish-white. Underside pale ochreous-brown, with dusky broad medial transverse indistinct fascia. Thorax and head hoary ; palpi ochreous, black-speckled ; legs black, banded with white ; abdomen brown.

Expanse $1\frac{5}{10}$ inch.

Hab. Kashmir (Margan Pass) ; Pir Pinjal. In coll. Dr. Staudinger and F. Moore.

HECATERA MODESTA, n. sp.

Male. Fore wing black, interspersed with white speckles ; basal, discal, and marginal sinuous lines, and orbicular and reniform spots, blacker, with white-speckled borders : hind wing uniformly dusky brown. Underside dusky brown. Thorax and head hoary grey ; palpi brown, speckled with white beneath ; legs brown, with white bands ; abdomen brown.

Expanse $1\frac{3}{10}$ inch.

Hab. Kashmir (Margan Pass, Sonamurg). In coll. Dr. Staudinger and F. Moore.

Allied to, but very distinct from, *H. magnolii* and *H. xanthocyanea.*

Genus EUPLEXIA, *Stephens.*

EUPLEXIA DISTORTA.

Euplexia distorta, Moore, P. Z. S. 1881, p. 354, pl. xxxviii. fig. 18.

Male and female. Fore wing black, with transverse distorted white bands, including an outer marginal sinuous-bordered band : hind wing pale blackish-cinereous externally and whitish basally ; cilia white. Thorax black, with white tegulæ, crest, and collar ; abdomen blackish at the tip ; palpi black, tipt with white ; legs black, with white streaks and tarsal bands.

Expanse $1\frac{3}{4}$ inch.

Hab. Darjiling. In coll. Dr. Staudinger.

EUPLEXIA SINUATA, n. sp. (Plate IV. fig. 25.)

Female. Fore wing dark brown, the cell including costal border and the outer margin below apex purple-brown ; a prominent white antemedial transverse deeply-sinuous black-bordered line, and a postmedial slightly sinuous line, the latter curved inwards towards the costa; a straight marginal sinuous line; some basal irregular white streaks; orbicular and reniform spots white bordered; discal area beyond the cell and outside the postmedial line ochreous, and an ochreous patch near the base: hind wing pale brown, with narrow white sinuous marginal band from anal angle, with a few black and white flecks up the two lower median veins. Body, palpi, and legs dark purple-brown, the latter with whitish tarsal bands.

Expanse 1½ inch.

Hab. Darjiling (April). In coll. Dr. Staudinger.

NIKARA, nov. gen.

Female. Fore wing—costa nearly straight, exterior margin oblique, slightly convex in the middle, posterior margin short; second subcostal emitted before end of the cell, trifurcate; third and fourth at equal distances from base of second; fifth curving upward from end of the cell and touching the third at nearly half distance between it and the fourth; radial from end of the cell; discocellular concave; two upper median branches from angles above end of the cell, third at its lower end, fourth some distance before the end. Hind wing —costa nearly straight; exterior margin convex, abdominal margin short; two subcostal branches from end of the cell; discocellulars concave, upper slightly longest; radial from angle of the middle; two upper medians from lower end of the cell, third at some distance before its end. Body short; palpi pilose, third joint minute; legs pilose beneath; antennæ filiform.

NIKARA CASTANEA, n. sp.   (Plate IV. fig. 24.)

Female. Fore wing chestnut-brown, the discoidal area and the cell darker brown, and bordered externally by a black waved line; orbicular and reniform spots, streaks between the base of median veins, two lines down the exterior margin, and the costal and posterior margins, speckled with greyish-white: hind wing brown, with an indistinct darker discal fascia. Underside pale brown, palest on basal areas; both wings with a transverse discal brown line. Thorax chestnut-brown, speckled with white; abdomen brown; legs dark brown, speckled with white.

Expanse 1₁⁵₂ inch.

Hab. Darjiling. In coll. Dr. Staudinger.

Genus APPANA, *Moore*.

Appana, Moore, P. Z. S. 1881, p. 355.

Fore wing somewhat short, costa straight, apex slightly pointed, exterior margin oblique, very slightly waved and convex; first subcostal branch emitted from half length of the cell, second at one fourth, trifurcate, third at one fourth, and fourth at one half from base of second ; fifth from end of the cell, projecting upward and touching the third near its base ; cell long ; upper discocellular very short, lower bent in the middle ; upper radial from angle near subcostal, lower radial and upper median from angles immediately above end of the cell, middle median from its end, lower median from one third before its end ; submedian slightly recurved. Hind wing triangular, exterior margin convexly oblique and waved ; two subcostal branches emitted from a short distance beyond end of the cell ; discocellular obliquely concave, radial from near its lower end ; two upper median branches from end of cell ; submedian and internal veins long. Body moderate ; abdomen extending beyond hind wing ; palpi short, stout, squamose, third joint very short ; antennæ setose ; legs pilose beneath.

Allied to *Habryntis*, Lederer (*H. scita*, Hübner).

APPANA INDICA.

Phlogophora indica, Moore, P. Z. S. 1867, p. 57.
Appana indica, Moore, P. Z. S. 1881, p. 355.

Fore wing pale purplish-ochreous, with a medial transverse broad band of chestnut-brown, palest on the costa, and enclosing a paler orbicular and reniform spots, which are confluent below the cell ; two black-speckled transverse streaks and a contiguous black spot at base of the wing ; a pale yellow submarginal line ; veins speckled with black and white : hind wing paler, with an indistinct darker discal and marginal band. Thorax chestnut-brown, with white-fringed collar ; abdomen paler ; palpi and legs ochreous-brown.

Expanse $1\frac{4}{10}$ inch.

Hab. Darjiling. In coll. F. Moore and Dr. Staudinger.

Genus EUROIS, *Hübner.*

EUROIS MAGNIFICA, n. sp.

Male. Fore wing purple-red, veins grey-speckled ; crossed by a yellow subbasal, an antemedial zigzag black-speckled bordered line, a postmedial dentate line (the outer points of which on the veins are white-and-black tipt), and a submarginal black-bordered dentate line ; the costal border and interspaces of the veins from base below the cell to postmedial line yellow-and-black speckled, the speckles most thickly disposed on each side of postmedial line ; orbicular and reniform marks yellow-lined, and with a grey-speckled central streak ; three yellow dots on costal border before the apex ; a marginal black dentate lunular line.

Female. Differs in having the markings nearly obsolete along the posterior border, the interspaces between orbicular and reniform spots black ; the postmedial and submarginal dentate lines not so widely apart, and the latter with more uniform black dentate inner border : hind wing cupreous-brown, paling to slaty-brown towards the base ; cilia greyish-white. Thorax and head ochreous-yellow, with a red-brown collar on outer edge of tegulæ ;

inner edge black-speckled; palpi red-brown, third joint and tip of second yellow; fore and middle tibiæ red in front; hind legs red-brown, with pale tarsal bands.

Expanse, ♂ 2⅗, ♀ 2⅞ inches.

Hab. Darjiling. In coll. Dr. Staudinger and F. Moore.

Allied to European *E. occulta* and to the Japanese *E. virens.*

Genus BERRILÆA.

Berrhæa, Walker, Catal. Lep. Het. B. M. xv. p. 1721 (1858); Moore, P. Z. S. 1881, p. 356.

Wings rather narrow: fore wing elongate; costa straight; apex slightly pointed; exterior margin oblique, scalloped; posterior margin recurved; costal vein extending to two thirds the margin; first subcostal branch emitted at half length of the cell, second trifurcate, emitted at one fourth before end of the cell, third at one fourth, and fourth at one half from base of second, fifth from end of the cell, curving upward and touching third near its base; discocellular slightly angled close to each end, deeply concave in middle; radials from its upper and lower angle; cell long, narrow; two upper median branches from angles at end of the cell, lower at one fourth before the end; submedian recurved. Hind wing long, exterior margin convex, waved; abdominal margin short; costal vein extending to apex; two subcostal branches from end of the cell; discocellular oblique, concave, radial from near its lower end; two upper median branches from end of the cell, lower at one third before its end; submedian and internal veins straight. Body stout; abdomen extending beyond hind wing; palpi stout, broad, ascending, densely clothed with long scales; third joint short, squamose; legs densely pilose; antennæ very minutely pectinate in male.

Type *B. aurigera.*

Allied to *Trachea (T. atriplicis).*

BERRILÆA AURIGERA.

Berrhæa aurigera, Walker, Catal. Lep. Het. B. M. xv. p. 1721 (1858); Moore, P. Z. S. 1881, p. 356.

Fore wing ochreous-brown, crossed by a pale-ochreous duplex black-bordered zigzag antemedial line and a sinuous postmedial line; some black-bordered ochreous basal streaks, others on the costal border and middle of posterior border; a submarginal zigzag interrupted ochreous and brown fascia; a black marginal lunular line; orbicular and reniform marks ochreous, large and widely separated at their anterior ends, but joined by a lower streak, which extends below the cell and runs into a whitish-ochreous quadrate spot: hind wing æneous-brown, the base being whitish-cinereous. Thorax ochreous-brown, abdomen cinereous-brown; palpi brown; legs brown with ochreous tarsal bands.

Expanse, ♂ 1⅞, ♀ 2 inches.

Hab. Darjiling (*Atkinson*). In coll. Dr. Staudinger.

BERRHÆA MEGASTIGMA.

Hadena megastigma, Walker, Catal. Lep. Het. B. M. xxxiii. p. 738 (1865).
Berrhæa megastigma, Moore, P. Z. S. 1881, p. 356.

Hab. Darjiling.

BERRHÆA ALBINOTA.

Hadena albinota, Moore, P. Z. S. 1867, p. 58.
Berrhæa albinota, Moore, P. Z. S. 1881, p. 356.

Hab. Darjiling.

BERRHÆA OLIVACEA.

Berrhæa olivacea, Moore, P. Z. S. 1881, p. 357.

Allied to *B. megastigma*; comparatively smaller. Fore wing similarly marked, but with less prominent ochreous and black streaks, the submarginal zigzag ochreous line slender throughout its length, the orbicular and reniform marks slightly smaller: hind wing cupreous-brown, slightly paler at the base.

Expanse, ♂ 1⅜, ♀ 1⅝ inch.

Hab. Darjiling. In coll. F. Moore and Dr. Staudinger.

Genus DRYOBATA.

DRYOBATA LEUCOSTICTA, n. sp. (Plate IV. fig. 22.)

Fore wing dark olive-brown, with a broad irregular transverse subbasal and a submarginal whitish band, both bordered by a sinuous black line; orbicular and reniform spots and a patch below the cell white, orbicular spot with a blackish centre; some white streaks on costal border; a marginal row of black spots bordered by a white lunule: hind wing pale brown. Thorax, palpi, and legs dark olive-brown, white-streaked; abdomen brown.

Expanse 1⅞ inch.

Hab. Tonglo, Darjiling, 10,000 feet. In coll. Dr. Staudinger.

Allied to *D. roboris* of Europe.

HYADA, nov. gen.

Male. Wings somewhat short: fore wing with the costa almost straight, exterior margin oblique and convex, posterior margin short; first subcostal emitted at two fifths before end of cell, second at one fifth, trifurcate, fifth curving up from end of the cell and touching third near its base; radial from end of the cell; discocellulars concave; two upper medians from angles above end of the cell, third at its end, fourth opposite the first subcostal. hind wing—costa long, exterior margin slightly sinuous and obliquely convex, abdominal margin short; two subcostal branches from beyond end of the cell; discocellular angled in the middle; radial from their angle; two upper medians from end of the cell. Body

u

moderate; thorax laxly clothed; palpi porrect, very laxly clothed, third joint long and slender; legs laxly pilose; antennæ finely bipectinate.

Allied to *Dasypolia*.

HYADA GRISEA, n. sp.   (Plate IV. fig. 26.)

Male. Fore wing ochreous-grey, brown-speckled; crossed by a sinuous subbasal and three discal brown lines, and a marginal row of broad dentate marks; a whitish orbicular and less distinct reniform spot: hind wing uniform ochreous-brown. Thorax grey; palpi, legs, and abdomen greyish-brown. Underside pale uniform ochreous-grey.

Expanse 1¾ inch.

Hab. Alutong, Sikkim, 15,000 feet (October).   In coll. Dr. Staudinger.

Genus HADENA, *Treits.*

HADENA CONSTELLATA, n. sp.   (Plate IV. fig. 21.)

Male and female. Fore wing dark ferruginous-brown ; crossed by an indistinct black subbasal, an antemedial and a postmedial double sinuous line, and a submarginal pale ochreous lunular line bordered on both sides by black points ; a marginal line of black dots and pale cilial spots ; a prominent black quadrate mark below the orbicular spot ; orbicular spot oblique and black-lined, reniform spot black-lined and with an internal pure white spot (lunular in male, quadrate in female), and upper and lower contiguous white dots: hind wing and abdomen paler ferruginous-brown. Thorax, palpi, and legs darker; palpi and legs black-speckled, tarsi black-banded.

Expanse 1¾ inch.

Hab. Darjiling.   In coll. Dr. Staudinger.

Allied to *H. satura* and *H. confundens.*

HADENA DISTANS, n. sp.

Near to *H. indistans.* Wings brighter ferruginous: fore wing with a transverse short subbasal, an antemedial and a postmedial sinuous double black line ; a black quadrate mark below the cell ; a pale-centred orbicular and reniform spots, the interspaces between them and the subbasal line very dark ferruginous-brown ; a submarginal transverse pale zigzag line with dark-brown inner dentate points and outer marginal border; hind wing and thorax dark ferruginous-brown ; abdomen paler.

Expanse 1¼ inch.

Hab. Darjiling ; Khasia Hills.   In coll. F. Moore and Dr. Staudinger.

HADENA HASTATA, n. sp.   (Plate IV. fig. 20.)

Male and female. Fore wing ferruginous, crossed by a subbasal zigzag grey-bordered prominent black line, which is double inwardly below the cell ; a recurved discal narrow sinuous pale-bordered black line ; outwardly below the cell is an elongated lanceolate black mark ; orbicular and reniform spots black-lined ; a marginal row of black points : hind wing pale whitish-ferruginous, with a slight brown discocellular spot. Thorax, palpi, and legs ferruginous ; palpi and legs brown-speckled ; tarsi with brown bands.

Expanse, ♂ 1⅛, ♀ 1¾ inch.

Hab. Darjiling. In coll. Dr. Staudinger.

CHUTAPHA, nov. gen.

Fore wing long, narrow ; costa straight ; apex acute ; exterior margin very oblique, posterior margin nearly straight ; third subcostal emitted at one third before end of the cell, second at one fourth, third at one fourth from second, fourth at one third from the third, fifth from end of the cell and touching the third near its base ; discocellular angled close to each end, concave in the middle ; radials from upper and lower angles ; two upper median branches from angles at end of the cell, lower at one third before its end ; submedian straight. Hind wing somewhat long, narrow ; exterior margin very oblique, abdominal margin short ; costal vein straight, extending to apex ; two subcostal branches from end of cell ; discocellular concave, slightly bent near lower end ; radial from the angle ; two upper median branches from end of the cell, lower at one third ; submedian and internal vein straight. Thorax moderately stout ; abdomen slender ; palpi porrect ; second joint pilose, third joint short, cylindrical ; femora pilose beneath ; antennæ minutely pectinate to tip.

CHUTAPHA COSTALIS, n. sp.

Male and female. Fore wing yellowish-ochreous, costal border whitish-ochreous ; with a transverse basal, antemedial, and a postmedial whitish-bordered indistinct black sinuous band, and a submarginal zigzag fascia ; orbicular and reniform marks whitish-ochreous, with a continuous quadrate whiter black-bordered spot below the cell : hind wing yellowish-cinereous, with a very indistinct brownish-white sinuous-bordered submarginal fascia ; lower median vein and cilial border brown-speckled. Thorax ochreous ; abdomen cinereous ; palpi and legs above ochreous-brown.

Expanse 1½ inch.

Hab. Darjiling. In coll. F. Moore and Dr. Staudinger.

Family X Y L I N I D Æ.

Genus CUCULLIA, *Ochs.*

CUCULLIA ATKINSONI, n. sp.

Fore wing purplish-brown ; veins darker ; the posterior and discal areas longitudinally streaked with purplish-grey ; a zigzag black streak between the median and submedian veins, a lengthened streak extending along the posterior margin to near the angle, and then ascending in a duplex zigzag line a short distance up the disk; some linear streaks at posterior angle, and a marginal lunular line; orbicular and reniform marks with grey inner border: hind wing cinereous, with broad brown outer band; cilia cinereous-white. Body grey; band on collar, dorsal tufts, front of head, palpi, and streaks on legs above dark brown.

Expanse 2⅓ inches.

Hab. Darjiling. In coll. Dr. Staudinger.

JARASANA, nov. gen.

Female. Fore wing elongated, narrow; costa straight; exterior margin oblique, scalloped; first subcostal emitted at nearly one half before end of the cell, second at one sixth, trifid, third at one sixth from its base, fourth at one third from below the third, fifth from end of cell and slightly touching third near its base; discocellular bent at each end, concave in middle; radials from the angles; two upper medians from angles at end of the cell, lower at one fifth before the end. Hind wing narrowed, scalloped; two subcostals from end of the cell; discocellular slightly bent inward at the middle; radial from near the lower end; two upper medians from acute end of the cell, lower close to the end. Body stout; palpi ascending, squamose; third joint long, slender, extending above vertex; legs thick; tibiæ pilose; antennæ simple.

JARASANA LATIVITTA, n. sp.

Female. Fore wing greyish-ochreous; some short indistinct outwardly oblique brownish streaks on the costa, one from the apex directed towards the disk, and a more distinct blackish grey-bordered streak from exterior margin above posterior angle, and terminating sinuously on the posterior margin, where it is diffusely extended; an indistinct black-speckled reniform mark, and a marginal row of dots: hind wing purplish-white, with a broad black apical transverse band; cilia white. Underside white; both wings with a blackish subapical transverse band. Body, palpi, and legs greyish-ochreous, sparsely brown-speckled.

Expanse 1⅔ inch.

Hab. Benares. In coll. Dr. Staudinger.

Genus CALOPHASIA, *Stephens*.

CALOPHASIA CASHMIRENSIS.

Calophasia cashmirensis, Moore, P. Z. S. 1881, p. 358.

Fore wing pale whitish-ochreous, with a pale yellowish-ochreous medial transverse band bordered on both sides by an indistinct black sinuous double line, and medially traversed below the cell by a more distinct black lunular fascia; costal border blackish; orbicular and reniform marks blackish, and lined with pale ochreous; base of wing black-speckled, and longitudinally streaked near posterior margin; some black dentate discal marks, a patch above posterior angle, and a marginal row of white-bordered black points; hind wing dusky-white, with a very pale dusky-black border, and a distinct black lunular marginal line. Body pale ochreous; palpi and legs black-speckled; legs with black tarsal bands.

Expanse $1\frac{1}{12}$ inch.

Hab. Changas, Cashmir. In coll. Dr. Staudinger.

Near to the European *C. linariæ*, Fabr.

Family HÆMEROSIIDÆ.

BAORISA, nov. gen.

Male. Fore wing with almost straight costa, apex pointed; exterior margin oblique, convex, even; posterior margin slightly convex towards the base; first subcostal emitted at half length of the cell and extending to near apex, second at one sixth before the end, bifid, third extending to the apex, fourth and fifth from a foot-stalk at half distance between base of third and end of the cell, fourth anastomosed to third near its base; upper discocellular short, straight, lower concave; radial from their angle; two upper median branches at equal distances from above end of the cell, third from its end, lower at one fourth before the end; submedian recurved at some distance from margin. Hind wing triangular; apex very convex; exterior margin oblique, slightly waved; anal angle acute, abdominal margin short; costal vein extending to apex; cell short; two subcostal branches from end of the cell; disco-cellular very obliquely concave; radial from angle close to lower end of the cell; two upper median branches from acute lower end of the cell, lower branch from one fourth before the end; submedian and internal vein nearly straight, the latter terminating at anal angle. Body robust, extending beyond hind wing; palpi very short; legs stout, squamose, slightly pilose at the side; antennæ filiform.

Allied to *Apsarasa*.

BAORISA HIEROGLYPHICA, n. sp. (Plate IV. fig. 14.)

Male. White: fore wing with six metallic-blue streaks obliquely on base of the costal border, the sixth being the longest and extending across end of the cell; a purple-black mark like the letter A beyond, closely followed by a blue letter-V mark, and then some black radiating lines from its lower point, two other short blue lines ascending from posterior angle, and two small blue spots at equal distances on the submedian vein; below the cell is a light-yellow fascia, beyond which and between the letter V and the streaks from the posterior angle is a patch of greyish-ochreous bordered by an outer bright red line: hind wing roseate-white, unmarked. Underside with less distinct black exterior marks, the red line slightly visible: hind wing roseate-white, with a small blackish spot on middle of costa, and two near the apex. Body white; thorax at the side yellowish; front of head with a black band ; legs, including tarsi, broadly banded with black; abdomen beneath with lateral black bands; antennæ white.

Expanse 2 inches.

Hab. Darjiling. In coll. Dr. Staudinger.

Family ACONTIIDÆ.

Genus NARANGA, *Moore*.

Naranga, Moore, P. Z. S. 1881, p. 359.

Wings small—fore wing elongated, narrow, acute at the apex; exterior margin oblique: hind wing slightly elongated and narrow. Veins similar to those in *Xanthodes*. Palpi small, smooth, slightly ascending; third joint minute, slender, short; legs slender, smooth. Type *N. diffusa* *.

NARANGA QUADRIVITTATA, n. sp.

Male and female. Fore wing dark yellow, with two purple-brown oblique transverse outer irregular bands, the inner indistinctly abbreviated on the costa before the apex, the outer slightly interrupted in the middle; two minute black discocellular spots: hind wing yellow, discal area suffused with purple-brown. Body and legs yellow.

Expanse, ♂ $1\frac{9}{10}$, ♀ $1\frac{1}{2}$ inch.

Hab. Calcutta. In coll. Dr. Staudinger and F. Moore.

A darker-winged species than *N. diffusa*, with two prominent oblique bands.

NARANGA FERRUGINEA, n. sp.

Male and female. Fore wing purplish-ferruginous; with two medial oblique yellow fasciæ: hind wing cupreous-brown. Cilia yellow. Thorax ferruginous-yellow; abdomen brown; palpi and legs yellow, with brown tarsal bands.

Expanse $1\frac{8}{10}$ inch.

Hab. Calcutta. In coll. Dr. Staudinger and F. Moore.

HICCODA, nov. gen.

Fore wing lengthened, narrow; costa slightly arched at base, exterior margin oblique, convex; costal vein extending two thirds of the margin; first subcostal branch emitted at one half before end of the cell, second trifurcate, emitted from end of the cell, third and fourth at equal distances from base of second, fifth also from end of the cell and touching third near its base; discocellular very slender, radials from near upper and lower end;

* *Xanthodes diffusa*, Walker, Catal. Lep. Het. B. M. xxxiii., Suppl. p. 779 (1865).

middle median one sixth before end of the cell, lower at one third before its end; submedian curved. Hind wing triangularly ovate; costal vein straight, extending to apex; two subcostals emitted beyond end of the cell; discocellular concave anteriorly and angled near lower end, radial from the angle; two upper medians from some distance beyond end of the cell, lower at one third before its end; submedian and internal vein straight. Body small; palpi small, porrect, laxly squamose, second joint projecting beyond the head, third joint short: legs squamose; antennæ setulose.

HICCODA DOSAROIDES, n. sp.

Pale ochreous-yellow: fore wing with a large quadrate dark-brown discocellular spot, a costal streak above it, some speckled spots at its base, a curved duplex streak below the cell, and a transverse discal sinuous pale-yellow line with brown points; a slight brown fascia below the apex; a marginal row of brown dots: hind wing cinereous-grey. Body pale ochreous. Underside ochreous-yellow, numerously covered with brown speckles.

Expanse $\frac{18}{24}$ inch.

Hab. Calcutta; Bombay; Ceylon. In coll. Dr. Staudinger and F. Moore.

Has much the appearance of *Dosara*, a genus of Pyrales.

Genus ACONTIA, *Ochs.*

ACONTIA VIALIS, n. sp.

Hydrelia acontioides, Van M. d. Rioy, MS.

Male and female. Pale ochreous-yellow, suffused with purplish-brown: fore wing with an outer oblique wavy-bordered dark-clouded purple-brown band, which is crossed from the costa to posterior margin by a white waved line; a marginal line of black and white dots; cilia purple-brown: hind wing suffused with pale purple-brown; cilia purplish-cinereous. Body ochreous-yellow; palpi brown-speckled; fore legs in front with brown bands.

Expanse $\frac{10}{12}$ inch.

Hab. Dharmsala (*B. Powell & Hocking*); Darjiling (*Atkinson*). In coll. F. Moore, Rev. H. Hocking, and Dr. Staudinger.

Allied to *A. Inda,* Felder.

Genus CHURIA, *Moore.*

Churia, Moore, P. Z. S. 1881, p. 359.

Fore wing elongated, narrow, rectangular; costal vein extending to two thirds the margin; first subcostal emitted at one half and second at one eighth before end of the cell, second trifurcate, third and fourth at equal distances from base of second, fifth from end of the cell and slightly touching third at its base; discocellular bent at its upper and lower end and very convex in the middle, radials from the angles; upper median branch from end of the cell, middle branch from one eighth and lower from beyond one third before end of the cell; submedian slightly curved at the base. Hind wing short; costal vein straight, extending to apex; two subcostal branches from end of the cell; discocellular obliquely concave; radial from its lower end immediately above angle of the cell; two upper median branches from beyond end of the cell, lower from one third before the end; submedian and internal veins straight. Body stout, abdomen long; palpi porrect, second joint laterally broad at the tip, clothed with coarse lax scales, third joint short, thick, half length of the second; legs stout, squamose; antennæ setose.

Type *C. nigrisigna.*

CHURIA NIGRISIGNA.

Churia nigrisigna, Moore, P. Z. S. 1881, p. 360, pl. xxxvii. fig. 13.

Male and female. Upper side pale brownish-ochreous: fore wing with a small black spot on middle of the discocellular veinlet. Cilia ochreous-white. Underside paler along the posterior border of fore wing and on the hind wing. Palpi and legs above pale brownish-ochreous.

Expanse, ♂ $\frac{9}{10}$, ♀ $1\frac{1}{10}$ inch.

Hab. Calcutta. In coll. Dr. Staudinger.

CHURIA OCHRACEA.

Churia ochracea, Moore, P. Z. S. 1881, p. 360.

Male. Upper side paler ochreous than in *C. nigrisigna.* No black spot on the fore wing. Underside pale ochreous. Thorax, palpi, and legs above ochreous.

Expanse $\frac{8}{10}$ inch.

Hab. Calcutta. In coll. Dr. Staudinger.

Family HELIOTHIDÆ.

Genus ADISURA, *Moore.*

Adisura, Moore, P. Z. S. 1881, p. 367.

Fore wing comparatively short and broad, triangular, costa nearly straight, apex very acute, exterior margin oblique and even, posterior margin short; first subcostal branch emitted from half length of the cell, second near the end, trifurcate, the third and fourth at equal distances from base of second, fifth curved upwards from end of the cell and touching the third close to its base; discocellular concave, slightly bent near its lower end; upper radial from end of the cell, lower from angle of discocellular; upper median branch from angle above end of the cell, middle branch from the end, lower at some distance before its end. Hind wing triangular, rather broad and short; costa nearly straight, exterior margin convex and much waved, abdominal margin short; two subcostal branches emitted from end of the cell;

discocellular very slender, slightly concave ; radial extremely slender, emitted from middle of discocellular; two upper median branches from slightly beyond end of the cell, lower from one third before its end. Body short, stout; palpi stout, laxly squamose, apical joint thick, short; legs laxly pilose; antennæ minutely pectinated in male.

Type *A. Atkinsoni*.

ADISURA ATKINSONI.

Adisura Atkinsoni, Moore, P. Z. S. 1881, p. 368, pl. xxxvii. fig. 6.

Male and female. Fore wing pale purplish brownish-ochreous, with a suffused paler fascia from base through the cell to the apex ; indistinctly speckled with minute black scales, which are most apparent on the costal border ; a curved discal transverse recurved series of minute black points, which are less distinct in the male : hind wing pale ochreous-yellow, with a slight purplish-brown submarginal fascia ; median and submedian veins lined with darker brown scales. Underside of both wings uniformly pale ochreous, with a few brown speckles along the costal border. Thorax brownish-ochreous; pectus, palpi at the side, and legs brighter ochreous, fore and middle femora with a brown streak ; abdomen above brown-speckled.

Expanse, ♂ $1\frac{2}{12}$, ♀ $1\frac{3}{12}$ inch.

Hab. Darjiling. In coll. Dr. Staudinger.

ADISURA MARGINALIS.

Anthophila marginalis, Walker, Cat. Lep. Het. B. M. xii. p. 830 (1857).
Adisura marginalis, Moore, P. Z. S. 1881, p. 368.

Male and female. Pale gamboge-yellow : fore wing with a pale pink band along the costa and a similar band above the posterior margin ; extreme edge of the costa yellow : cilia pale pink, edged with white : hind wing paler yellow at the base, with a slight ochreous outer border ; cilia edged with white.

Expanse, ♂ $\frac{1}{12}$, ♀ 1 inch.

Hab. Calcutta. In coll. F. Moore and Dr. Staudinger.

ADISURA DULCIS.

Adisura dulcis, Moore, P. Z. S. 1881, p. 368, pl. xxxvii. fig. 20.

Male and female. Fore wing golden-yellow, with a prominent purplish-pink band along the costal, exterior, and posterior margins ; cilia entirely pink : hind wing pale yellow, with a distinct dusky-brown marginal band ; cilia yellowish-white. Underside pale yellow ; fore wing with the veins broadly suffused with dusky-black, and a blackish discal fascia. Body pale brownish-ochreous ; thorax, palpi, and legs above ochreous-brown.

Expanse 1 inch.

Hab. Darjiling. In coll. Dr. Staudinger.

u

ADISURA SIMILIS.

Adisura similis, Moore, P. Z. S. 1881, p. 369.

Allied to *A. marginalis*. Differs from it in being of a pale ochreous-yellow; the fore wing with similar marginal bands and cilia, but with the disk crossed by an indistinct recurved row of brown speckles, some speckles also being present towards the base of hind margin: hind wing whitish at the base.

Expanse, ♂ ♀ 1½ inch.

Hab. Calcutta. In coll. F. Moore and Dr. Staudinger.

Genus PRADATTA, *Moore.*

Pradatta, Moore, P. Z. S. 1881, p. 364.

Fore wing comparatively short and broad; costa slightly depressed in the middle, exterior margin oblique, posterior margin convex near the base; cell long; first subcostal branch rather short, second emitted from near end of the cell, trifurcate, third and fourth close together, fifth curved abruptly upward from end of the cell and anastomosed to third near its base; discocellular very slender, bent close to each end, concave in middle, upper and lower radial from the angles; upper median branch emitted from angle above end of the cell, middle branch from the end, lower from some distance before the end. Hind wing comparatively long and narrow; two subcostal branches from end of the cell; discocellular bent in middle, a very slender radial from the middle; cell long; two upper median branches from end of the cell, lower from some distance before the end. Body long, slender; thorax laxly pilose; palpi pilose, small, apex very short; legs slightly pilose beneath; fore tibiæ in male armed with a long and a short black spine in front; antennæ in male minutely pectinate.

Type *P. Beatrix.*

PRADATTA BEATRIX.

Synia Beatrix, Van M. D. Roy, MS.
Pradatta Beatrix, Moore, P. Z. S. 1881, p. 365.

Male and female. Fore wing pale pink, with a longitudinal pale-yellow fascia extending from the base of the cell to exterior margin, and a similar fascia below the cell spreading below the median vein ⸱to the outer margin; some specimens have the median vein to end of the cell tinged with black: hind wing white, with pink lining to the veins and outer border. Underside—fore wing with paler costal and outer border than above: hind wing with pink costal border. Thorax pinkish-brown, palest on tegulæ; abdomen pale yellowish above, pinkish beneath; palpi and legs above ochreous-red; fore tibial claws black.

Expanse 1½ inch.

Hab. Canara, S. India; Dharmsala, N.W. Himalaya; Saidabad, Cashmir. In coll. F. Moore, Dr. Staudinger, and Lord Walsingham.

PRADATTA DECORATA.

Pradatta decorata, Moore, P. Z. S. 1881, p. 365.

Male and female. Fore wing pale yellow; with a broad triangular pale-crimson band, extending from base through and below the cell, and thence obliquely upward across the disk to the apex; a recurved series of six white spots on the discal portion of the band, one on middle of median vein, and one on submedian vein; cilia crimson: hind wing paler yellow; cilia whitish, slightly tinged with crimson at the apex. Body pale yellow; thorax ochreous; front of head, tip of palpi, and legs above crimson; fore tibial claws black.

Expanse $\frac{3}{4}$ to 1 inch.

Hab. Deccan (*Dr. Day*); Manpuri, N.W. Provinces (*Horne*); Allahabad (*Hellard*); Sind Valley and Saidabad, Cashmir (*Atkinson*). In coll. F. Moore and Dr. Staudinger.

Family ANTHOPHILIDÆ.

Genus HYDRELIA, *Guén.*

HYDRELIA CONJUGATA.

Hydrelia conjugata, Moore, P. Z. S. 1881, p. 369.

Male and female. Fore wing dark umber-brown; costal border pale ochreous-brown, with darker streaks between the costal and subcostal veins; a white-bordered brown elongated outwardly-oblique orbicular mark, which is confluent hindward with a similar upright reniform mark, an oblique pale streak below the cell in a line with the orbicular mark. These markings indistinct in the female. Hind wing ochreous-brown; thorax, palpi, and legs above brown-speckled.

Expanse $1\frac{1}{4}$ inch.

Hab. Darjiling. In coll. Dr. Staudinger.

Genus THALPOCHARES, *Lederer.*

(*Micra*, Guén.)

THALPOCHARES TRIFASCIATA.

Thalpochares trifasciata, Moore, P. Z. S. 1881, p. 370, pl. xxxviii. fig. 21.

White: fore wing with two oblique transverse narrow basal, and a broad outer lilac-grey band, each thickly studded with ochreous-brown scales; a slight apical red patch, bordered by a few black dots, which continue hindward indistinctly to the angle: hind wing thickly studded with ochreous-brown scales on posterior area. Body brown-scaled; fore tarsi with brown bands.

Expanse $\frac{8}{10}$ inch.

Hab. Calcutta. In coll. Dr. Staudinger.

THALPOCHARES QUADRILINEATA.

Thalpochares quadrilineata, Moore, P. Z. S. 1881, p. 370, pl. xxxviii. fig. 14.

Fore wing pale ochreous, irrorated with minute brown scales, these scales darkest along inner border of four transverse equidistant pale lines, and also on the costa before the apex, the basal line very indistinct; a white streak from the apex, followed by a recurved row of indistinct black speckles; outer border bright ochreous: hind wing ochreous-white, with ochreous marginal line. Body, palpi, and legs above ochreous.

Expanse $\frac{1}{2}$ inch.

Hab. Calcutta. In coll. Dr. Staudinger.

THALPOCHARES RIVULA, n. sp.

Roseate-white: fore wing with an oblique medial transverse ochreous-brown band, and a submarginal recurved white line, the discal interspace being very pale purple, the outer margin pale ochreous-brown, with a minute black apical dot, and another dot above the posterior angle: hind wing with a pale ochreous-brown outer margin. Body ochreous-white; palpi and legs above pale ochreous.

Expanse $\frac{1}{2}$ inch.

Hab. Calcutta. In coll. Dr. Staudinger and F. Moore.

Genus ACANTHOLIPES, *Lederer.*

ACANTHOLIPES HYPENOÏDES.

Acantholipes hypenoïdes, Moore, P. Z. S. 1881, p. 372.

Male and female. Upperside—fore wing greyish ochreous-brown, numerously covered with dark-brown speckles; with a transverse lower discal blackish-brown band, which is broadest in the female, bordered outwardly by a slender yellowish line indistinctly angled at its upper end and bent inward to the costa; a dark-brown waved fascia below the apex; the outer margin with a pale line below the apex, and some pale speckles at end of the costa: hind wing pale ochreous-brown. Underside pale ochreous-brown; both wings slightly speckled with darker brown along the costal border: hind wing with indistinct transverse brown-speckled line. Body, palpi, and legs above greyish-brown.

Expanse $\frac{9}{10}$ to 1 inch.

Hab. Darjiling. In coll. Dr. Staudinger.

Family ERASTRIIDÆ.

Genus ERASTRIA, *Ochs.*

ERASTRIA PALLIDISCA.

Erastria pallidisca, Moore, P. Z. S. 1881, p. 372, pl. xxxvii. fig. 1 *b*.

Male and female. Smaller than *E. albiorbis*: fore wing paler ferruginous-brown; the transverse markings smaller, with the medial area brownish-white, and irregularly speckled hindward; the orbicular and reniform spots indistinct, smaller, and of the same colour as the discal area.

Expanse 1 inch.

Hab. Darjiling. In coll. Dr. Staudinger.

ERASTRIA MARGINATA.

Erastria marginata, Moore, P. Z. S. 1881, p. 372, pl. xxxvii. fig. 21.

Fore wing dark greyish ferruginous-brown, with a broad pale ferruginous band along the hind margin, and extending two thirds up the outer margin, where it is slightly whitish and black-streaked; an indistinct whitish discal transverse sinuous line; orbicular and reniform spots indistinct: hind wing cinereous-brown. Thorax pale ferruginous; palpi and legs dark ferruginous-brown, with pale bands.

Expanse $1\frac{2}{12}$ inch.

Hab. Darjiling. In coll. Dr. Staudinger.

ERASTRIA ALBIORBIS, n. sp.

Male and female. Fore wing ferruginous-brown, medial area slightly paler; crossed by a subbasal pale-bordered prominent black elbowed line, an outwardly-oblique antemedial slightly waved line, a postmedial recurved sinuous line, and a submarginal pale sinuous line with slight black inner points; orbicular and reniform spots white, the latter with brown-speckled centre: hind wing white, with a brown-speckled marginal and cilial line. Thorax, head, palpi, and legs dark brown; legs with pale bands; abdomen pale brown.

Expanse, ♂ $1\frac{1}{12}$, ♀ $1\frac{3}{12}$ inch.

Hab. Darjiling. In coll. Dr. Staudinger and F. Moore.

ERASTRIA FUSCA, n. sp.

Male and female. Fore wing ferruginous-brown; medial area slightly palest; the transverse lines as in *E. albiorbis*; orbicular and reniform spots pale ferruginous and black-lined: hind wing brownish white, with brown-speckled marginal and cilial line.

Expanse, ♂ $1\frac{1}{10}$, ♀ $1\frac{3}{10}$ inch.

Hab. Darjiling. In coll. Dr. Staudinger and F. Moore.

Nearest allied to *E. fusca*.

Male and female. Fore wing much darker ferruginous-brown, the basal and outer area more clouded; transverse black lines less distinct; orbicular and reniform spots smaller, pale ferruginous: hind wing dark cinereous; outer margin brownish.

Expanse 1$\frac{1}{10}$ inch.

Hab. Darjiling. In coll. Dr. Staudinger and F. Moore.

Allied to *E. fuscula*, of Europe.

Fore wing dark ferruginous-brown, the basal area slightly brighter; transverse markings similar; the orbicular and reniform spots grey-brown; the outer border has the white confined along the lower part of the discal sinuous line, and in the female is present only as a narrow border to it: hind wing pale cinereous-brown.

Expanse 1$\frac{1}{10}$ inch.

Hab. Darjiling. In coll. Dr. Staudinger and F. Moore.

Genus PHOTHEDES, *Lederer*.

Phothedes bipars, Moore, P. Z. S. 1881, p. 373, pl. xxxviii. fig. 7.

Allied to the European *P. captiuncula*, Zeller. Fore wing with the basal half dark brown, enclosing a white-lined narrow reniform spot; outer half pale brown, the margin and cilia speckled with dark brown: hind wing and abdomen pale brown. Thorax and fore legs above dark brown; collar, front of head, and palpi greyish-brown.

Expanse 1$\frac{9}{10}$ inch.

Hab. Cherra Pungi, Assam. In coll. Dr. Staudinger.

Genus BANKIA, *Guén.*

Bankia abnormis, Van M. d. Rioy, MS.

Fore wing pale greenish-brown; with an antemedial transverse oblique white line, from which longitudinal streaks less distinctly extend to the base of the wing; a postmedial transverse similar white line, which is interrupted by a prominent white-bordered black-lined reniform mark, the upper end of the line extending obliquely outward from the costa; a short white line from the apex and some longitudinal looped lines on the disk, the white lines and streaks with dark-brown-speckled borders: hind wing cinereous-brown. Cilia cinereous-white. Body pale green-brown; thorax with white bands; abdomen dark-speckled.

Expanse $\frac{7}{8}$ to $\frac{8}{8}$ inch.

Hab. Dharmsala (*B. Powell, Hocking*); Calcutta (*Atkinson*). In coll. F. Moore, Rev. H. Hocking, and Dr. Staudinger.

BANKIA BASALIS, n. sp.

Fore wing pale ochreous-brown, basal third of the wing obliquely white; a postmedial oblique white line terminating in two black dots forming the reniform mark, above which is a short white outwardly-oblique streak from the costa; on the disk from the apex to posterior angle are some short longitudinal black streaks which are interruptedly bordered with white; a marginal black line with white inner border: hind wing cinereous-brown. Cilia cinereous-white. Thorax white, collar brown-streaked; abdomen cinereous-white, brown-speckled; palpi speckled and tarsi banded with brown.

Expanse $\frac{8}{10}$ inch.

Hab. Darjiling (*Atkinson*); Shanghai (*Pryer*). In coll. Dr. Staudinger and F. Moore.

BANKIA OBLIQUA, n. sp.

Bankia obliqua, Van M. D. Rioy, MS.

Male and female. Fore wing cupreous-brown, basal third and an oblique band from middle of the costa to posterior angle, pure white; a black dot at base of wing: hind wing cinereous-brown. Thorax and head white; front and collar brown-speckled; palpi brown; tarsi with brown bands; abdomen cinereous-brown.

Expanse, ♂ $\frac{7}{5}$, ♀ $\frac{9}{4}$ inch.

Hab. Dharmsala (*Hocking*), Kaschmir, Changra (*Atkinson*). In coll. Rev. H. Hocking and Dr. Staudinger.

Family ERIOPIDÆ.

Genus CALLOPISTRIA.

Callopistria, Hübner, Verz. bek. Schmett. p. 216 (1816).
Lagopus*, Latr. N. Dict. H. N. xvii. p. 199 (1816?).
Eriopus, Treitschke, Schmett. Eur. v. 1, p. 365 (1825).

Fore wing acuminate at the apex; exterior margin oblique, angular in the middle; scalloped; first subcostal emitted at half length of the cell, second close to end of the cell, third and fourth at nearly equal distances apart from base of second, fifth from end of the

---

* Previously used as a genus of Birds.

cell and anastomosing with third for a short distance above its base ; radial from angle close to subcostal ; discocellular slightly concave ; two upper median branches from angles above end of the cell, third from its end, fourth at one third before its end. Hind wing longer than broad, exterior margin convex, scalloped ; two subcostal branches from end of the cell, upper discocellular slightly concave, lower short, oblique, nearly straight ; radial from their angle ; two median branches from end of the cell, third at about one fourth before its end. Thorax laxly pilose ; antennæ minutely bipectinate, more or less distorted and bent at half its length in the male ; palpi slightly ascending, second joint laterally broad, widest and pointed in front, pilose, third joint long, slender, cylindrical ; femora, tibiæ, and tarsi densely and laxly clothed with hair in the male, less so in the female.

Type *C. pteridis*. (*Hab.* Europe.)

CALLOPISTRIA RECURVATA, n. sp.

Fore wing rufous-brown, veins lined with ochreous-white ; some silvery-white-bordered black subbasal transverse zigzag lines, a duplex antemedial straight line which is bent outward at the median vein ; lunate orbicular and reniform marks ; a recurved postmedial line, and zigzag short streaks below the apex ; the basal, costal, and apical interspaces blackish-brown, the discal area brightest and traversed by a black lunular line ending in a white spot ; two black-bordered white zigzag lines below the apical streak : hind wing brown. Thorax rufous-brown, collar black-streaked ; abdomen brown.

Expanse 1¼ to 1¾ inch.

Hab. Darjiling (*Atkinson*) ; Calcutta ; Ceylon. In coll. Dr. Staudinger and F. Moore.

C. exotica, Guén., which also occurs at Darjiling, is a smaller insect, and differs in having the antemedial transverse line entirely convex, and the postmedial line more boldly recurved. *C. repleta*, also from Darjiling, has the antemedial line bent at the submedian vein, and the postmedial line is almost straight.

Genus PHALGA, Moore.

Phalga, Moore, P. Z. S. 1881, p. 375.

Fore wing—costa almost straight, apex pointed, exterior margin oblique and scalloped, angular in the middle ; first subcostal branch emitted at one third before end of the cell, second at one sixth before its end, third at one eighth from below base of second, fourth at three fourths from third ; fifth from end of the cell, bent obliquely upward and slightly touching third near its base ; discocellular extremely slender, slightly bent at each end, convex in the middle, radials from upper and lower angles ; upper median branch emitted from angle above end of the cell, middle branch from its end, lower at nearly one half before its end ; submedian recurving from the base. Hind wing narrow, exterior margin convex, slightly scalloped ; abdominal margin short ; costal vein extending to apex, two subcostal branches emitted from end of the cell ; discocellular extremely slender, radial from its lower end ; cell very short ; upper median branch from angle above end of the cell, middle branch from its end, lower at one third before the end; submedian and internal vein recurved. Body moderate, abdomen laterally tufted; palpi ascending, not extending above the head, second joint stout, third slender; fore tibia laxly tufted; antennæ filiform.

Allied to *Lineopalpa*, Guénée.

PHALGA SINUOSA.

Phalga sinuosa, Moore, P. Z. S. 1881, p. 375, pl. xxxvii. fig. 7.

Fore wing pale dull brownish-ochreous, with a very indistinct black-speckled-bordered pale zigzag subbasal transverse line, a more distinct black treble discal acute-angled zigzag line, a submarginal single line, and less distinct marginal lunular line; a pale-yellowish reniform mark: hind wing ochreous-brown; cilia brownish-ochreous; a slender black marginal lunular line and streaks above anal angle. Body brownish-ochreous; thorax, palpi, and fore legs ochreous-brown.

Expanse 1¼ inch.

Hab. Darjiling. In coll. Dr. Staudinger.

LUGANA, n. g.

Male. Fore wing somewhat short, costa slightly curved; exterior margin oblique, slightly convex; posterior margin convex towards the base; costal vein recurved, extending to two thirds the margin; first subcostal emitted at nearly one half before end of the cell, second at one eighth before the end, third and fourth at nearly equal distances from base of second, fifth from end of the cell and curved upward and touching the third near its base; discocellulars oblique, deeply concave in the middle, bent at each end, radials from the angles; cell long, extending rather more than half the wing; upper median from angle above end of the cell, middle median from very near its end, and lower at nearly one half before its end; submedian curved near its base. Hind wing short; exterior margin convex, abdominal margin short; costal vein slightly waved; subcostal straight, its two branches from one third beyond end of the cell; discocellular slender, obliquely concave, radial from its middle; cell broad; two upper medians from end of the cell, lower at one third before the end; submedian and internal veins straight. Body moderately stout, abdomen long, with dorsal and lateral tufts; palpi laxly squamous, second joint very long, ascending to vertex, third joint short and thick; fore and middle tibiæ pilose above, hind femora and tibiæ densely tufted; antennæ grooved and twisted from the base, and dilating into a broad rounded cavity in the middle, and from the cavity to the tip slender, its base finely serrate-pectinate, with longer and broader serrations from the cavity and thence decreasing to the tip.

Type *L. antennata*.

I

Lugana antennata, n. sp.

Fore wing chestnut-brown, the basal area black-speckled, the outer area brightest and slightly grey-speckled; an indistinct small pale orbicular spot and a figure-of-8 reniform spot with dark centre: hind wing cinereous-brown. Body, palpi, and legs chestnut-brown; antennæ chestnut-brown, with the dilated cavity black. Underside uniform pale cinereous-brown.

Expanse 1 inch.

Hab. Darjiling. In coll. Dr. Staudinger and F. Moore.

Lugana renalis, n. sp.

Allied to *L. antennata.* A third less in size. Fore wing uniformly darker brown, with a pale transverse discal recurved sinuous band, and prominent figure-of-8 white-bordered black-centred reniform mark: hind wing and underside pale pinkish cinereous-brown. Body, palpi, and legs above dark brown; dilated cavity of antennæ black.

Expanse $\frac{18}{20}$ inch.

Hab. Calcutta (*Atkinson*); Andaman Isles (*Roepstorff*). In coll. Dr. Staudinger and F. Moore.

Genus ÆGILIA, *Walker.*

Ægilia obscura, n. sp.

Male and female. Fore wing dark brown, washed with purplish-grey; with very indistinct ochreous streaks between the lower veins, an indistinct postmedial and an antemedial transverse waved sinuous pale-bordered blackish lines, a more distinct ochreous subapical waved line, a marginal row of dots, and a reniform mark: hind wing dusky brown, palest at the base. Body, palpi, and legs dark brown; legs with pale bands.

Expanse $1\frac{1}{4}$ inch.

Hab. Darjiling. In coll. Dr. Staudinger.

Specimens taken by Mr. Wallace at Sarawak are undistinguishable from the above.

Ægilia angulata, n. sp.

Female. Fore wing ochreous-brown, with indistinctly paler streaks between the veins; an indistinct antemedial and a postmedial transverse waved brown double line, the outer line bent inward to middle of the posterior margin; a pale waved apical streak, and a black spot at upper end of the cell: hind wing paler brown, with slight ochreous streak from anal angle. Body, palpi, and legs dark brown, legs with grey bands.

Expanse 1 inch.

Hab. Darjiling. In coll. Dr. Staudinger.

Family EURHIPIDÆ.

Genus EUTELIA, *Hübner*.

EUTELIA SICCIFOLIA.

Eutelia siccifolia, Moore, P. Z. S. 1881, p. 375.

Greenish-ochreous, numerously covered with short indistinct dusky strigæ: fore wing crossed by five or six irregular waved indistinct blackish lines and an oblique subapical line, the costal border clouded with brown, the edge at the apex black-speckled: hind wing suffused with purplish-brown on exterior border; with irregular transverse indistinct blackish lines; a slender semidiaphanous white discocellular streak. Front of thorax, head, palpi at the side and in front, black; legs above black.

Expanse $1\frac{1}{10}$ inch.

Hab. Darjiling. In coll. Dr. Staudinger.

Allied to *E. viridatrix.*

EUTELIA INEXTRICATA, n. sp.

Male and female. Fore wing dark rufous-brown, varied with greyish-brown across the disk; veins from the base lined with white; the basal area crossed by several irregular white lines; a white streak within the cell and a large ragged spot at its end; a transverse discal waved purple-red line ending in a blue-grey spot on posterior margin, the line angled in front of upper and lower end of the cell, and outwardly bordered by three white-and-brown lines; the apex whitish, with a white interrupted streak curving from costa before the apex to posterior angle; a marginal row of black dots and white-bordered lines: hind wing whitish, with broad brown marginal band traversed by a white lunular streak from the anal angle and a lunular line from the apex; some black streaks also above anal angle. Cilia white, bordered with rufous-brown. Body rufous-brown; collar and anterior segments of abdomen white-margined.

Expanse $1\frac{3}{8}$ inch.

Hab. Darjiling; Cherra Punji. In coll. Dr. Staudinger and F. Moore.

Genus CHILUMETIA, *Walker*.

CHILUMETIA ALTERNANS, n. sp.

Fore wing dark purplish-brown, with six transverse pale ochreous bands, and alternate brown intervening bands traversed by a grey streak, the discal band being most prominent; orbicular and reniform marks defined by a slight black-speckled line; outer border broadly speckled with greyish-white, the outer margin with white lunular line: hind wing pale dull brown, with a marginal row of darker brown white-bordered lunules; cilia greyish-white, lined with brown. Body and abdomen brown ; palpi grey at the tip, third joint with a brown band.

Expanse 1 inch.

Hab. Darjiling. In coll. Dr. Staudinger.

A larger insect than *C. gutticentris*, with broader transverse alternate bands.

Genus VARNIA, *Walker.*

VARNIA FENESTRATA.

Varnia fenestrata, Moore, P. Z. S. 1881, p. 376.

Deep dull chocolate-red, washed with chalybeate-grey, marked with very indistinct blackish confluent strigæ ; fore wing with transverse very indistinct black lines, those on the basal half waved, the discal and subapical lines being oblique, straight, and joined together on the interdiscal space bordering these two lines brighter red ; some pale yellow spots on the costal edge : hind wing with a large irregular quadrate semidiaphanous white discocellular spot, and some contiguous pale yellow streaks. Body with red dorsal streaks ; palpi black laterally.

Expanse $1\frac{3}{16}$ inch.

Hab. Darjiling. In coll. Dr. Staudinger.

Family PLUSIIDÆ.

Genus ABROSTOLA, *Ochs.*

ABROSTOLA ANOPHIOIDES, n. sp.

Fore wing umber-brown, basal area and exterior border greyish ochreous-brown ; crossed by an antemedial and a postmedial narrow black waved line, both outwardly margined by a pale-bordered brown line, the postmedial line sinuous and paler at the anterior end ; a black-lined broad orbicular and reniform marks, below which is a short streak and a bifid mark ; outer border with waved pale-bordered brown fasciæ ; hind wing pale brownish-white, with broad brown marginal band in male, and more uniformly brown in female. Body dark brown : crest and abdominal tufts pale ochreous-brown ; fore and middle legs dark brown.

Expanse, ♂ $1\frac{4}{8}$, ♀ $1\frac{5}{8}$ inch.

Hab. Darjiling. In coll. Dr. Staudinger and F. Moore.

Nearest allied to *A. asclepiadis.*

Genus PLUSIA, *Ochs.*

PLUSIA RETICULATA, n. sp.

Fore wing dark cupreous-brown, brightest on exterior border ; washed with purplish chalybeate-grey, crossed by a very slender waved basal, subbasal, a sinuous discal, and two submarginal golden-yellow lines; an oblique orbicular spot, a contiguous mark below it, and a reniform mark, each formed by a golden-yellow line: hind wing and body pale brown.

Expanse 1¾ inch.

Hab. Darjiling. In coll. Dr. Staudinger and F. Moore.

PLUSIA PANNOSA, n. sp.

Fore wing purple greyish-brown; with a transverse subbasal and discal brown-bordered pale straight line, the interdiscal space below the cell glossy cupreous-brown, containing a white loop and unattached pendent round spot; orbicular spot minute, bordered with dark brown and white, reniform spot broken, dark brown, lower portion white-circled, upper portion indistinct: a short streak at base of the costa, a cupreous-brown waved fascia below the apex, a marginal greyish line terminating in a triangular spot at posterior angle: hind wing and abdomen pale cupreous-brown. Thorax, palpi, and legs purple-grey-brown.

Expanse 1¼ inch.

Hab. Darjiling; Khasia Hills. In coll. Dr. Staudinger and F. Moore.

PLUSIA CONFUSA, n. sp.

Fore wing glossy greyish-purple-brown, with a transverse basal, subbasal, and a discal waved slender black-bordered pale line, the interdiscal area clouded with darker brown; a submarginal darker brown waved zigzag fascia, a marginal row of slender black points; orbicular mark pale, oblique, reniform mark indistinct; a golden-yellow oblique loop with an attached pendent spot beneath the cell: hind wing brown, paler at the base. Thorax purple-brown, grey-speckled; abdomen brown; palpi and legs brown; tarsi with pale bands. Near to *P. permissa.*

Expanse 1½ inch.

Hab. Darjiling. In coll. Dr. Staudinger and F. Moore.

PLUSIA ARGYROSIGNA, n. sp.

Fore wing glossy purplish ochreous-brown; a lower subbasal oblique pale line and a less distinct discal line, the interdiscal area and base of costa and a zigzag submarginal fascia dark olivaceous ochreous-brown, discal silvery-white lobed mark curved and entire, and with a slight streak above it: hind wing pale brown, with indistinct paler transverse fascia. Thorax dark ochreous-brown, grey-speckled, front tinged with red; abdomen brownish-ochreous; palpi and legs above dark ochreous-brown.

Expanse 1¾ inch.

Hab. Kashmir (Sind Valley); Dalhousie. In coll. Dr. Staudinger and F. Moore.

Allied to *P. iota* and to *P. v-aureum.* Differs in the brown colour of the fore wings, and in the discal mark being entirely white and unbroken.

Genus PLUSIODONTA, *Guén.*

PLUSIODONTA AURIPICTA, n. sp.

Fore wing deep chestnut-red, with a purplish-grey line enclosing a broad angular space at base of the costa, the hindward angle of the line extending to the submedian vein, beyond which are two or three waved similar-coloured lines across the middle, and followed by an oblique discal double acutely zigzag line which is slightly blackish anteriorly; reniform mark distinct, composed of a purplish-white line and anterior streak; discal area with a metallic golden lunular black-lined bordered waved streak from the apex, and a broader irregular constricted streak from posterior angle, the lower interspace between which and the discal purplish-lined; the lobe of hind margin and a quadrate spot near base of costa is also of a metallic golden colour; a slight fascia from middle of exterior margin to apical streak, and a marginal lunular line purple-grey: hind wing and abdomen ænescent brown; cilia cinereous. Thorax, head, palpi, and legs above chestnut-red.

Expanse, ♂ 1₁⁴₂, ♀ 1₁⁷₂ inch.

Hab. Darjiling; Cherra. In coll. Dr. Staudinger and F. Moore.

A larger insect than the Ceylonese *P. conducens*, Walker.

Genus EUCHALCIA, *Hübner.*

EUCHALCIA CASHMIRENSIS.

Euchalcia cashmirensis, Moore, P. Z. S. 1881, p. 378.

Fore wing brownish olive-green, with a transverse olive-white basal line, an antemedial line curving below the cell, an undulated postmedial line, and two submarginal lines; orbicular and reniform marks formed by a similar olive-white line; the outer border of the pale lines tinged with cupreous-brown: hind wing pale purplish-brown; cilia ochreous. Thorax brownish-olive; abdomen ochreous, dorsal tufts bright ochreous; palpi and legs pale ochreous; tarsi and antennæ brighter ochreous.

Expanse 1⅔ inch.

Hab. Sind Valley, Cashmir. In coll. Dr. Staudinger.

Allied to *E. uralensis* and *E. modesta.*

Family CALPIDÆ.

Genus CULASTA, *Moore.*

Culasta, Moore, P. Z. S. 1881, p. 376.

Fore wing elongate; costa nearly straight; apex acute; exterior margin convex towards the posterior angle, posterior margin very convex near the base; first subcostal branch emitted at nearly one half before end of the cell, second at one fifth, third from near base of second, and fourth from near the apex; fifth from end of the cell, ascending to but not touching the third near its base; discocellular bent near each end, concave and very slender in the middle, radials from the angles; cell long, extending more than half length of the wing; upper median branch from angle above end of the cell, middle branch from its end, lower at one third before the end ; submedian recurved. Hind wing somewhat short and broad, exterior margin waved, convex ; costal vein nearly straight ; two subcostal branches from end of the cell; discocellular bent inward in the middle, radial from its lower end ; cell broad, short; two upper median branches emitted from angle at end of the cell, lower at one fourth before its end ; submedian and internal vein slightly curved. Body stout ; head flat above ; palpi large, thick, pointed at the tip, ascending to the apex, and then projecting out in front ; legs moderately long, laxly squamose ; antennæ filiform.

CULASTA INDECISA.

Culasta indecisa, Moore, P. Z. S. 1881, p. 377.

Fore wing pale greyish-ochreous, greyish externally ; with an indistinct oblique grey streak ascending from middle of posterior margin to below the apex, the streak bordered on the inner side by a contiguous brown line, which is broken and diffused at the apex, and on the outer side by broader suffused brown lines ; a minute brown dot at lower end of the cell, and a row of dots on outer margin : hind wing whitish-ochreous ; cilia white. Body, palpi, and legs pale ochreous.

Expanse 1½ inch.

Hab. Madras ; Bombay (*Dr. Leith*); Benares (*Atkinson*). In coll. F. Moore and Dr. Staudinger.

Genus CALPE, *Treits.*

CALPE FASCIATA, n. sp.

Larger than *C. Thalictri* : the fore wing comparatively longer and narrower, the exterior margin being more oblique and less convex, the fascia more distinct and oblique, the line traversing the disk from the apex nearly straight, the discocellular streak with a blackish outer border ; a distinct submarginal series of black speckles : hind wing uniformly ochreous-brown. Underside—fore wing uniformly dusky brownish-ochreous : hind wing throughout of a pale ochreous, with a slightly darker discocellular lunule.

Expanse 2⅜ inches.

Hab. Darjiling. In coll. Dr. Staudinger and F. Moore.

Family HEMICERIDÆ.

NAGASENA, n. g.

Fore wing elongated ; costa arched at base ; exterior margin short, obliquely concave ; posterior margin convex towards the base ; first subcostal emitted at nearly one half before end of the cell, second from end of the cell, trifid, third at one fourth from below base of second, fourth at one fourth from below base of third, and terminating below the apex ; fifth from end of the cell, extending parallelly alongside base of second and touching the third at its base ; discocellular bent at each end, concave in middle, radials from the angles ; middle median at one sixth and lower at one half before end of the cell : submedian

slightly curved towards the base. Hind wing short, convex externally; costal vein slightly
recurved; cell half length of the wing; two subcostals emitted from end of the cell; discocellular obliquely concave, bent at lower end, radial from the angle; middle median at one
sixth and lower at one third before end of the cell; submedian and internal straight.
Thorax moderately stout, abdomen slender; palpi slender, squamose, ascending to vertex,
third joint half length of second; femora laxly clothed; antennæ minutely bipectinate in
male.

Allied to *Westermannia.*

NAGASENA ALBESCENS, n. sp.

Silky purplish greyish-white: fore wing very minutely and indistinctly brown-speckled;
with a curved ochreous-brown line extending from apex to base of posterior margin; the
speckles somewhat clustered across the disk and forming a very ill-defined sinuous fascia:
hind wing paler. Thorax white, minutely brown-speckled; palpi and fore legs above pale
brown. Underside—fore wing pale brownish-white, hind wing white.

Expanse 1¼ inch.

Hab. Darjiling. In coll. Dr. Staudinger and F. Moore.

Family HYBLÆIDÆ.

Genus PHYCODES, *Guén.*

PHYCODES MINOR.

Phycodes minor, Moore, P. Z. S. 1881, p. 378.

Fore wing cupreous-grey, with a slender cupreous-brown medial transverse band: hind
wing greyish-cupreous; cilia white. Underside uniformly brown. Body cupreous-grey;
second joint of palpi white, third joint black; legs cupreous-brown above, femora beneath
and bands above white.

Expanse ⅞ inch.

Hab. N.W. India. Caragola, Bengal (*Atkinson*). In coll. F. Moore and Dr. Staudinger.

PHYCODES MACULATA.

Phycodes maculata, Moore, P. Z. S. 1881, p. 378.

Fore wing cupreous-brown, very indistinctly speckled with minute grey scales; with
several golden-yellow spots on the basal and medial area, and longitudinal streaks on the
exterior border: hind wing with a pale yellow linear streak from the base, a slender streak
above the anal angle, and three spots on the upper part of the disk; cilia pale cinereous-
yellow. Body cupreous-black, abdomen with slight yellow segmental bands; palpi black
above, pure white at the side; legs black, femora golden-yellow beneath, tarsi with yellow
bands; antennæ black.

Expanse 1 1/13 inch.

Hab. Darjiling. In coll. F. Moore and Dr. Staudinger.

Family GONOPTERIDÆ.

Genus GONOTIS, *Guén.*

GONOTIS BRUNNEA, n. sp.

Male. Brownish-ochreous: fore wing with the interdiscal area dusky, a transverse antemedial and a postmedial zigzag indistinct blackish-speckled line, and a submarginal zigzag row of blackish points; cilia slightly black-streaked: hind wing dusky ochreous; cilia paler. Legs with pale ochreous terminal bands; antennæ rather broadly bipectinate.

Expanse 1½ inch.

Hab. Calcutta. In coll. Dr. Staudinger and F. Moore.

COARICA, n. g.

Fore wing elongated; costa slightly arched at the base and apex, angle acute; exterior margin waved, slightly produced in the middle; first subcostal emitted at one half before end of the cell, second at one fifth, trifid, third at one fifth from its base, fourth at one fourth before the apex, fifth from end of the cell and slightly touching third near its base; disco-cellulars bent at each end, concave in the middle, radials from the angles; two upper median veins from angles at end of the cell, lower at one third before the end. Hind wing long, narrow; exterior margin convex, waved; two subcostals from end of the cell; discocellular obliquely concave, radial from near lower end; two upper medians from lower angle of the cell. Body long; palpi long, slender, compactly squamose; femora stout, pilose; antennæ long, simple.

COARICA FASCIATA, n. sp. (Plate V. fig. 1.)

Male. Fore wing pale ochreous-brown, indistinctly marked with short paler strigæ between the veins; crossed by a broad postmedial dark brown band formed of three divisions, each slenderly margined by a white line and bordered outwardly by a broad grey-white shade; a slight black streak below the cell, and an interrupted, submarginal, sinuous black lunular line, with whitish outer border; a small yellow orbicular spot: hind wing pale cinereous-brown, with a white-bordered black lunule from anal angle. Thorax ochreous-brown; front of thorax and vertex dark brown; palpi and legs above dark brown; tarsi with pale bands; abdomen pale cinereous-brown.

Expanse 1¾ inch.

Hab. Darjiling. In coll. Dr. Staudinger.

FALANA, n. g.

Fore wing somewhat short, broad anteriorly, costa straight, apex pointed; exterior margin irregularly scalloped, angular in the middle; first subcostal branch emitted at nearly one half before end of the cell, second from close to the end, third from very near base of second, fourth from near end of third and terminating at the apex, fifth from end of the cell and touching third near its base; discocellular slightly bent near each end, slightly concave

K

in the middle, radials from the angles; upper and middle median branches from angles at end of the cell, lower at nearly one half before its end; submedian nearly straight. Hind wing somewhat small, short, and narrow; exterior margin slightly uneven; costal vein nearly straight; two subcostal branches from end of the cell; discocellular slender, straight, inwardly oblique, radial from near its lower end; cell short; two upper median branches from beyond end of the cell, lower from close before its end; submedian and internal vein straight. Body moderate, abdomen extending beyond hind wings; palpi ascending, second joint squamose, extending to vertex, third joint of nearly equal length, slender; legs long, tibiæ tufted, hind femora tufted beneath; antennæ setose.

FALANA SORDIDA, n. sp.

Dull brownish-ochreous: fore wing with an indistinct basal and medial transverse darker band, the latter inwardly bordered by a slight black sinuous line and outwardly by a less distinct duplex sinuous line, the latter ending on the costa in a pale streak; the discal and marginal area black-speckled, with a discal series of indistinct speckled spots, which are most prominent at apical end; an indistinct black orbicular dot and an elongated pale-centred black reniform mark: hind wing ochreous-brown; cilia ochreous. Body ochreous; palpi and legs above ochreous-brown, tuft on hind femora white.

Expanse $1\frac{3}{8}$ inch.

Hab. Cherra Punji, Assam (*Austen*). In coll. F. Moore and Dr. Staudinger.

Genus THALATTA, *Walker.*

Thalatta, Walker, Catal. Lep. Het. B. M. xiii. p. 996.

THALATTA FASCIOSA, n. sp. (Plate V. fig. 2.)

Fore wing chestnut-brown, washed by three or four transverse waved purplish-grey fasciæ, crossed by a postmedial, reddish-bordered, waved line and a submarginal greyish-white sinuous line terminating in two small black dots at apical end: hind wing paler brown. Thorax, palpi, and legs above dark brown.

Expanse $1\frac{2}{6}$ inch.

Hab. Cherra Punji. In coll. Dr. Staudinger.

Family AMPHIPYRIDÆ.

Genus AMPHIPYRA, *Ochs.*

AMPHIPYRA CORVUS.

Amphipyra corvus, Motsch.

Allied to the European *A. livida.* Larger in size: fore wing much darker cupreous-black, the hind wing also of a deeper copper-red and darker apical border. Underside also of a much deeper tint.

Expanse $1\frac{3}{4}$ inch.

Hab. Khasia Hills. In coll. Dr. Staudinger.

Identical with specimens from China, Shanghai, and Japan.

AMPHIPYRA CUPREIPENNIS, n. sp.

Male. Fore wing very dark velvety coppery-brown, with a very indistinct paler post-medial transverse waved line, and which is seen only in certain lights: hind wing coppery-red, with broad brown anterior margin. Female darker; fore wing uniformly coloured throughout. Thorax, palpi, and legs above coppery-brown; abdomen brown. Underside as above, but paler and more glossy.

Expanse, ♂ 2⅛, ♀ 2⅜ inches.

Hab. Darjiling. In coll. Dr. Staudinger.

TAMBANA, n. g.

Fore wing long, narrow; costa almost straight, apex pointed, exterior margin oblique and slightly convex; costal vein extending two thirds the margin; first subcostal branch emitted at nearly one half before end of the cell, second at one eighth, third at one third from below second, fourth at one half from third, fifth from end of the cell and touching third near its base; discocellular bent near each end, concave in the middle, radials from the angles; cell long, extending to more than half the wing; upper median emitted from end of the cell, middle branch from near the end, lower at nearly one half before the end; submedian slightly recurved. Hind wing somewhat long and narrow; apex slightly convex, exterior margin obliquely convex, abdominal margin short; costal vein slightly waved; two subcostal branches emitted from beyond end of the cell; discocellular very slightly concave, radial from near its lower end; cell not extending to half the wing; two upper median branches from end of the cell, lower at one third before its end; submedian and internal vein straight. Body moderately stout; abdomen with slight dorsal and lateral tufts; palpi stout, thick, laxly squamose, ascending but not reaching beyond vertex, third joint short, conical; legs long, tibia pilose above; antennæ minutely pectinated in male.

Type *T. variegata.*

TAMBANA VARIEGATA, n. sp.

Fore wing dark glossy chestnut-brown; crossed by two indistinct white-speckled zigzag antemedial lines and two similar postmedial lines; the basal area white-speckled, the outer discal area with white-speckled longitudinal streaks and marginal speckled-lunular line; interspaces between the markings transversely fasciated with darker brown; some white spots on costal edge near apex: hind wing glossy yellowish-cupreous, with brown marginal band. Cilia white-spotted. Body yellowish-cupreous; thorax in front, head, dorsal and anal tufts darker; palpi and legs above dark chestnut-brown, banded and speckled with white; head, thorax, and abdominal tufts white-fringed, a white spot on tegulæ. Underside of both wings glossy yellowish-cupreous.

Expanse 2⅗ inches.

Hab. Darjiling. In coll. Dr. Staudinger and F. Moore.

K 2

TAMBANA CATOCALINA, n. sp.   (Plate V. fig. 3.)

Female. Fore wing olive-brown, crossed by a straight antemedial and a waved post-medial grey-bordered blackish line; medial interspace olivaceous-grey, crossed by a central waved olive-brown narrow fascia and enclosing a large reniform spot; exterior border olivaceous-grey, divided from the postmedial line by a zigzag olive-brown fascia, the middle of which extends obliquely outwards; a marginal row of brown dots: hind wing ochreous-yellow, with a pale ochreous-brown marginal band. Thorax, palpi, and legs ochreous-grey; abdomen ochreous. Underside ochreous-yellow, with ochreous-grey borders and a dusky-brown discocellular spot on fore wing.

Expanse 1¼ inch.

Hab. Darjiling. In coll. Dr. Staudinger.

Genus PERINÆNIA, *Butler.*

Perinænia, Butler, Ann. N. H. 1878, p. 289.

This genus is allied to *Scotophila* (*S. tragopogonis*) of Europe.

PERINÆNIA ACCIPITER.

Spintherops accipiter, Felder, Novara-Reise, Lep. iv. pl. 111. fig. 29.

Fore wing umber-brown, sparsely speckled with minute black scales, numerously covered with short transverse paler strigæ; an indistinctly-indicated pale-bordered darker sinuous discal line; veins externally from the discal line indistinctly blackish; a marginal row of dots; a small white orbicular dot and a blackish streak extending to the discal line, the streak encompassing a reniform cluster of whitish speckles: hind wing ochreous-brown, with darker marginal band; cilia pale ochreous-brown. Thorax, palpi, and legs above umber-brown, sparsely black-speckled; abdomen paler.

Expanse 1½ inch.

Hab. Pangi (*Stoliczka*); Darjiling (*Atkinson*). In coll. Dr. Staudinger.

Allied to *P. lignosa*, Butler, Types of Heterocera in Brit. Mus. ii. p. 37, pl. 32. fig. 7, from Japan.

MITHILA, n. g.

Fore wing elongated, narrow, costa much arched at base, exterior margin slightly oblique, convex, posterior margin lobed towards the base; costal vein extending to four fifths the margin; first subcostal emitted at one half before end of the cell, second trifurcate, emitted at one twelfth before the end, third at one fourth from base of second, fourth at one half from third, fifth from end of the cell and joined by a short spur to second below base of the third; cell extending beyond half length of the wing; discocellulars angled close to each end, concave in the middle, radials from upper and lower angle; middle median emitted at one twelfth before end of the cell, lower at one third before the end; submedian straight, extending some distance above posterior margin. Hind wing long, narrow; anterior margin somewhat longer than posterior margin of fore wing, apex very convex, abdominal margin very short; costal vein straight, extending to apex; two subcostals emitted from immediately beyond end of the cell; discocellular very concave, radial from slight angle near its lower end; middle median emitted close before end of the cell, lower at one third before the end; submedian and internal vein slightly curved. Body very stout, abdomen extending half beyond the hind wings; thorax and abdomen laxly crested; palpi long, porrect, second joint extending to front of head, clothed beneath with thick long lax scales, third joint long, squamose; legs densely pilose; antennæ setose.

MITHILA LICHENOSA, n. sp.

Fore wing greyish olive-brown, crossed by a basal, a subbasal, a discal, and two submarginal delicate white speckled-bordered indistinct black sinuous lines; orbicular and reniform spots white-speckled with black centre: hind wing brown, palest at base. Thorax olive-brown, white-speckled; abdomen brown; palpi and legs brown, tarsi with pale bands.

Expanse $1\frac{3}{10}$ to $1\frac{6}{10}$ inch.

Hab. Darjiling. In coll. Dr. Staudinger and F. Moore.

Genus BLENINA, *Walker.*

BLENINA PANNOSA, n. sp. (Plate V. fig. 4.)

Fore wing crossed with four very indistinct slender, brown, irregular sinuous white-bordered lines, the basal interspace olivaceous, the antemedial and upper postmedial interspaces grey, the upper medial space olivaceous-brown, the lower medial part broadly dark brown, this latter colour extending to the posterior margin; a black dot within the cell, and a distinct white-bordered streak at the end: hind wing ochreous-yellow, with a broad cupreous-brown outer border; cilia yellow. Underside paler yellow: fore wing with a slender dusky black transverse medial band, and a confluent broad outer dusky ochreous-brown band: hind wing with a narrow marginal band. Thorax, palpi, and legs above olivaceous-grey; abdomen yellow.

Expanse $1\frac{1}{2}$ inch.

Hab. Calcutta. In coll. Dr. Staudinger.

Allied to *B. donans*, of Ceylon.

BLENINA VARIEGATA, n. sp.

Fore wing crossed with four nearly equidistant slender sinuous black lines, with slight whitish sinuous borders, the borders of the two inner lines and lower medial interspace in some specimens quite white, the basal, medial, and outer interspaces olivaceous, the whole surface of the wing being black-speckled; a black dot within the cell and a streak at the end: hind wing dusky black, with bright ochreous medial band and cilia. Underside ochreous: fore wing with transverse confluent medial and broad outer dusky black band: hind wing with similar but narrower medial and outer bands. Thorax olivaceous, grey-speckled; abdomen dusky ochreous-brown.

Expanse $1\frac{4}{10}$ to $1\frac{7}{10}$ inch.

Hab. Darjiling. In coll. Dr. Staudinger and F. Moore.

BLENINA QUINARIA, n. sp. (Plate V. fig. 5.)

Fore wing crossed by four nearly equidistant slender black irregular sinuous pale-bordered lines, the basal, medial, and outer spaces olivaceous and clouded with brown speckles, the other two interspaces purplish-grey; a slender black streak at end of the cell, and an imperfectly-formed black irregular streak from base of the cell to above posterior angle: hind wing dusky-brown, with a medial indistinct pale ochreous waved band; cilia pale ochreous. Underside with a broad medial and an outer dusky black band. Thorax olivaceous, streaked with grey; palpi and legs black-speckled.

Expanse 1 1/16 inch.

Hab. Darjiling. In coll. Dr. Staudinger.

Allied to *B. senex* (*Danduca senex*, Butler, Types Het. Lep. B. M. iii. pl. 44. fig. 6), from Japan.

AMBELLA, n. g.

Male. Fore wing rectangular, costal and posterior margins curved from the base; exterior margin short, scalloped, and angular in the middle; costal and subcostal veins equally wide apart from the costa; first subcostal emitted at one half before end of the cell, second at one fifth, third at one third from base of second, fourth from end of the cell and anastomosing with third for a short distance near its base, fifth thrown off below the juncture; discocellular angled at both ends, middle very convex, radials from the angles; upper median from angle above end of the cell, second at end of the cell, third at nearly one half before the end. Hind wing broad; exterior margin waved, angular in the middle; costal vein straight; two subcostals from end of the cell; discocellular concave, radial from slight angle at lower end; two upper medians from end of the cell, lower from near the end; submedian and internal vein nearly straight. Cilia of both wings broad and composed of spatular scales. Body stout; palpi stout, terminal joint flattened, fusiform; fore and middle legs somewhat pilose; antennæ finely bipectinate to tip.

Allied to *Eliocroca*, Walker.

AMBELLA ANGULIPENNIS, n. sp. (Plate V. fig. 6.)

Fore wing olive-green, crossed by four irregular indistinct, slender, black, sinuous greyish-ochreous-bordered lines, a cluster of black speckles in middle of the cell, and two black dots on the costa near the apex; cilia greyish-ochreous: hind wing greyish-ochreous, with a prominent dusky-black outer band, the basal area also suffused with pale dusky ochreous. Underside ochreous, with a narrow medial and broad outer dusky black band. Thorax, palpi, and legs above olive-green; abdomen greyish-ochreous.

Expanse 1 9/16 inch.

Hab. Darjiling. In coll. Dr. Staudinger.

Family TOXOCAMPIDÆ.

Genus TOXOCAMPA.

TOXOCAMPA CUCULLATA, n. sp.

Male. Fore wing pale brownish-ochreous, brightest on exterior border; numerously covered with dark-brown- and black-speckled strigæ, which on the disk form a slight transverse indistinct fascia; reniform mark ochreous on the lower lobe, with dark brown border and black outer contiguous spots; a marginal row of black dots: hind wing and abdomen ochreous-brown, darkest on outer border; cilia pale brownish-ochreous. Thorax brownish-ochreous, black-speckled, a broad black collar on front of thorax; palpi, front of head, and legs above dark brown. Female darker coloured, the fore wing with two transverse subbasal indistinct fasciæ and a more distinct discal fascia; reniform mark much blacker.

Expanse 2 inches.

Hab. Nynee Tal, N.W. Himalayas. In coll. Dr. Staudinger and F. Moore.

Allied to the European *T. Lusoria.*

Family POLYDESMIDÆ.

BAMRA, n. g.

Fore wing elongated, triangular; costa straight; exterior margin very oblique, even, and slightly convex; posterior margin short; first subcostal emitted at nearly half length of the cell, second at one fifth before its end, trifid, third emitted from below end of second at one sixth its length, fourth from the third at one third before the apex, fifth ascending from end of cell and joined to third near its base; upper radial from upper end of the cell; discocellular slightly concave; lower radial and upper median from angles above end of the cell, second at its end, lower at nearly one half before the end. Hind wing short, exterior margin very convex; costal vein nearly straight; cell very short; two subcostals from end of cell; upper discocellular very obliquely concave, long, lower very short, radial from their juncture; two upper medians from acute lower end of cell, third at some distance before its end and opposite base of subcostals. Body stout, squamous; palpi large, ascending, first joint pilose, second squamous, third long and cylindrical; legs pilose beneath; antennæ filiform.

Type *B. discalis.*

BAMRA DISCALIS.

Agriopis discalis, Moore, P. Z. S. 1867, p. 57, pl. vii. fig. 2.

Hab. Bengal.

BAMRA ALBICOLA.

Felinia albicola, Walker, Catal. Lep. Het. B. M. xiv. p. 1515 (1858).

Hab. India. In coll. Saunders (Oxford Museum).

Male and female. Fore wing grey ; crossed by indistinct brownish lunular bands, two subbasal and a discal transverse irregular zigzag black line, and a submarginal black line curving from before apex to second median branch, along which it runs to the outer margin, another black line also extending along the lower median vein to the outer margin ; a black streak from the discal line at end of cell running inward to the costa : hind wing white, with pale brown marginal band. Thorax and abdomen grey ; collar, lateral band on palpi, and bands on tarsi dark brown.

Expanse, ♂ 1⅝, ♀ 1⅞ inch.

Hab. Darjiling. In coll. Dr. Staudinger and F. Moore.

OROMENA, n. g.

Fore wing elongated, triangular ; costa slightly depressed in the middle, apex pointed, exterior margin oblique and convex ; costal vein extending two thirds the margin ; first subcostal emitted at half length before end of the cell, second at one sixth before the end, trifid, third at nearly one third from base of second, and terminating at the apex, fourth at nearly one half from below third, fifth from end of cell, ascending and touching third close to its base ; cell short, not extending to half length of the wing ; discocellular slightly bent near each end, radials from the angles ; upper median from angle above end of the cell, middle branch from the end, lower from one fifth before the end ; submedian slightly concave near base. Hind wing small, exterior margin very obliquely convex, slightly scalloped ; abdominal margin short ; costal vein straight, extending to apex ; two subcostal branches from end of the cell ; discocellular very convex, oblique, radial from near its lower end ; cell short, one third length of the wing ; two upper medians from acute lower angle of the cell, lower branch from near the end ; submedian and internal straight. Body stout ; abdomen extending beyond hind wings ; palpi robust, extending to vertex, second joint thick, laxly squamous, third joint cylindrical, one third length of second in male and nearly equal in female ; legs stout, tibia pilose, tufted ; antennæ pectinated, the pectinations finely plumose in both sexes.

Type *O. reliquenda.*

OROMENA RELIQUENDA.

Briada reliquenda, Walker, Catal. Lep. Het. B. M. pt. xv. p. 1802 (1858), ♂.

Male. Fore wing golden olivaceous-brown ; crossed by a subbasal, an antemedial, and a postmedial black white-bordered sinuous line, their interspaces more or less black-speckled, the reniform mark being darker black, the postmedial line bordered by a broad black irregular band which is slightly white-speckled ; a submarginal row of slender black denticulate marks with white inner borders, and a marginal row of less distinct similar-coloured lunules : hind wing ænescent-yellow, with a broad outer and a slender waved discal ænescent-brown band, the former with an indistinct marginal row of whitish-bordered black lunules. Female with more prominent and broader white borders to the transverse sinuous lines, the outer black irregular band whiter-speckled and almost white at the costal end. Cilia with pale points at end of the veins. Thorax, palpi, and legs golden olivaceous-brown; legs with darker brown bands; abdomen ænescent-brown.

Expanse, ♂ 1¾, ♀ 2¼ inches.

Hab. Darjiling. In coll. Brit. Mus., Dr. Staudinger, and F. Moore.

DONDA, n. g.

Fore wing elongated, less triangular than in *Oromena*: hind wing somewhat shorter, exterior margin more convex; venation similar. Body stouter; palpi compactly clothed, terminal joint more slender; antennæ simple in both sexes. Also allied to *Belciana*.

Type *D. eurychlora* (*Dandaca eurychlora*, Walk. Catal. Lep. Het. B. M. p. 1670).

DONDA THORACICA, n. sp. (Plate V. fig. 7.)

Female. Fore wing dark cupreous-brown, crossed by a basal, a broad antemedial, and a postmedial olive-grey irregular band interruptedly traversed by black sinuous lines, a submarginal slender zigzag olive-grey fascia, and a cilial row of minute dots; orbicular and reniform marks defined by lower white spots; some white spots also on costal edge: hind wing cupreous-brown, crossed by a slender pale discal waved line terminating in a white streak above anal angle; a submarginal row of minute white dots and a streak at anal angle; cilia edged with white. Thorax and head above olive-yellow; front of thorax, palpi, and legs above chestnut-brown, grey-speckled; legs with white bands.

Expanse 2¼ inches.

Hab. Darjiling. In coll. Dr. Staudinger.

Family HYPOGRAMMIDÆ.

Genus CALLYNA, *Guén.*

CALLYNA SEMIVITTA, n. sp.

Allied to *C. jaguaria.* Of larger size: fore wing with the costal and posterior borders of a much paler chestnut-red colour, the middle of the wing only being longitudinally fasciated with dark chestnut-brown; veins speckled with grey; the transverse medial greyish-ochreous sinuous-bordered band is confined to the anterior half of the wing, not extending beyond the lower median vein; the orbicular and reniform marks are white-lined; apical and submarginal markings similar.

Expanse, ♂ 1¾, ♀ 2¼ inches.

Hab. Darjiling. In coll. Dr. Staudinger and F. Moore.

L

Family CATEPHIDÆ.

Genus ANOPHIA, *Guén.*

ANOPHIA PERDICIPENNIS, n. sp.   (Plate V. fig. 18.)

Fore wing grey, crossed by a short basal costal curved duplex black line, a transverse antemedial sinuous duplex black line, and a postmedial irregularly-undulated less-distinctly duplex line; an orbicular and a reniform indistinct black-lined grey mark, and a curved streak below the cell; a black streak between the two upper median veins and some less-defined streaks below the apex; the transverse duplex lines bordered with chestnut-brown: hind wing with white basal area and broad greyish cupreous-brown marginal band; cilia white-streaked.   Thorax, head, palpi, and legs above chestnut-brown, grey-speckled; fore tarsi with black bands; abdomen greyish-brown.

Expanse 1⅜ inch.

Hab. Darjiling.   In coll. Dr. Staudinger.

ZARIMA, n. g.

Male. Fore wing elongated, costa slightly arched at the base, apex slightly pointed; exterior margin obliquely convex, waved; posterior margin convex towards the base; costal vein recurved, extending two thirds the wing; first subcostal emitted at one half before end of the cell, second trifid, emitted at one third before the end, third at one fourth from base of second, fourth at one fourth from base of third, fifth from end of the cell and slightly touching third near the base; discocellular concave, slightly bent near its lower end; upper radial from end of the cell, lower from angle of discocellular; upper median from angle above end of the cell, middle branch from the end, lower from one third before the end; submedian slightly curved near the base.   Hind wing small; exterior margin convex, waved; abdominal margin short; costal vein nearly straight, extending to apex; two subcostals from end of the cell; discocellular obliquely convex, radial from slight angle near its lower end; cell short, broad; two upper medians from end of the cell, lower at one fourth before the end; submedian and internal veins straight.   Body moderately robust; abdomen extending much beyond hind wing, laterally tufted; thorax tufted above; palpi ascending to level of vertex, laxly squamose, second joint stout, third cylindrical, fully half length of second; legs tufted; antennæ somewhat broadly pectinated in male, minutely in female.

Allied to *Anophia.*

ZARIMA DENTIFERA, n. sp.   (Plate V. fig. 19.)

Male. Fore wing greyish ochreous-brown, crossed by a subbasal and antemedial partly white- and black-bordered zigzag line, and an oblique postmedial lunular black line, with grey-and-brown outer border, the black lunular line acutely indented between the middle and lower median veins; a submarginal indistinct pale line, and a marginal row of black dots; an ochreous streak on posterior border; orbicular and reniform marks defined by a black-speckled line, with a quadrate mark extending below the cell, forming a continuous quadrate mark, reniform mark with greyish centre and white-speckled outer border : hind wing white, with broad brown marginal band ; cilia edged with grey. *Female* darker : fore wing vinous-brown, with pale transverse markings, the lower medial area darkest ; the postmedial transverse lunular line lobed outward between the lower discocellular vein and the middle median branch, and not indented as in male ; outer half of reniform mark yellow. Body, palpi, and legs ochreous-brown ; tarsi with darker bands.

Expanse, ♂ 1⅜, ♀ 1⅝ inch.

Hab. Darjiling. In coll. Dr. Staudinger.

VAPARA, n. g.

Fore wing comparatively broad, exterior margin even, posterior margin convex in the middle ; first subcostal emitted at one third before end of the cell, second bifid, emitted at one fifth, fourth also bifid, emitted from end of the cell, curving upward and touching third near its base, the fifth being thrown off above end of the cell and just below the point of juncture ; discocellular slightly oblique and convex ; upper radial from end of the cell in a line with subcostal ; lower radial and upper median from slight angles above end of the cell, middle median close to its end, lower at one third before the end ; submedian slightly curved. Hind wing short, broad ; costal vein straight ; two subcostals from end of the cell ; discocellular slightly oblique and convex, radial from near its lower end ; two upper medians from end of cell, lower at one third before the end ; submedian and internal vein nearly straight. Body moderate, abdomen of male extending beyond hind wing, anal appendages tufted ; tibiæ compactly pilose ; palpi porrect, laxly squamose, stouter and shorter than in *Erygia* ; antennæ minutely pectinate in male. Allied to *Erygia*.

Type *V. fasciata.*

VAPARA INDISTINCTA, n. sp.

Fore wing brownish-ochreous, with two indistinct transverse basal, two medial, and two discal slender dusky-brown sinuous lunular bands, and a marginal row of points ; an orbicular dot and pale reniform mark : hind wing paler. Body brownish-ochreous, front of thorax and bands on fore legs ochreous-brown.

Expanse, ♂ 1⅖, ♀ 1⅝ inch.

Hab. Darjiling. In coll. Dr. Staudinger and F. Moore.

VAPARA FASCIATA, n. sp.

Fore wing brownish-ochreous, with a broad black subbasal transverse sinuous-lined band and two discal zigzag speckled fasciæ, the latter slightly bordered with white speckles at the costal end ; interspaces between the markings sparsely black-speckled ; a small black reniform mark ; cilia with black spots : hind wing ochreous-brown, with indistinctly darker discocellular streak and outer band ; cilia ochreous. Body brownish-ochreous, abdomen brown, bands on palpi and tarsi blackish.

Expanse 1⅖ inch.

Hab. Darjiling. In coll. Dr. Staudinger and F. Moore.

Genus STICTOPTERA, *Guén.*

STICTOPTERA OLIVASCENS, n. sp.

Male and female. Fore wing olive-grey, outer border brownish, traversed by grey lunular lines, two distinct dentate spots from apex, some indistinct brown streaks on costa, a black-lined distinct reniform mark, an orbicular black dot, and some black dots below the cell : hind wing diaphanous-white, with a broad brownish marginal band, veins brown. Body and legs greyish, legs brown-streaked.

Expanse, ♂ $1\frac{1}{16}$, ♀ $1\frac{2}{16}$ inch.

Hab. Khasia Hills. In coll. Dr. Staudinger and F. Moore.

Allied to *S. denticulata.*

Genus STEIRIA, *Walker.*

STEIRIA VARIABILIS, n. sp.

Male. Fore wing dark brown, with an indistinct greyish-brown duplex waved streak curving upward from base of posterior margin to the costa, the space between it and the lower part of the cell more or less dark black ; a recurved discal lunular similar line, and a dusky-brown-bordered submarginal zigzag fascia, the latter longitudinally streaked with black below the apex ; a marginal row of grey-bordered black lunules ; an indistinct black dot at base of the costa, an orbicular spot and reniform mark : hind wing semitransparent, pale purplish-blue, with broad dusky-brown marginal band. Body dark brown, thoracic crest fringed with black, base ochreous.

Variety. With the markings less distinct, the ground-colour dusky black, the portion of the markings on posterior border and before the apex bordered with pale ochreous.

Female brownish-ochreous, brightest on basal area, with black longitudinal streaks from base, the lines on posterior margin indistinct, the discal area from middle of costa to below the apex dusky-black, the streaks at apex and posterior angle bordered with ochreous. Body ochreous.

Expanse $1\frac{3}{8}$ inch.

Hab. Darjiling. In coll. Dr. Staudinger and F. Moore.

Allied to *S. cucullioides*, Guén.

SADARSA, n. g.

Fore wing very long and narrow ; costa slightly arched towards the end, apex acute, exterior margin very oblique, posterior margin slightly convex at the base ; first subcostal emitted at more than one half before end of the cell ; second at one eighth before the end, trifid ; fifth from end of the cell, curving upward and slightly touching third near its base ; cell extending three fifths of the wing ; discocellular concave, slightly bent near lower end ; upper radial from end of the cell, lower from angle of discocellular ; two upper medians from angles at end of the cell, lower from one fourth before the end. Hind wing short, broad ; exterior margin oblique, waved ; cell very short, extending only to one third the wing ; two subcostals from end of the cell; discocellular slightly bent towards lower end; radial and two upper medians from lower end of the cell, lower median at nearly one half before end of the cell. Thorax robust; abdomen long, somewhat slender; palpi somewhat porrect, compactly squamose, long; second joint extending to front of the head, third joint long; legs long, tibiæ laxly squamose; antennæ simple. Allied to *Gyrtona*, Walker.

Type *S. longipennis.*

SADARSA LONGIPENNIS, n. sp. (Plate V. fig. 14.)

Male. Fore wing pale violet-grey, with indistinct brown lunular spots between the veins, which are mostly apparent obliquely from the apex to posterior margin; a blackish dot below the cell: hind wing pale brown, the basal area paler and almost transparent. Body, palpi, and legs brown.

Expanse 1⅜ inch.

Hab. Darjiling. In coll. Dr. Staudinger.

SADARSA TENUIS.

Cucullia tenuis, Moore, P. Z. S. 1867, p. 60.

Hab. Darjiling (*Atkinson*). In coll. Dr. Staudinger and F. Moore.

Genus GYRTONA, *Walker.*

GYRTONA ALBODENTATA, n. sp.

Fore wing dark purple-brown, with indistinct transverse discal ochreous lunular-sinuous line, an orbicular and a reniform spot, a distinct submarginal curved row of minute white dentate spots, a cluster of grey scales on middle of posterior margin: hind wing pale brown, basal area whitish. Body, palpi, and legs above brown.

Expanse 1¹⁄₁₀ inch.

Hab. Cherra Punji, Khasia Hills. In coll. Dr. Staudinger and F. Moore.

Genus MELIPOTIS, *Hübner.*
Ercheia, Walker.

MELIPOTIS STRIGIPENNIS, n. sp.

Fore wing deep bright umber-brown; veins indistinctly lined with black, and with intervening longitudinal narrow ochreous streaks; an interrupted zigzag black ochreous-bordered apical streak; a double sinuous black line obliquely from exterior margin below the apex to posterior margin, and a sinuous ochreous line between them and the posterior angle; an indistinct double sinuous black transverse antemedial line; a small ochreous-bordered black orbicular and reniform mark, and a black-centred mark below the cell; posterior end of the streaks whitish-bordered in the male, and the posterior margin of the wing in the female longitudinally streaked with ochreous. Hind wing dusky umber-brown, with a paler waved discal streak and an outer spot. Body dark brown.

Expanse, ♂ 1⅞, ♀ 2¼ inches.

Hab. Darjiling. In coll. Dr. Staudinger and F. Moore.
Nearest allied to the Japanese *E. umbrosa*, Butler.

MELIPOTIS COSTIPANNOSA, n. sp. (Plate V. fig. 8.)

Male and female. Fore wing umber-brown, with a blackish-brown sinuous-bordered patch on base of the costa, and an elongated black-streaked apical patch; a slender black line from lower angle of the former to posterior margin, and a double sinuous line recurved from the latter; a black-lined mark between the median and submedian veins, and a slender whitish sinuous submarginal line; a black dot within the cell, and two white dots at the end: hind wing dusky brown, with two white spots, and two white streaks on the cilia.

Expanse, ♂ 2, ♀ 2⅛ inches.
Hab. Darjiling. In coll. Dr. Staudinger.

Family CATOCALIDÆ.

Genus CATOCALA, *Ochs.*

CATOCALA TAPESTRINA, n. sp. (Plate V. fig. 13.)

Fore wing chestnut-brown; veins across the discal area broadly speckled with purple-grey; a transverse basal short sinuous pale streak, an antemedial waved narrow black band with greyish-inner-bordered line, a postmedial deeply sinuous black line with greyish-yellow-outer-bordered line, the lines bent inward to the costa between the discocellular veins; a discal yellowish zigzag fascia, and a marginal row of black-bordered yellow spots; a grey-speckled discocellular lunule, and an auriform mark below the angle of the cell composed of a black-bordered greyish-yellow line with chestnut-brown centre; the basal, medial, and discal areas with black longitudinally suffused streaks: hind wing deep yellow, with broad outer waved band, and a narrower medial irregular brown band; a marginal row of small yellow spots; abdominal margin broadly paler brown. Thorax, head, and palpi chestnut-brown; tegulæ and palpi grey-fringed; abdomen above pale brown; legs grey.

Expanse 3 inches.
Hab. Darjiling In coll. Dr. Staudinger.

Family PHYLLODIDÆ.

Genus PHYLLODES, *Guén.*

PHYLLODES ORNATA, n. sp.

Female. Fore wing ochreous-red, the posterior and exterior borders suffused reddish-brown; numerously covered with blackish strigæ; crossed by five dusky-black bent strigate lines; a blackish streak extending from upper end of the cell to exterior margin below the apex; two pure white black-bordered spots at end of the cell, the lower spot triangular in form; exterior margin slightly white-speckled: hind wing dark purplish-brown, with a broad ochreous-yellow marginal band decreasing in width to near anal angle; cilia dark brown. Thorax and palpi ochreous-red; legs purple-brown; abdomen purplish-brown.

Expanse 4 inches.

Hab. Darjiling. In coll. Dr. Staudinger and F. Moore.

Family EREBIIDÆ.

Genus SYPNA, *Guén.*

SYPNA PLANA, n. sp. (Plate V. fig. 24.)

Female. Bright uniform ochreous-brown, crossed by several indistinct dusky-black-speckled sinuous lines, a more distinct submarginal zigzag line, and marginal lunules: hind wing dark ochreous-brown, with a broad apical yellow band, narrow sinuous anal streaks, and lower marginal black lunules. Underside with three transverse blackish bands, comparatively narrower than in *S. floccosa.* Near to *S. apicalis.*

Expanse 2¼ inches.

Hab. Cherra Punji. In coll. Dr. Staudinger.

SYPNA REPLICATA.

Elpia replicata, Felder et Rogenh. Novara Reise, Lep. iv. pl. 117. f. 25, ♂ (1874).

Male. Pale umber-brown, the basal half darker brown and traversed by two pale zigzag lines, followed by a medial club-shaped fascia, which is indistinctly marked by paler streaks; a submarginal zigzag black fascia and marginal pale-bordered black lunules: hind wing dusky brown, with a broad yellowish-ochreous apical band, and narrower less distinct lower marginal streaks to anal angle, and black lunules.

Expanse 2 inches.

Hab. Darjiling. In coll. Dr. Staudinger.

Felder gives British Guiana as the habitat of this insect. This is evidently incorrect.

SYPNA FLOCCOSA, n. sp. (Plate V. fig. 23.)

Allied to *S. replicata. Female.* Fore wing uniformly darker umber-brown, with a broad medial transverse greyish-white floccose patch: hind wing paler, with shorter apical yellow band and less distinct lower streaks. Underside very pale brownish-ochreous, with three transverse narrow black bands, the outer band on hind wing the broadest and confluent at the upper end with the inner band.

Expanse 2⅜ inches.

Hab. Darjiling. In coll. Dr. Staudinger.

SYPNA BRUNNEA, n. sp.

Allied to *S. cyanivitta.* Comparatively smaller, and with shorter wings. Fore wing with similar transverse medial lines, the interspace being entirely dark brown in both sexes. Underside with more prominent medial transverse fascia.

Expanse, ♂ 1⅝, ♀ 2⅛ inches.
Hab. Darjiling.   In coll. Dr. Staudinger and F. Moore.

SYPNA ALBOVITTATA, n. sp.   (Plate V. fig. 25.)

Allied to *S. tenebrosa*. *Male.* Pale umber-brown : fore wing crossed by a broad sub-basal and an antemedial dark-brown-speckled band, each band outwardly bordered towards the costa by a broad white zigzag band ; a dark brown zigzag submarginal fascia and a distinct series of submarginal white-bordered black lunules ; orbicular spot black-bordered, reniform mark indistinct : hind wing with indistinctly darker medial transverse narrow line, and zigzag streaks from anal angle, a white-bordered black lunular marginal line.   Underside with three transverse fasciæ, the inner indistinct.
Expanse 2⅛ inches.
Hab. Darjiling.   In coll. Dr. Staudinger.

SYPNA PANNOSA, n. sp.   (Plate V. fig. 12.)

Male. Lilacine ochreous-brown.   Fore wing with a broad black basal oblique band, traversed by a brown sinuous line, and outwardly bordered by a dark brown line ; a medial transverse, very indistinct, waved darker brownish narrow fascia, and submarginal zigzag fascia ; a black triangular costal patch at the apex, and a constricted patch at posterior angle : hind wing with a dusky brown marginal fascia merging into black at the anal angle, and above which are some short blackish sinuous streaks ; apical border and cilia brownish-ochreous, traversed by black lunules.   Thorax and tip of abdomen blackish.   Underside paler, with two narrow inner and a broad outer dusky black fascia, the latter angulated outward.
Expanse 2⅛ inches.
Hab. Khasia Hills.   In coll. Dr. Staudinger.

Family HYPOPYRIDÆ.

Genus SPIRAMA, *Guén.*

SPIRAMA MODESTA, n. sp.

Allied to *S. triloba* (*Hypopyra mollis,* Guén., ♀).   *Male.* Upperside olivaceous brownish-ochreous, with the lower medial area darker brown ; markings similar, the trilobate spot and discocellular lunule prominently black and slenderly bordered with grey : hind wing with the basal area brown ; a medial and a discal darker straight-bordered band. *Female* paler : fore wing with less distinct markings, the inner discal lines being most prominent towards the posterior margin, and the outer line with dark points : hind wing with the inner discal lines and outer denticulated line prominent, the submarginal line pale ochreous and straight.
Expanse, ♂ 2⅝, ♀ 2⅞ inches.
Hab. Silhet ; Darjiling.   In coll. Dr. Staudinger and F. Moore.

HYPOPYRA DISTANS, n. sp.

Male. Upperside very pale uniform olivaceous-ochreous, the outer border very slightly tinged with brighter olive: fore wing with a very indistinct oblique medial brown fascia bent inward to the costa, a contiguous sinuous line, an outer pale sinuous line, and apical fascia; three distinct black discal spots, the lowest large, the others minute, and each white-bordered: hind wing with pale-brown transverse fascia and indistinct outer sinuous pale fasciæ. *Female* of a brighter ochreous tint, and not suffused with olive; markings more indistinct; discal spots on fore wing indistinctly replaced by whitish-ochreous spots. Underside ochreous-yellow; bands not prominent.

Expanse $2\frac{7}{8}$ inches.

Hab. Bombay (*Dr. Leith*). In coll. F. Moore.

Genus HAMODES, *Guén.*

HAMODES MARGINATA, n. sp.

Male and female. Bright brownish-ochreous; a pale-bordered dark brown line extending from apex of fore wing to near anal angle: fore wing with three dusky-brown oblique streaks on the costa, the streaks very indistinctly bent and continued obliquely across to the posterior margin; a dusky brown reniform spot, and a broad yellowish irregular-bordered indistinct marginal band: hind wing with a more distinct yellow irregular-bordered marginal band. Sides of thorax, tip of palpi, and legs above dusky-black. Underside brighter ochreous; both wings with two transverse dusky brown lines, the inner line slender.

Expanse $2\frac{1}{2}$ inches.

Hab. Darjiling. In coll. Dr. Staudinger.

Family OPHIUSIDÆ.

Genus OPHIODES, *Guén.*

OPHIODES ADUSTA, n. sp. (Plate VI. fig. 11.)

Nearest to *O. trapezium* (*cognata*, Walk.). Wings shorter: fore wing pale purplish sienna-red; transverse bands similarly disposed but darker, and the interspace between the medial and submarginal of a much deeper colour; reniform spot very pale: hind wing very pale reddish-ochreous, with broad dusky brown outer band.

Expanse $2\frac{1}{16}$ inches.

Hab. Cherra Punji. In coll. Dr. Staudinger.

OPHIODES INDISTINCTA, n. sp. (Plate VI. fig. 12.)

Fore wing brownish-ochreous, indistinctly brown-speckled; with a subbasal and a medial transverse very indistinct pale line, the subbasal line angled outward at the submedian vein, the medial line waved and black-speckled at the costal end; a submarginal duplex sinuous slightly blackish-bordered grey line; reniform mark indistinct: hind wing with very broad dark dusky-brown band extending to near the base. Underside: both wings with a dusky outer band.

Expanse $1_{1}^{9}_{0}$ inch.

Hab. Khasia Hills. In coll. Dr. Staudinger.

DORDURA, n. g.

Fore wing elongated, triangular; costa straight, exterior margin very oblique; costal vein long, extending four fifths the margin; first subcostal emitted at one half, and second at one eighth before end of the cell, third at one fifth from below base of second, fourth at two thirds from base of third, fifth from end of the cell, ascending and touching third near its base; discocellular convex in middle, bent at each end, radials from the angles; cell extending to half the wing; upper median from angle above end of the cell, middle median from immediately before the end, lower at one half before the end; submedian curved near base. Hind wing triangular, apex slightly convex, exterior margin oblique, somewhat angular near anal end, costal vein nearly straight, extending to apex, two subcostals from end of the cell; discocellular obliquely concave, radial from its lower end; cell short; two upper medians from acute angles at lower end of the cell, lower branch at one third before the end; submedian and internal vein straight. Body moderately stout; abdomen of male longer than hind wing; palpi ascending to level of vertex, flattened, second joint broadest at tip, third joint slender, half length of second; antennæ finely pectinated in male; femora slightly pilose beneath, tibiæ pilose above.

Allied to *Hypætra.*

Type *D. apicalis.*

DORDURA APICALIS, n. sp. (Plate V. fig. 20.)

Fore wing uniform brownish-ochreous, very sparsely speckled with minute brown scales, which are slightly clustered on lower discal area; two very indistinct medial transverse oblique brown lines, the outer line sometimes ending in a very prominent dark brown pale-bordered triangular costal patch before the apex; a marginal row of minute black dots, and an indistinct brown-speckled orbicular and reniform spot: hind wing paler at the base, with a brownish outer border, which is traversed by a submarginal dusky-brown fascia extending broadly from the apex, and is sinuous and darkest at the anal angle. Body brownish-ochreous; palpi at the tip, and legs with terminal brown bands.

Expanse $1\frac{3}{8}$ inch.

Hab. Calcutta. In coll. Dr. Staudinger.

DORDURA ALIENA.

Hypætra aliena, Walker, Catal. Lep. Het. Brit. Mus. pt. xxxiii. p. 964 (1865).

Hab. Calcutta (*Atkinson*).

NOTE.—*Ophiusa subcostalis,* Walker, Catal. Lep. Het. B. M. pt. xxxiii. p. 969, from N. China, will also come into the genus *Dordura.*

Genus OPHIUSA, *Guén.*

OPHIUSA FALCATA, n. sp. (Plate VI. fig. 14.)

Fore wing dark brown, with a short basal, a recurved outwardly-oblique duplex subbasal, and a curved discal purplish-lilac slender line, the latter acutely angled towards the apex, near its costal end, and joins a similar dark-bordered straight line to the apical angle, giving the wing a falcate appearance; the basal area to middle of the wing purplish-lilac, the outer space to the curved discal line being very dark brown; a very indistinct submarginal pale sinuous line with lilac points: hind wing paler brown, the basal and outer area suffused with dull purplish-lilac.

Expanse 2¼ inches.

Hab. Khasia Hills. In coll. Dr. Staudinger.

PASIPEDA, n. g.

Fore wing with the costa slightly arched at the base and before the end, apex acute; exterior margin oblique, convex hindward; cell short, less than half the wing; first subcostal emitted at one half before end of the cell, second at one fifth, trifid, the third from below second at one sixth from its base above end of the cell, fourth at one sixth before the apex, fifth from end of the cell and touching third close to its base; discocellular bent inward at each end, concave in the middle, radials from the angles; two upper medians from angles obliquely above end of the cell, lower at one half before the end; submedian slightly curved downward from the base. Hind wing short, somewhat quadrate; exterior margin convexly angular in the middle; cell very short; costal vein nearly straight, extending to apex; two subcostals from end of the cell; discocellular concave, radial from near its lower end; two upper medians from end of the cell, lower at one third before the end; submedian and internal vein slightly recurved. Body stout; palpi ascending, flattened, squamose, second joint long, third joint slender, two thirds length of second; legs short, thick; antennæ simple.

PASIPEDA PALUMBA.

Hulodes palumba, Guénée, Noct. iii. p. 211 (1852).
Remigia colligens, Walker, Catal. Lep. Het. B. M. pt. xxxiii. p. 1019 (1865).

Hab. Java; Singapore; Silhet; Cherra (*Atkinson*); S. India; Ceylon.

Family EUCLIDIIDÆ.

Genus CHALCIOPE, *Hübn.*

Chalciope, Hübner, Verz. bek. Schmett. p. 268 (1818-25).
Trigonodes, Guén. Noct. vii. p. 281 (1852).

CHALCIOPE DISJUNCTA, n. sp.

Allied to *C. Cephise* (Cram. Pap. Exot. pl. 227. fig. C). A much smaller insect: fore wing with a lengthened black fusiform lower basal streak and an entirely separated triangular

M 2

discal patch, the latter with slightly concave margins, and both with slender white-bordered line : hind wing with a slight dusky outer border.

Expanse 1½ inch.

Hab. Parisnath, Bengal (*Atkinson*); Bombay (*Dr. Leith*).   In coll. Dr. Staudinger and F. Moore.

Family POAPHILIDÆ.

Genus POAPHILA.

POAPHILA QUADRILINEATA, n. sp.   (Plate V. fig. 22.)

Fore wing lilac-brown, costal edge ochreous ; crossed by an antemedial and a postmedial ochreous-bordered brown line ; some brown speckles at end of the cell, and a row of speckles midway between the outer line and the margin : hind wing pale brownish ochreous, with a darker brownish outer band.   Underside ochreous : fore wing with dusky outer band and small discocellular spot.   Front of thorax, head, palpi, and legs above bright ochreous ; hind part of thorax and abdomen lilac-brown.

Expanse, ♂ 1⁴⁄₁₀, ♀ 1²⁄₁₀ inch.

Hab. Darjiling ; Cherra Punji.   In coll. Dr. Staudinger.

POAPHILA OCULATA, n. sp.   (Plate V. fig. 11.)

Fore wing ochreous, crossed by three medial irregular brown lines, the outer line suffused ; a large prominent greyish-centred reniform spot : hind wing paler ochreous, with a dusky-brown outer band, the basal area brownish-speckled.   Underside paler ochreous ; fore wing with a small brownish discocellular spot, and both wings with a brownish outer band.   Front of thorax, head, palpi, and legs above bright ochreous ; lower part of thorax and abdomen brownish ochreous.

Expanse 1 inch.

Hab. Bengal.   In coll. Dr. Staudinger.

POAPHILA PALLENS, n. sp.   (Plate V. fig. 9.)

Fore wing whitish-ochreous, with a pale brownish suffused lunular submarginal band, and a marginal row of spots : hind wing pale brownish-ochreous.   Body and underside pale brownish-ochreous.

Expanse 1⅛ inch.

Hab. Calcutta.   In coll. Dr. Staudinger.

POAPHILA UNIFORMIS, n. sp.   (Plate V. fig. 10.)

Male. Pale lilac greyish-ochreous.   *Female.* Pale brownish-ochreous : fore wing with a nearly straight submarginal pale line, which is slightly brownish-bordered externally : a marginal row of minute brownish dots, a dot at end of the cell, and one in its middle.

Expanse 1 inch.

Hab. Parisnath ; Calcutta.   In coll. Dr. Staudinger.

Genus BORSIPPA, *Walk.*

BORSIPPA MARGINATA, n. sp.

Fore wing pale ochreous-brown; crossed by two straight subbasal, a medial, and a discal slightly curved dark brown lines, each line with a slight pale border; a broad blackish-brown irregular-bordered marginal band: hind wing pale dull brown. Underside paler than above.

Expanse $1\frac{4}{10}$ inch.

Hab. Darjiling. In coll. Dr. Staudinger and F. Moore.

Allied to *B. quadrilineata*, Walk.

NASAYA, n. g.

Wings short, broad: fore wing with nearly straight costa, apex acute; exterior margin nearly erect, slightly convex hindward; posterior margin slightly convex towards base; first subcostal at one half before end of the cell; second at one fifth, trifid, third at one fifth from base of second, fourth at one half from third; fifth from end of the cell, ascending and slightly touching third near the base; discocellular bent at each end, concave in the middle, radials from the angles; two upper medians from angles at end of the cell, lower at nearly one half before the end. Hind wing convex externally; abdominal margin short; two subcostals from end of the cell; discocellular concave, radial from near lower end; two upper medians from end of the cell, lower from one third before the end. Body slender; palpi ascending, second joint compactly squamose, extending to vertex, third joint long, slender; antennæ setose; femora pilose beneath; tibiæ thickish, compactly pilose.

NASAYA HEPATICA, n. sp.

Fore wing chestnut-brown, with numerous transverse olive-grey strigæ, which are clustered thickly from the apex to end of the cell: hind wing dusky brown. Thorax, palpi, and legs above chestnut-brown; abdomen dusky-brown. Underside uniform dusky brown.

Expanse $1\frac{3}{10}$ inch.

Hab. Darjiling. In coll. Dr. Staudinger and F. Moore.

Genus DIERNA, *Walker.*

DIERNA MULTISTRIGARIA, n. sp.

Ochreous-grey; fore wing numerously covered with transverse continuous purplish-brown slender waved strigæ, and a prominent oblique discal duplex darker brown line which is broadly bordered outwardly by a brown indistinct shade, and followed by a submarginal indistinct sinuous fascia: hind wing crossed by numerous shorter brown strigæ; outer border dusky, with pale sinuous streaks from anal angle.

Expanse $1\frac{5}{8}$ inch.

Hab. Bombay; Cherra (*Atkinson*). In coll. F. Moore and Dr. Staudinger.

Genus PHURYS, *Guén.*

PHURYS FASCIOSA, n. sp. (Plate VI. fig. 6.)

Female. Pale greyish brownish-ochreous: fore wing crossed by an oblique straight antemedial purplish-brown line, and a discal duplex line slightly curving from apex to near posterior angle, the interspaces from the base to the discal line traversed by parallel waved less distinct purple-brown fasciæ; a submarginal less distinct sinuous fascia; a discocellular pale lunular mark: hind wing crossed by less distinct continuous brown lines.

Expanse 1⅝ inch.

Hab. Darjiling. In coll. Dr. Staudinger.

PHURYS SIMILIS, n. sp. (Plate VI. fig. 5.)

Female. Allied to *P. fasciosa*; of a more brownish-ochreous colour: fore wing crossed by a similar antemedial and discal line, the interspaces from the base each with only two intervening parallel lines, and the outer margin with a more distinct pale line; a black dot in middle of the cell.

Expanse 1⅔ inch.

Hab. Darjiling. In coll. Dr. Staudinger.

PHURYS DISSIMILIS, n. sp.

Allied to *P. similis*. Pale yellowish-ochreous: fore wing more acutely pointed at the apex; crossed by six ochreous-brown slightly obliquely-curved fasciæ, the fifth most prominent and suffusedly bordered on its outer margin; a submarginal similar fascia, and a marginal row of dots: hind wing paler, with indistinct similar fasciæ.

Expanse 1¼ inch.

Hab. Cherra Punji. In coll. Dr. Staudinger and F. Moore.

Genus ILUZA, *Walker.*

ILUZA TRANSVERSA, n. sp. (Plate VI. fig. 13.)

Male. Pale greyish-ochreous, indistinctly speckled with minute black scales; crossed by a prominent slender black oblique line, extending from apex of fore wing to anal angle, the line slenderly bordered within by pale ochreous and without by reddish-ochreous; the disk crossed by an indistinct sinuous speckled fascia, and an outer marginal row of dots; an indistinct sinuous subbasal fascia and discocellular mark on fore wing.

Expanse 1⅞ inch.

Hab. Darjiling. In coll. Dr. Staudinger.

ILUZA DUPLEXA, n. sp. (Plate VI. fig. 7.)

Female. Purplish brownish-ochreous: fore wing crossed by a slender chestnut-brown subbasal line and two oblique discal contiguous lines, the latter continued across the lower part of hind wing to the anal angle, the outer discal line being broadest and extending to apical angle ; a submarginal sinuous chestnut-brown slender zigzag line on both wings, and a dusky round discocellular spot on fore wing: hind wing paler-coloured.

Expanse 1¾ inch.

Hab. Khasia Hills. In coll. Dr. Staudinger.

TOCHARA, n. g.

Fore wing triangular, apex acute, exterior margin very oblique, posterior margin short ; first subcostal emitted at three fourths and second at one fourth before end of the cell, second trifid, third and fourth at equal distances ; fifth from end of the cell, bent upward and touching third near its base ; discocellular bent at each end, concave in the middle, radials from the angles ; middle median at one eighth, lower at one third before end of the cell. Hind wing short ; exterior margin convex, somewhat truncated at anal angle ; cell short, one third the wing ; two subcostals from end of the cell ; discocellular concave, radial from its lower end ; middle median from near end of the cell, lower at one half before the end. Body long, abdomen slender ; palpi ascending, compactly squamose, second joint broad, extending to vertex, third joint two thirds length of second, cylindrical ; legs thickly pilose ; antennæ very long, two thirds length of the costa, bipectinate to tip.

Allied to *Iluza*.

Tochara obliqua, n. sp. (Plate VI. fig. 27.)

Pale ochreous-brown, indistinctly grey-speckled : fore wing crossed by an indistinct dusky-brown zigzag subbasal line, a medial, and a more irregular discal line, beyond which is a very prominent oblique, straight, slender whitish-bordered outer line, extending from apex to anal angle of hind wing, the outer border of this line with an indistinctly brighter brown sinuous-margined fascia, the sinuous margin also slightly brown-speckled, and on the hind wing with a central discal distinct dusky-black spot ; an indistinct dusky-brown orbicular spot and reniform lunule, the latter with a whitish central spot : both wings also with an indistinct slender dusky scalloped marginal line. Female brighter-coloured ; marked as in male.

Expanse, ♂ 1¾, ♀ 2 inches.

Hab. Khasia Hills ; Cherra. In coll. Dr. Staudinger.

Family THERMESIIDÆ.

Genus SANYS, *Guén.*

Sanys flexus, n. sp.

Allied to *S. angulina* ; smaller. Upperside similarly coloured : fore wing with an indistinct antemedial pale-bordered waved line, a more prominent postmedial line, the latter bent inward to the costa ; two small indistinct black-speckled dots at end of the cell, and one in the middle ; a submarginal sinuous black-speckled line and a less distinct marginal row of dots: hind wing with a medial transverse pale-bordered brown line, and a marginal row of black dots.

Expanse 1 inch.

Hab. Cherra Punji.　In coll. Dr. Staudinger and F. Moore.

Genus THERMESIA, *Hübn.*

THERMESIA OBLITA, n. sp.

Brownish-ochreous, sparsely speckled with minute blackish scales: fore wing crossed by three indistinct subbasal darker oblique fasciæ, a pale-bordered sinuous line, and an outer discal slightly black-speckled similar line; a small black dot in middle of the cell: hind wing with indistinct subbasal transverse darker sinuous fasciæ, and a more distinct straight submarginal narrow pale band.　Underside paler, with darker marginal band and narrow discal fascia.　Front of thorax, head, palpi, and legs above dark umber-brown.

Expanse $1\frac{1}{4}$ inch.

Hab. Parisnath, Bengal.　In coll. Dr. Staudinger and F. Moore.

Genus CAPNODES, *Guén.*

CAPNODES PALLENS, n. sp.　(Plate V. fig. 21.)

Upperside pale brownish-ochreous.　Markings similar to those of *C. trifasciata* but more distinctly black, the white spots on the points of the outer sinuous line less prominent.　Underside very pale brownish-ochreous.

Expanse $1\frac{3}{10}$ inch.

Hab. Calcutta.　In coll. Dr. Staudinger.

DURDARA, n. g.

Fore wing elongated, narrow; costa slightly arched towards the end, apex very acute; exterior margin short, obliquely convex in the middle; posterior margin long; first subcostal emitted at one half before end of the cell, second bifid near its base and terminating at apex, second and fourth from near end, and fifth from end of the cell; discocellular bent at each end, concave in the middle; radials from the angles; cell long, extending half the wing; middle median from near end of the cell, lower at one half before the end.　Hind wing short, exterior margin very obliquely convex, abdominal margin short; first subcostal at one half before end of the cell; discocellular slightly oblique and concave, radial from its lower end; cell very short; middle median from very near end of the cell, lower at one third before the end; submedian and internal veins wide apart.　Body short, stout; palpi ascending slightly above the vertex, thick, second joint compactly squamose, third joint cylindrical, about half length of second; legs laxly squamose, tibiæ thick; antennæ simple.

Type *D. myrtæa.*

DURDARA MYRTÆA.

Phalæna (Noctua) myrtæa, Drury, Exot. Ins. ii. pl. 2. fig. 3 (1773).
Thermesia myrtæa, Walker, Catal. Lep. Het. B. M. xv. p. 1575.

Red; both wings crossed by numerous indistinct black linearly-disposed strigæ; an indistinct dusky spot at end of each cell. Underside paler, marked as above; the fore wing also having a slight short blackish fascia below the apex. Cilia edged with grey.

Expanse, ♂ 1₁³₂, ♀ 1₁⁵₃ inch.

Hab. Bombay (*Wilkinson*); Darjiling (*Atkinson*). In coll. Dr. Staudinger and F. Moore.

DURDARA PYRALIATA, n. sp.

Dull purplish-red, crossed by several linearly-disposed indistinct brown strigæ: fore wing with a lower discal trilobate spot, the upper and lower lobes diaphanous-white, the middle lobe yellowish-streaked, the upper lobe also divided across the middle. Underside purplish-grey, crossed by red strigæ; diaphanous spots as above.

Expanse ¾ inch.

Hab. Calcutta. In coll. Dr. Staudinger.

Has somewhat the appearance of *Pyralis Eloralis.*

DURDARA LOBATA, n. sp. (Plate V. fig. 16.)

Purplish-grey: both wings crossed by several indistinct delicate brown strigæ, less numerous than in *D. pyraliata*: fore wing with the trilobed discal spot smaller than in *D. pyraliata*, and narrowly black-bordered, the upper lobe being smaller, obliquely oval and entire.

Expanse 1 inch.

Hab. Khasia Hills. In coll. Dr. Staudinger.

RAPARNA, n. g.

Wings small, short, broad; fore wing triangular, apex pointed; exterior margin short, very slightly convex; posterior margin long; first subcostal emitted at one half before end of the cell, second at one fifth, trifid, third at one fifth from base of second, fourth at one fourth from base of third, fifth from end of the cell and touching third close to its base; discocellular bent at each end, concave in middle, radials from the angles; two upper medians from angles at end of the cell, lower at nearly one half before the end, submedian curved. Hind wing short, exterior margin very convex; cell one third the wing; two subcostals from end of the cell; discocellular slightly concave, radial from near lower end; two upper medians from end of the cell, lower at nearly one half before the end; submedian and internal straight. Body moderate; palpi ascending, squamose, second joint broadest at apex, extending to vertex, third joint two thirds its length, slender, pointed; legs slender; antennæ very minutely bipectinated.

Type *R. ochreipennis.*

N

RAPARNA OCHREIPENNIS, n. sp.   (Plate VI. fig. 8.)

Bright ochreous-yellow, with numerous very indistinct darker strigæ: fore wing with a purple-brown transverse antemedial and a postmedial sinuous line; a spot at base and middle of costa, a dot at end of the cell, and a very indistinct transverse submarginal sinuous fascia: hind wing with a less distinct transverse medial and discal sinuous line. Palpi and fore legs above pale brown.

Expanse $\frac{7}{8}$ inch.

Hab. Parisnath, Bengal (April). In coll. Dr. Staudinger and F. Moore.

RAPARNA TRANSVERSA, n. sp.

Ochreous-yellow, numerously covered with broad purplish-red transverse strigæ: fore wing with a straight purplish-brown-speckled narrow postmedial band, which is also continued across the middle of hind wing.

Expanse $\frac{6}{8}$ inch.

Hab. N.W. Himalaya; Parisnath, Bengal (*Atkinson*). In coll. F. Moore and Dr. Staudinger.

RAPARNA UNDULATA, n. sp.

Pale ochreous-yellow, with very indistinct and sparsely-disposed darker strigæ: fore wing with an antemedial and postmedial transverse purple-brown waved narrow band, and hind wing with a medial waved band; a small brown spot at lower end of each cell.

Expanse $\frac{8}{8}$ inch.

Hab. Dharmsala (*Hocking*); Calcutta (*Atkinson*), June. In coll. Rev. H. Hocking and Dr. Staudinger.

Genus SELENIS, *Guén.*

SELENIS RETICULATA, n. sp.   (Plate VI. fig. 9.)

Pale brownish-ochreous: fore wing with a slender white subbasal line extending along the median vein and forming a reniform mark, from the top of which it curves outward below the apex and runs to the posterior angle, and then across the hind wing to the anal angle; a slender line also running from lower end of the cell to posterior margin; the base, costal and apical borders whitish; some blackish dots disposed at the base and below the apex, and some white dots on costa before the apex: hind wing also with a basal and medial transverse white line, a mark within the cell, and a suffused marginal sinuous line. Thorax and head whitish; abdomen, palpi, and legs ochreous.

Expanse $\frac{9}{10}$ inch.

Hab. Darjiling. In coll. Dr. Staudinger.

SELENIS OBSCURA, n. sp.   (Plate VI. fig. 10.)

Brownish-ochreous: fore wing with an indistinct pale slender subbasal, a discal, and a submarginal transverse sinuous line; a pale yellowish streak at the apex, and a slight patch on posterior border; some yellow spots on costa before the apex: hind wing with a yellowish costal border, and two sinuous medial transverse lines. Body ochreous-brown ; thorax with a pale yellowish band.

Expanse 1¼ inch.

Hab. Darjiling. In coll. Dr. Staudinger.

Genus MESTLETA, *Walker.*

MESTLETA ANGULIFERA, n. sp.

Pale purplish-greyish-ochreous: fore wing crossed by an oblique medial narrow brown band, which is acutely bent inward to the costa near its upper end, an antemedial and a postmedial slender sinuous brown line, both bent inward at the costal end ; a slight brown pale-bordered oblique streak from the apex, which is continued hindward in a blackish sinuous line ; a slender pale-bordered brownish marginal line : hind wing with a narrow brown medial transverse band, and two indistinct outer sinuous lines. Underside pale brownish-ochreous. Thorax, palpi, and legs ochreous-brown ; abdomen and band on the tarsi greyish-ochreous.

Expanse ⅞ inch.

Hab. Calcutta. In coll. Dr. Staudinger and F. Moore.

" Larva feeds in the pods of *Jonesia asoka.* June."—*Atkinson.*

MESTLETA ACONTIOIDES, n. sp. (Plate V. fig. 15.)

Fore wing ochreous-white, with three minute brown streaks on middle of the costal edge ; from the third streak a slender fine line runs to the disk ; an oblique brown fascia extending from apex to middle of posterior margin, the fascia sharply defined on its outer border and black-marked at its apical end ; below the apex is a slight sinuous blackish line : hind wing whitish, with a brown short fasciated streak from above anal angle. Front of thorax, head, palpi, and legs above brown ; abdomen pale brownish-ochreous.

Expanse ¹⁰⁄₁₆ inch.

Hab. Calcutta. In coll. Dr. Staudinger.

SONAGARA, n. g.

Fore wing triangular ; costa nearly straight, apex acute, exterior margin oblique and slightly convex, posterior angle pointed ; first subcostal emitted at two thirds before end of the cell, second, third, and fourth near end of the cell and at equal distances from the fifth, third terminating at the apex ; discocellular bent at each end, concave in the middle, radials from the angles ; middle median at one fifth and lower at more than one half before end of the cell. Hind wing broadly triangular ; exterior margin convex ; first subcostal at nearly one half before end of the cell ; discocellular slightly oblique, bent inward at lower end, radial from the angle ; middle median very near end of the cell, lower at one third before the end ; submedian and internal vein at equal distances from lower median. Body short, stout ; palpi ascending and convergent at the tip, second joint stout, laxly squamose, third joint very short, conical; legs stout, tibiæ thick, laxly squamose; antennæ simple in both sexes.

Type *S. strigipennis.*

SONAGARA STRIGIPENNIS, n. sp.

Ochreous-red, with a narrow dusky-brown band straight from apex to middle of abdominal margin: fore wing with a slender slightly waved line extending from the base below the apex to posterior angle; some outwardly-oblique waved very slender brown strigæ from the costa, and some longitudinally on the outer border: hind wing also with a recurved slender brown line extending from costa before the apex to above anal angle, the interspaces from the base with very slender brown transverse strigæ. Cilia edged with grey. Underside marked as above, the fore wing also having a slight streak at end of the cell. Body reddish-ochreous.

Expanse, ♂ $1\frac{4}{10}$, ♀ $1\frac{5}{10}$ inch.

Hab. Darjiling. In coll. Dr. Staudinger and F. Moore.

SONAGARA STRIGOSA, n. sp. (Plate V. fig. 17.)

Brownish-ochreous. Both wings with inwardly transverse uniformly disposed, slender black continuous strigæ, and a more distinct streak from apex to middle of abdominal margin; an ill-defined reniform mark on fore wing. Underside marked as above.

Expanse $\frac{1}{1}\frac{1}{1}$ inch.

Hab. Calcutta. In coll. Dr. Staudinger.

Allied to, but quite distinct from, *S. reticulata* (*Thermesia reticulata*, Walker).

HINGULA, n. g.

Fore wing elongated, narrow, apex acute; exterior margin short, waved, convex hindward; cell extending more than half the wing; first subcostal emitted at more than one half before end of the cell, second at one fourth, trifid, third from below second immediately above end of cell, fourth from below third close to apex, fifth from end of the cell, bent upward and touching third near its base; discocellular bent at each end, radials from the angles; middle median close to end of the cell, lower at one half before the end; submedian nearly straight. Hind wing short; exterior margin convex, waved; cell short, extending one third the wing; two subcostals from end of the cell; discocellular slightly concave, bent near lower end, radial from the angle; two upper medians from end of the cell, lower at one half before the end; submedian and internal straight. Body short, stout; palpi ascending, second joint extending half beyond the head, broad at anterior end, squamose, third joint one third its length, slender, pointed; antennæ minutely pectinated; legs squamose.

Near to *Daxata*, Walk. (Catal. Lep. Het. Brit. Mus. xxxiii. p. 1108), and to the American genus *Dagassa.*

HINGULA ALBOLUNATA, n. sp.

Greyish ochreous-brown : fore wing with an indistinct black lunular subbasal transverse line, and a pale-bordered discal line curving outward from middle of the costa to below end of the cell and then extending to posterior margin ; a submarginal grey-pointed sinuous fascia, and a marginal grey-pointed slender lunular line ; a black-bordered white lunule in middle of the cell and a reversely-joined double lunule at the end : hind wing with a transverse medial double black pale-bordered lunular line, a pale-bordered sinuous fascia, and a marginal slender lunular line ; a whitish lunule in middle of the cell, and two dots at the end.

Expanse ⅞ inch.

Hab. Nilgiris ; Cherra Punji (*Atkinson*). In coll. F. Moore and Dr. Staudinger.

HINGULA CERVINA, n. sp.

Purplish greyish-brown : fore wing with an indistinct pale-bordered darker brown transverse subbasal lunular line, and a similar irregular discal line, both darkest at costal end ; marginal lunular line indistinct ; a white lunule in middle and a reversely-joined double lunule at end of the cell : hind wing with a transverse medial brown lunular line ; a white lunule in middle and two spots at end of the cell.

Expanse ⅞ inch.

Hab. Manpuri (*Horne*); Calcutta (*Atkinson*). In coll. F. Moore and Dr. Staudinger.

Family FOCILLIDÆ.

Genus ZETHES, *Ramb.*

ZETHES AMYNOIDES, n. sp. (Plate VI. fig. 2.)

Female. Umber-brown : both wings crossed by an antemedial and a postmedial sinuous darker brown line, with an intermediate transverse indistinct fascia, and a less distinct submarginal sinuous line, each line greyish-ochreous at the costal end ; veins slightly grey-speckled ; a small greyish-ochreous spot at lower end of the cell. Underside brighter brown, with a grey-speckled bordered medial darker brown sinuous line ; discocellular lunule, and patch of grey speckles on outer margin.

Expanse 1 5/11 inch.

Hab. Calcutta. In coll. Dr. Staudinger.

Genus THYRIDOSPILA, *Guén.*

THYRIDOSPILA FASCIATA, n. sp. (Plate VI. fig. 20.)

Greyish umber-brown : fore wing with an antemedial and postmedial transverse pale-bordered brown curved waved line, a medial dark-brown fascia terminating in a broad quadrate patch on the costa, and a submarginal zigzag brown line bordered by a brown patch below the apex ; some white dots on the costal edge : hind wing with a discal pale-bordered transverse brown line, inner dark-brown fascia, and submarginal zigzag brown-bordered line. Underside with similar transverse lines: fore wing with a white orbicular spot and black-centred white-bordered reniform mark, and a cluster of grey speckles before the apex: hind wing also with a black discocellular lunule.

Expanse 1 ¹⁄₈ inch.

Hab. Darjiling. In coll. Dr. Staudinger.

Genus SARACA, *Walker.*

SARACA PANNOSA, n. sp.

Purplish greyish-brown: fore wing crossed by an antemedial grey-bordered brown zigzag line, a postmedial straight line bent inward at the costal end, and a submarginal sinuous line, the postmedial line inwardly bordered by a broad darker brown fascia, which is bent inward to the costa and thence curved upward to the apex, the curved costal inter-space being grey; an indistinct grey orbicular and reniform spot: hind wing with a broader grey-bordered double brown curved line, suffused on each side with darker brown; a sub-marginal sinuous brown line. Underside ochreous-brown, with indistinct transverse discal sinuous brown line, and discocellular lunule.

Expanse 1½ inch.

Hab. Cherra; Darjiling. In coll. Dr. Staudinger and F. Moore.

Allied to *S. trimantesalis* (*Egnasia trimantesalis*), Walk. Catal. Lep. Het. B. M. xvi. p. 220.

HARMATELIA, n. g.

Fore wing elongated; costa slightly arched near end, apex pointed; exterior margin short, slightly convex and almost angular in the middle; first subcostal at nearly one half before end of the cell, second at one fourth, trifid, third from below second above end of the cell, fourth near the apex, fifth from end of the cell, bent upward and touching third very near its base; cell extending more than half the wing; discocellulars bent at each end, concave in the middle, radials from the angles; upper median from angle above end of the cell, middle median from its end, lower at one half before the end; submedian slightly recurved. Hind wing short, convex externally; cell one third the wing; costal vein extending to apex; two subcostals from end of the cell; discocellular concave, bent near lower end, radial from the angle; two upper medians from acute lower end of the cell, lower median from one third before the end. Body stout; palpi ascending, squamose, second joint extending to vertex, third joint of nearly equal length, cylindrical; antennæ finely bipecti-nated; legs thickly pilose beneath.

Type *H. basalis.*

HARMATELIA BIPARTITA, n. sp.

Fore wing purplish chocolate-brown, the outer area washed with grey, the dark portion confined to half the wing, the dividing line being acutely angular at end of the cell and on lower median vein; a discal ochreous-bordered brown sinuous fascia, an indistinct darker brown waved subbasal line : hind wing and abdomen paler brown. Body, palpi, and legs dark brown.

Expanse 1½ to 1¾ inch.

Hab. Khasia Hills; Cherra Punji. In coll. Dr. Staudinger and F. Moore.

HARMATELIA BASALIS, n. sp. (Plate VI. fig. 13.)

Fore wing with a transverse discal slender purple-grey undulated line, the inner area to the base of wing dark chocolate-brown, the outer area pale chocolate-brown ; an indistinct black orbicular dot, and a submarginal sinuous pale-bordered fascia, which is darkest at the apex and grey-speckled at posterior end ; some grey dots on costa before the apex : hind wing and abdomen pale chocolate-brown. Thorax, head, palpi, and legs dark brown.

Expanse 1¾ inch.

Hab. Cherra (*Austen*); Darjiling (*Atkinson*). In coll. F. Moore and Dr. Staudinger.

Genus RILÆSENA, *Walker.*

Rhæsena, Walker, Catal. Lep. Het. B. M. pt. xxxv. p. 1973.

RILÆSENA OBLIQUIFASCIATA, n. sp.

Fore wing deep brownish-ochreous, with a pale-bordered dark-brown waved transverse antemedial line, a short upper median line, and a submarginal line ; a more prominent dark-bordered pale outwardly-oblique medial waved fascia ; some pale spots on costa before the apex : hind wing pale ochreous-brown. Body brownish-ochreous.

Expanse 1 inch.

Hab. Bombay (*Leith*); Calcutta (*Atkinson*). In coll. F. Moore and Dr. Staudinger.

"Larva green ; with twelve legs. Feeds on *Tragea.* April. Imago emerged Nov. 23." —*Atkinson, MS. note.*

Genus CULTRIPALPA, *Guén.*

CULTRIPALPA INDISTINCTA, n. sp.

Female. Greyish-ochreous, crossed by a slender white-bordered transverse medial waved line, which extends from the costa, encircling a large reniform mark, and then across both wings to above anal angle ; the basal area indistinctly fasciated with white and brownish sinuous lines, the outer areas numerously brown-speckled ; a small brown dot at base of the cell, the reniform mark also brown-bordered on each side, and a brown spot below it : hind wing also with a slender dentate discocellular white-bordered brown streak ; both wings with a slender white-bordered brown marginal lunular line.

Expanse 1¾ inch.

Hab. Calcutta (August to October). In coll. Dr. Staudinger and F. Moore.

CULTRIPALPA TRIFASCIATA, n. sp. (Plate VI. fig. 1.)

Female. Brownish-ochreous, sparsely covered with delicate darker brown strigæ: fore wing crossed by two subbasal and a medial zigzag brown fascia, the outer fascia joined to a transverse irregular sinuous whitish-bordered black line, which less distinctly crosses the hind wing; an indistinct greyish orbicular lunular mark, some brown spots before the apex, and some white dots on the costa.

Expanse 1⅔ inch.

Hab. Cherra Punji (October, November). In coll. Dr. Staudinger and F. Moore.

Genus EGNASIA, *Walker.*

EGNASIA KHASIANA, n. sp.

Near to *E. ephyridalis.* Differs in being of a brighter ochreous-yellow colour and the outer borders brownish-ochreous; the diaphanous spots on both wings similar, the two outer transverse black lines not sinuous, the inner one being broader and suffused, the outer even and more distinct.

Expanse, ♂ 1⅔, ♀ 1⅘ inch.

Hab. Khasia Hills (*Austen & Atkinson*). In coll. F. Moore and Dr. Staudinger.

EGNASIA SINUOSA, n. sp.

Allied to *E. accingalis.* Both wings similarly marked. Differs in the transverse lines being regularly sinuous throughout, the orbicular and reniform spots smaller, the reniform spot represented only by two minute spots.

Expanse 1⅔ inch.

Hab. Calcutta; Moulmein. In coll. Dr. Staudinger and F. Moore.

EGNASIA COSTIPANNOSA, n. sp.

Vinous greyish-brown: fore wing with a transverse medial dark-brown fascia, followed by an indistinct double sinuous line and a very prominent discal pale-bordered blackish irregular sinuous line, terminating at the costal end in a broad large patch; a slender marginal blackish sinuous line; a black orbicular dot and reniform mark: hind wing with an irregular waved transverse discal pale-bordered black band, and slender marginal sinuous line; both wings with an indistinct short dark-brown fascia from middle of outer band to the extreme margin.

Expanse 1⅗ inch.

Hab. Darjiling. In coll. Dr. Staudinger and F. Moore.

EGNASIA CASTANEA, n. sp.

Purplish chestnut-red; both wings crossed by a narrow grey-centred dark-brown double waved discal line, the line twice angled near costal end; a medial dark-brown fascia, and a submarginal white-speckled acute-pointed sinuous black line; a small grey orbicular spot, a less distinct reniform mark, and a more distinct trilobed spot below the latter on the fore wing ; a grey orbicular spot, and on the hind wing a short black grey-bordered sinuous streak above anal angle: fore wing also with a subbasal sinuous double line, each of the lines bent inward to the costa and grey-bordered at that end. Underside paler, grey-speckled, the discal line on fore wing straight and bent inward near the costa ; orbicular and reniform mark black-lined ; other markings as above.

Expanse 1¼ to 1½ inch.

Hab. Khasia Hills ; Darjiling. In coll. Dr. Staudinger and F. Moore.

EGNASIA MOROSA. (Plate VI. fig. 4.)

Allied to *E. castanea.* Umber-brown : fore wing with similar transverse zigzag lines, but less prominent, the reniform mark and lower trilobate spot indistinct: markings on hind wing also similar, but the discal line less waved and terminating at the anal angle. Underside pale brown, the discal line on the hind wing terminating at the anal angle.

Expanse 1½ inch.

Hab. Darjiling. In coll. Dr. Staudinger.

ACHARYA, n. g.

Male. Fore wing elongate, narrow ; costa almost straight, apex pointed ; exterior margin deeply scalloped, somewhat angular in the middle ; first subcostal emitted about one half and second at one sixth before end of the cell, second trifid, third at a short distance from base of second, fourth near the apex, fifth from end of the cell, ascending and touching third near its base ; discocellular bent at each end, concave in the middle, radials from the angles; middle median from near end of the cell, lower at one half before the end. Hind wing deeply scalloped and somewhat angular on the exterior margin ; two subcostals from beyond end of the cell ; discocellular very obliquely concave, radial from near its lower end ; two upper medians from end of the cell, lower at one half before the end ; submedian and internal wide apart. Body long, moderately slender ; palpi long, large, ascending, first and second joints flat, compactly squamose, third joint of the same length as second, thickly tufted throughout its length ; legs long, squamose, tibiæ slightly pilose ; antennæ bipectinated, the pectinations on the outer side long and thick to near the tip, those on the inner side very delicate and indistinct, midrib incrassated and distorted near the base.

Allied to *Egnasia.*

ACHARYA CRASSICORNIS, n. sp. (Plate VI. fig. 3.)

Male. Dark chestnut-brown ; the fore wing and basal area of hind wing indistinctly crossed with short grey and black strigæ: fore wing with an elongated subcostal ochreous streak extending from base to apex ; some short oblique ochreous streaks on the costa, a subbasal transverse slender zigzag yellowish line, and a similar black-bordered medial lunular zigzag line, a small diaphanous-white spot at end of the cell; an indistinct submarginal sinuous slender ochreous line, clustered in the middle with ochreous speckles and at the apex with purple-grey speckles: hind wing with a transverse subbasal and a medial brown-bordered ochreous line, a diaphanous-white spot at end of the cell, and a submarginal sinuous ochreous line. Body dark brown; palpi and legs ochreous, with brown bands. Underside dark brown, mottled with ochreous.

Expanse 1½ inch.

Hab. Silhet. In coll. Dr. Staudinger.

Family HYPENIDÆ.

CORCOBARA, n. g.

Fore wing very long, narrow; costa slightly arched at the base, from whence it is straight to the end, apex very acutely pointed; exterior margin short, upper half erect, concave, and angled outward at end of upper median vein, very oblique hindward; posterior margin broadly convex; cell extending two thirds the wing; first subcostal emitted at one half before end of the cell, second at one fourth, trifid, fifth from end of the cell, bent upward and joined by a short spur to third near its base; discocellular bent at each end, very convex in middle, radials from the angles; middle median from close to end of the cell, lower at one half before the end; submedian slightly curved near base. Hind wing long, somewhat narrow; exterior margin very obliquely convex, and slightly angular in the middle; costal vein straight, extending to apex; cell nearly half the wing; two subcostals from end of the cell; discocellular concave, radial from near lower end; two upper medians from end of the cell, lower at one third before the end; submedian and internal slightly curved. Body moderately stout, abdomen rather long; palpi very long, three eighths of an inch in length, porrect, flat, very compactly squamose, second joint attenuating to anterior end and extending two thirds its length beyond the head, third joint more than half length of second, truncate at tip; legs compactly clothed; antennæ very minutely bipectinated.

Allied to *Anoratha*.

CORCOBARA ANGULIPENNIS, n. sp. (Plate VI. fig. 16.)

Fore wing purplish-brownish-ochreous, washed with glaucous-grey; crossed by numerous very indistinct purplish-grey rather long strigæ; the costal and posterior borders sparsely black-speckled; an ochreous reniform mark, with a black dot at its upper and lower end: hind wing bright ochreous-yellow, with a large black subapical spot. Thorax, palpi, fore and middle legs in front purplish-brownish-ochreous, front of thorax ochreous; abdomen bright ochreous-yellow. Underside ochreous-yellow; both wings with a blackish subapical spot, and black speckles on costal border.

Expanse 2 to 2⅛ inches.

Hab. Ceylon (*Mackwood*); Darjiling (*Atkinson*). In coll. F. Moore and Dr. Staudinger.

APANDA, n. g.

Fore wing elongated, triangular; costa very slightly arched at base and end, apex pointed; exterior margin slightly and obliquely convex; first subcostal emitted at two fifths before end of the cell, second at one fifth before the end, trifid; fifth from end of the cell, bent upward and touching third near its base; discocellular bent at each end, radials from the angles; cell nearly two thirds the wing; middle median close to end of the cell, lower at two fifths before the end. Hind wing somewhat triangular, exterior margin obliquely convex; costal vein nearly straight; cell short; two subcostals from end of the cell; discocellular concave, bent outward near lower end; radial from the angle; two upper medians emitted at some distance beyond end of the cell; lower at one third before the end. Body moderate; palpi ascending, second joint broadly flattened towards the apex, very laxly squamose, extending one third above vertex, third joint long, slender, laxly scaled; fore tibiæ compactly clothed, hind tibiæ laxly pilose; antennæ simple.

APANDA DENTICULATA, n. sp.   (Plate VI. fig. 24.)

Fore wing pale olive-brown, grey- and black-speckled; crossed by a black antemedial and postmedial zigzag line, and a submarginal white-pointed dentated line, the postmedial line indistinctly bordered outward by a greyish dentated fascia; a blackish sinuous streak below the apex, and an orbicular and reniform spot: hind wing cinereous-brown. Thorax grey-speckled; legs with pale bands.

Expanse $1\frac{4}{10}$ inch.

Hab. Darjiling. In coll. Dr. Staudinger and F. Moore.

HARITA, n. g.

Wings broad; fore wing short, triangular; apex pointed; exterior margin nearly erect, slightly convex; posterior margin long: first subcostal emitted at nearly one half before end of the cell, second at one fourth, trifid, third at one fifth and fourth at four fifths from base of second, fifth from end of the cell, bent upward and touching third near its base; discocellulars bent near each end, concave in the middle, radials from the angles; upper median from angle above end of the cell, middle median close to the end, lower at nearly one half before the end. Hind wing short; exterior margin very convex: two subcostals from end of the cell, discocellulars obliquely concave, radial from near lower end; two upper medians from acute lower end of the cell, lower branch at one third before the end. Body long, slender, shorter and robust in female; palpi long, slender, flat, very compactly clothed, second joint not extending beyond frontal tuft, third joint longer than second, broadest and blunt at apex; antennæ of male very finely bipectinated to tip; fore tibiæ tufted beneath, and densely clothed laterally with hair.

HARITA RECTILINEA, n. sp.   (Plate VI. fig. 23.)

Fore wing brownish-ochreous, crossed by a subbasal very indistinct brown waved line, and a medial very prominent grey-bordered dark-brown erect straight line; a submarginal row of indistinct white-pointed grey dots: hind wing and abdomen duller brownish-ochreous; thorax ochreous; frontal tuft and palpi dark brown.

Expanse $1\frac{1}{4}$ inch.

Hab. Khasia Hills. In coll. Dr. Staudinger and F. Moore.

MATHURA, n. g.

Allied to *Harita* : fore wing comparatively shorter, apex acute, exterior margin less convex ; hind wing broader, less convex externally. Body shorter, robust ; first and second joints of palpi laxly pilose beneath, second joint extending to end of frontal tuft, third joint half length of second, laxly and broadly tufted to two thirds its length, the tip being naked and pointed ; tibiæ more compactly clothed.

MATHURA ALBISIGNA, n. sp.

Ochreous-brown : fore wing crossed by a dark-brown irregularly waved antemedial line, a slightly waved postmedial line, and two submarginal very indistinct brown lunular lines ; a dark-brown dot at base of cell, and a prominent pure white spot in middle of the cell : hind wing duller. Thorax, palpi, and legs darker brown.

Expanse 1¼ inch.

Hab. Cherra Punji. In coll. Dr. Staudinger and F. Moore.

Genus RHYNCHINA, *Guén.*

RHYNCHINA ANGULIFASCIA.

Fore wing bright umber-brown, with a white straight streak from base below the cell to middle of the submedian vein, from whence it is bent upward and joins some short longitudinal oblique purple-white streaks to the apex ; a slender white waved submarginal line, bordered outwardly by purplish-white ; the upward part of the basal streak also bordered outwardly with purplish-white ; cilia pale brown, lined with white : hind wing pale brown ; cilia white. Underside pale brown. Body, palpi, and legs brown.

Expanse 1¼ inch.

Hab. Masuri, N.W. Himalaya (*Lang*); Kaschmir (*Atkinson*). In coll. F. Moore and Dr. Staudinger.

Genus HYPENA, *Schrank.*

HYPENA OCHREIPENNIS, n. sp.

Allied to *H. proboscidialis*. Brownish-ochreous : fore wing with numerous short transverse dark-brown strigæ, crossed by an indistinct curved antemedial and a distinct straight oblique postmedial dark-brown line : hind wing paler.

Expanse 1⅜ inch.

Hab. Darjiling. In coll. Dr. Staudinger and F. Moore.

HYPENA TORTUOSA, n. sp.

Purple-greyish-brown : fore wing with an indistinct dark-brown subbasal sinuous line, a prominent oblique tortuous postmedial transverse pale-bordered line, and two very indistinct pale-bordered sinuous submarginal lines, ending in a curved black apical streak, encompassing two black spots; a black dot within the cell, and a lunule at its end.

Expanse 1⅜ inch.

Hab. Deyra (*Austen*); Darjiling (*Atkinson*). In coll. F. Moore and Dr. Staudinger. Allied to *H. jocosalis.*

Hypena divaricata, n. sp.

Allied to *H. labatalis.* Dark ochreous-brown: fore wing with an indistinct subbasal purple-grey-bordered brown line, which is acutely angled outward below the cell; a similar oblique transverse discal line from the middle of posterior margin to costa, where it is dilated outward to the apex, and is there black-bordered; a black dot at base of the cell, another in the middle, and two at the end; the veins and discal area grey- and black-speckled.

Expanse 1¼ inch.

Hab. Khasia Hills; Darjiling. In coll. Dr. Staudinger and F. Moore.

Hypena ophiusoides, n. sp.

Ochreous-brown, washed with purplish-grey: fore wing crossed by an indistinct subbasal grey-bordered brown zigzag line, a very prominent medial biangulated line, the angles of the latter with broad inner black border; an indistinct submarginal pale-bordered lunular line; a minute grey orbicular spot.

Expanse 1¼ inch.

Hab. Khasia Hills. In coll. Dr. Staudinger and F. Moore.

Hypena mediana, n. sp.

Purplish-greyish-brown: fore wing with a broad medial transverse grey-bordered irregular angulated dark-brown band, which is narrowest at its lower end.

Expanse 1$\frac{1}{12}$ inch.

Hab. Parisnath, Bengal. In coll. Dr. Staudinger and F. Moore.

Hypena incurvata, n. sp.

Allied to *H. abducalis.* Differs in the medial longitudinal pale-bordered brown fascia being darker, less curved outward, and in extending to the posterior margin; the inner subbasal parallel line is curved and shorter; the pale-bordered brown outer marginal lines also extend to the posterior margin.

Expanse 1¼ inch.

Hab. Khasia Hills. In coll. Dr. Staudinger and F. Moore.

Hypena cidarioides, n. sp.

Purplish-greyish-brown: fore wing with the basal half purplish-brown, forming a prominent basal band, the outer border of which is obliquely waved and pale-lined; outer area crossed by brown lunular lines; a curved blackish streak below the apex; an indistinct darker sinuous subbasal transverse line, a spot in middle of the cell, and a distinct white lunule at the end: hind wing paler.

Expanse 1 inch.

Hab. Khasia Hills. In coll. Dr. Staudinger and F. Moore.

Allied to *H. tortuosa.*

HYPENA EXTERNA, n. sp.

Fore wing dark brown from base to beyond the cell, bordered by a waved pale-bordered line; the outer area pale brownish-grey, with a brown marginal and curved streak below the apex; a white spot at the apex, an indistinct subbasal pale sinuous transverse line, and two blackish spots within the cell: hind wing pale brownish-ochreous. Both wings with a marginal blackish lunular line.

Expanse 1 inch.

Hab. Darjiling. In coll. Dr. Staudinger and F. Moore.

HYPENA FLEXUOSA, n. sp.

Dark olive-brown, of a uniform tint throughout: fore wing crossed by an antemedial and a postmedial very slender crooked white line.

Expanse 1¼ inch.

Hab. Darjiling. In coll. Dr. Staudinger and F. Moore.

HYPENA GRISEIPENNIS, n. sp.

Fore wing brownish-grey, with an indistinct antemedial transverse pale slender angular grey line, and a more distinct postmedial sinuous line, the upper end of the latter bent inward to the costa; a pale oblique streak also from angle of the antemedial line to lower end of the postmedial line; a slight grey sinuous submarginal line, most distinct at apical end; two indistinct black spots within the cell: hind wing pale greyish-brown.

Expanse 1½ inch.

Hab. Cherra. In coll. Dr. Staudinger and F. Moore.

Near to *H. laceratalis* and *H. ignotalis.* Smaller than the latter, and may be distinguished by its less pointed fore wing, and by the postmedial sinuous line ending more obliquely inward at the costal end.

HYPENA LATIVITTA, n. sp.

Fore wing pale purplish-greyish-brown, with a prominent broad transverse subbasal dark-brown band, the outer border of which is waved, and the inner border acutely sinuous, the borders also slightly grey-speckled; a submarginal row of dentate dark-brown grey-bordered points, which are most distinct at the apical end: hind wing paler brown.

Expanse 1¾ inch.

Hab. Darjiling. In coll. Dr. Staudinger and F. Moore.

Allied to *H. tenebralis.* Distinguished from it in the total absence of the short transverse strigæ.

HYPENA MODESTA, n. sp.

Greyish-brown : fore wing with an indistinct brown pale-bordered irregular curved antemedial transverse line, and a nearly straight more distinct postmedial line ; a submarginal brownish fuscia, pale-bordered at apical end, and an indistinct brown spot in middle of the cell : hind wing paler.

Expanse $1\frac{1}{4}$ inch.

Hab. Cherra ; Darjiling. In coll. Dr. Staudinger and F. Moore.

Near to *H. therminalis.* Distinguished from it by the transverse lines being wider apart, and the postmedial line more erect.

HYPENA TRIANGULARIS, n. sp.

Pale brownish-ochreous-grey : fore wing with numerous extremely indistinct short brownish transverse strigæ ; a very prominent dark-brown triangular medial-costal band, and a less distinct sinuous apical streak ; a small blackish spot in middle of the cell, and one at its end.

Expanse $1\frac{1}{10}$ inch.

Hab. Khasia Hills. In coll. Dr. Staudinger and F. Moore.

HYPENA OCCATUS, n. sp.

Greyish-brownish-ochreous : fore wing with several very indistinct transverse brownish lunular lines, and crossed by an indistinct brown sinuous curved antemedial line, a slightly-waved oblique postmedial ochreous-bordered line, and a submarginal series of darker brown grey-bordered points, most distinct and sinuous below the apex, where also there is an inner terminating blackish spot ; a pale-bordered brown dot within the cell.

Expanse $1\frac{2}{3}$ inch.

Hab. Cherra Punji ; Khasia Hills. In coll. Dr. Staudinger and F. Moore.

HYPENA ADSIMILIS, n. sp.

Allied to *H. occatus.* Similar in colour : fore wing irrorated with irregularly-disposed dark-brown scales, the transverse lines being much less distinct, the antemedial line more erectly sinuous, and the postmedial more inwardly waved, the anterior end of the submarginal brown points with a straight lower apical streak ; two brown spots within the cell.

Expanse $1\frac{2}{3}$ inch.

Hab. Khasia Hills. In coll. Dr. Staudinger and F. Moore.

HYPENA STRIGIFASCIA, n. sp.

Violaceous-ochreous-brown : fore wing with numerous short blackish strigæ, which are most prominent and thick across the middle, where they form a more or less defined broad transverse slightly æneous-bordered band, and on the submargin, where they form a sinuous

192   HYPENIDÆ.—HERMINIIDÆ.

fascia and border an apical ochreous patch. Body, head, and palpi olive, with black strigæ: hind wing paler.

Expanse 1¾ inch.

Hab. Darjiling. In coll. Dr. Staudinger and F. Moore.

HYPENA SIMILATA, n. sp.

Allied to *H. iconicalis.* Differs in the indistinct subbasal angulated line being somewhat nearer the base, the discal line quite straight, and in having a black spot outside the latter at lower end of the cell.

Expanse 1⅛ inch.

Hab. Khasia Hills; Calcutta. In coll. Dr. Staudinger and F. Moore.

HYPENA UMBRIPENNIS, n. sp.

Dark umber-brown; fore wing minutely grey-speckled, crossed by an outwardly-curved waved antemedial line, an erect postmedial slender grey line, and an indistinct submarginal grey sinuous line; the grey speckles most thickly disposed outside the postmedial line and at the apex; a brown dot at base of the cell, and one within the middle. Cilia edged with grey. Head and palpi above grey-speckled.

Expanse 1¼ inch.

Hab. Khasia Hills. In coll. Dr. Staudinger and F. Moore.

Family HERMINIIDÆ.

Genus HERMINIA, *Latr.*

HERMINIA VIALIS, n. sp.

Male. Fore wing pale umber-brown, with a broad subbasal transverse whitish band, bordered by a double line of dark brown, the inner border of the band obliquely curved, the outer border extending below the cell, enclosing a large reniform mark, and terminating on the costa one third before the apex; a brown orbicular spot; a whitish slender zigzag submarginal line bordered with dark brown at its apical end: hind wing paler, with a slender sinuous darker-brown-bordered whitish streak from anal angle; an outer-marginal darker brown lunular line on both wings.

Expanse 1½ inch.

Hab. Cherra Punji. In coll. Dr. Staudinger and F. Moore.

Allied to *H. hadenalis.*

HERMINIA RESTRICTA, n. sp.

Allied to *H. hadenalis.*

Male and female. Fore wing with a similar medial transverse pale band, the outer border of which is very irregular and extends below end of the cell, thus restricting the posterior portion of the band to half the width of that in *H. hadenalis.*

Expanse 1¾ inch.

Hab. Darjiling. In coll. Dr. Staudinger and F. Moore.

HERMINIA LINEOSA, n. sp.

Male. Fore wing pale ochreous-brown, crossed by a dark-brown duplex antemedial and a postmedial line, a medial single line, and an irregular sinuous outer discal line; two pale spots at end of the cell: hind wing paler, indistinctly crossed by three darker fasciæ.

Expanse $1\frac{5}{16}$ inch.

Hab. Darjiling. In coll. Dr. Staudinger.

HERMINIA DUPLEXA, n. sp. (Plate VI. fig. 18.)

Male. Fore wing very pale brownish-ochreous, crossed by a basal, an antemedial, and a postmedial slender brown sinuous line, each bent inward at the costal end, and a more prominent submarginal straight duplex line; a distinct brown reniform mark: hind wing paler, with an angulated straight submarginal duplex line, and a slender sinuous discal line; a very indistinct discocellular spot.

Expanse 1 inch.

Hab. Darjiling. In coll. Dr. Staudinger.

Genus MADOPA, *Stephens.*

MADOPA QUADRILINEATA, n. sp.

Pale purplish-brownish-ochreous, brightest in female; a dark purple-brown straight oblique antemedial and a postmedial line crossing both wings, the former less distinct on the hind wings: fore wing with a parallel, very slender, brown zigzag subbasal and a discal line, and a slight orbicular dot and reniform mark.

Expanse $1\frac{1}{4}$ to $1\frac{1}{2}$ inch.

Hab. Darjiling. In coll. Dr. Staudinger and F. Moore.

Genus ZANCLOGNATHA, *Lederer.*

ZANCLOGNATHA ERECTA, n. sp.

Allied to *Z. tarsiplumalis* and to *Z. grisealis.* Fore wing pale umber-brown, crossed by a slender black erect basal and an antemedial line, and a recurved postmedial line, each line angled inward at the costal end, an outer discal erect darker and broader line, and a slight discocellular lunule: hind wing pale brownish-ochreous, with two indistinct slender medial transverse fasciæ.

Expanse, ♂ 1, ♀ $1\frac{2}{16}$ inch.

Hab. Darjiling. In coll. Dr. Staudinger and F. Moore.

ZANCLOGNATHA UNDULATA, n. sp.

Allied to *Z. erecta.* Pale dull brownish-ochreous: fore wing with a slender waved antemedial and a postmedial line, and a slightly more prominent pale-bordered waved outer discal line, and a slender discocellular lunule: hind wing with a pale outer transverse line.

P

Expanse, ♂ 1⁴⁄₁₆, ♀ 1 inch.
Hab. Darjiling. In coll. Dr. Staudinger and F. Moore.

Genus BERTULA, *Walker.*

BERTULA VIALIS, n. sp.

Male. Ochreous-brown: both wings with transverse narrow straight brown-bordered pale band and a less distinct submarginal sinuous line, the band extending across the disk of fore wing and middle of hind wing: fore wing with a small dark-brown orbicular spot and reniform mark. Body, palpi, and fore legs ochreous-brown; third joint of palpi three eighths of an inch in length, with a long ochreous brush-like tuft in front.
Expanse 1¼ inch.
Hab. Darjiling. In coll. Dr. Staudinger and F. Moore.

BERTULA PLACIDA, n. sp.

Allied to *B. chalybealis.* Differs on the fore wing in both the antemedial and post-medial transverse duplex lines being sinuous and less distinct; submarginal sinuous line less irregular, and the reniform mark lunate; the two discal duplex lines crossing the hind wing also sinuous.
Expanse 1½ inch.
Hab. Darjiling. In coll. Dr. Staudinger and F. Moore.

Genus AVITTA, *Walker.*

AVITTA FASCIOSA, n. sp.　(Plate VI. fig. 26.)

Differs from *A. subsignans,* Walker (Catal. Lep. Het. B. M. xv. p. 1675), in being larger, paler in colour, and of a greyish-ochreous-brown, the transverse markings more distinct; the antemedial, postmedial, and submarginal waved lines very slender, the two former comparatively nearer together; the intermediate fasciæ prominent; reniform mark with a pale centre. Underside with similar but much paler markings.
Expanse 1⁶⁄₈ inch.
Hab. Khasia Hills. In coll. Dr. Staudinger.

Genus BOCANA, *Walker.*

BOCANA RENALIS, n. sp.

Fore wing purple-brown, crossed by an indistinct black white-bordered sinuous ante-medial and a nearly erect postmedial line, the latter outwardly bordered by a chalybeate-white fascia, encompassing an indistinct white-lined reniform mark; a submarginal white-bordered sinuous black line; a black dot in middle of the cell: hind wing paler.
Expanse 1¼ inch.
Hab. Cherra; Khasia Hills. In coll. Dr. Staudinger and F. Moore.
Allied to *B. eagalis,* Walker.

BOCANA PICTA, n. sp. (Plate VI. fig. 21.)

Male. Fore wing with the basal half ochreous-red, bordered by a chalybeate-white line; crossed by a slender indistinct brown antemedial line, a broad dark-brown irregular discal fascia, and a marginal streak; a brown dot in middle of the cell, a white-lined slender discocellular lunule, and a marginal row of brown dots: hind wing pale brown. Thorax and head red; abdomen brown, with slight dorsal reddish crest; palpi, fore and middle legs dark brown, hind legs paler.

Expanse 1¼ inch.

Hab. Khasia Hills. In coll. Dr. Staudinger.

BOCANA MARGINATA, n. sp. (Plate VI. fig. 19.)

Allied to *B. Schaldusalis.*

Male. Pale brownish-ochreous: fore wing crossed by a submarginal curved prominent dark-brown band, the outer border of which is distinctly and sharply defined, and the inner border diffused; a minute discocellular brown dot and a marginal row of spots; outer margin of the wing much paler than the basal area: hind wing paler, with a similar band, which is angulated in the middle.

Expanse 1 1/12 inch.

Hab. Darjiling. In coll. Dr. Staudinger.

Genus AGINNA, *Walker.*

AGINNA SIMILIS, n. sp.

Male. Fore wing dull olivaceous-ochreous-brown, crossed by an antemedial and a postmedial very indistinct, brown, slightly sinuous line, and a submarginal, nearly erect, pale yellow line: hind wing pale cinereous-brown, with an indistinct submarginal whitish angular line.

Expanse 1 3/10 to 1 6/10 inch.

Hab. Darjiling. In coll. Dr. Staudinger and F. Moore.

Closely allied to *A. turpatalis* (*Bocana turpatalis*, Walk.). Differs in the transverse lines being less sinuous and more erect.

AGINNA SIMULATA, n. sp.

Allied to *A. turpatalis.* *Male* and *female* smaller: fore wing of a paler ochreous colour, with scarcely discernible transverse antemedial and postmedial sinuous lines, which are also less widely apart, the submarginal line paler and less prominent: hind wing pale whitish-ochreous, with very indistinct whitish submarginal line.

Expanse 1 inch.

Hab. Bombay (*Leith*); Calcutta (*Atkinson*). In coll. F. Moore and Dr. Staudinger.

CEPHENA, n. g.

Fore wing long, narrow, apex pointed; exterior margin short, scalloped hindward; posterior margin long; cell two thirds the wing; first subcostal emitted at one-half before end of the cell; second at one fourth, trifid; fifth from end of the cell, bent upward and touching third by a short spur near its base; discocellular bent at each end, concave in middle, radials from angles; middle median close before end of the cell, lower at two thirds before the end; submedian nearly straight. Hind wing long, exterior margin very obliquely convex, abdominal margin short; cell nearly one half the wing; two subcostals from end of the cell; discocellular concave, radial from near its lower end; two upper medians from acute angle beyond end of the cell, lower at one half before the end. Body long, abdomen extending beyond the wing; thorax slightly crested in front; head with a frontal tuft; palpi long, ascending, squamose, second joint broadly fusiform, extending half its length above the eyes, third joint nearly of the same length as second, slender, pointed; legs compactly pilose; antennæ serrated and pectinated to tip in male.

CEPHENA COSTATA, n. sp. (Plate VI. fig. 17.)

Fore wing very pale brownish-ochreous, with a longitudinal subcostal slender black fascia extending from base of costa along the cell to below the apex, the fascia with a short white streak along its middle; the costal border thickly black-speckled to two thirds the space, the line bordered below by a purple-brown shade, the veins beneath also slightly brown-speckled, and a small black spot on posterior margin near the angle: hind wing dusky ochreous-brown. Thorax white-speckled, crest ochreous-brown in front; frontal tuft white; palpi white at base and ochreous at the side; abdomen dusky-ochreous.

Expanse 1½ to 1¾ inch.

Hab. Khasia Hills; Darjiling.

ASTHALA, n. g.

Wings short: fore wing narrow, apex pointed, exterior margin obliquely convex; costal vein much recurved; first subcostal branch emitted at two thirds before end of the cell and nearly touching costal towards the end; second emitted at one third, bifid, the third being thrown off at one third from its base, fourth from end of the third, touching third near its base and anastomosing with it to one third its length; discocellular very slender, bent close to each end, concave in middle, radials from the angles; middle median from near end of the cell, lower at nearly one half before the end. Hind wing short, very broad, exterior margin convex and somewhat angular in the middle; two subcostals at some distance from beyond end of the cell; discocellular very slender and obliquely concave, radial from near its lower end; two upper medians from acute lower angle of the cell, lower median at nearly one half before the end. Body moderate; palpi ascending, first and second joints laxly pilose beneath, the hairs shortening to the tip, second joint slightly tufted above and ascending to a level with eyes, third joint as long as second, naked, very slender and pointed; antennæ very long, of the same length as costal margin, very delicately bipectinated; femora and tibiæ densely pilose.

Bocana silenusalis, Walker, Catal. Lep. Het. B. M. xvi. p. 179 (1858).

Hab. Borneo (*type*); Khasia Hills and Cherra (*Atkinson*). In coll. Dr. Staudinger and F. Moore.

Genus RIVULA, *Guén.*

RIVULA PALLIDA, n. sp.

Very pale brownish-ochreous : fore wing with a transverse medial and a discal very indistinct waved pale line, some short white streaks on costa near apex, and a marginal row of minute white dots ; two indistinct minute brown spots encircled by a pale line at end of the cell.

Expanse $\frac{8}{14}$ inch.

Hab. Calcutta. In coll. Dr. Staudinger and F. Moore.

Near to *R. bioeularis.* Wings larger, much duller-coloured, and more indistinctly marked.

PASIRA, n. g.

Fore wing more acuminate at the apex than in *Rivula* (*R. serieealis*), exterior margin more erect : hind wing shorter, and the male having a groove, which is set with short hairy scales on both sides, situated on the wing between the median and internal veins, extending from outer margin to below end of the cell, from whence it is bent below the median vein. Body slightly stouter than in *Rivula*; projecting front of the head, palpi, and legs similar; venation also similar.

PASIRA OCHRACEA, n. sp.

Male. Ochreous : fore wing with two indistinct minute black discocellular dots : hind wing externally suffused with pale purplish-ochreous ; abdominal margin and hairy scales along the groove ochreous-white. Underside paler ; discal area of fore wing in male suffused with brown.

Female. Both the fore and hind wings paler uniform ochreous. Body, palpi, and legs ochreous.

Expanse, ♂ $\frac{9}{12}$, ♀ $1\frac{2}{12}$ inch.

Hab. Calcutta. In coll. Dr. Staudinger.

BIBACTA, n. g.

Fore wing short, costa bent obliquely downward at two thirds from the base, apex pointed, exterior margin short and oblique ; a short fold composed of stout scales covering a tuft of long hair at base of the costa ; costal vein short, running close to the margin ; first subcostal very short, emitted immediately before end of the cell ; second short, trifid, emitted at half its length beyond the cell and terminating on the costa before the angle ; fourth from below the second beyond end of the cell and terminating at the apex; fifth from a very slight angle at upper end of the cell; discocellular extremely slender, concave; radial and two upper medians on a foot-stalk at some distance beyond end of the cell, lower median at one fourth before the end; submedian straight. Hind wing short, exterior margin convex; costal vein arched at the base, extending to apex; subcostal straight, its two branches emitted at one fourth beyond the cell; discocellular slender, concave, radial from near its lower end; cell short; two upper medians at one fifth beyond end of the cell, lower at one third before the end; submedian and internal slightly curved. Body short; palpi ascending, curved upward over the head, compactly squamose, second joint long, extending one third above the vertex, third joint half its length, laxly squamose above; antennæ very delicately bipectinated; legs compactly squamose above.

Allied to *Echana*, Walker.

BIBACTA TRUNCATA, n. sp. (Plate VI. fig. 25.)

Male. Fore wing umber-brown, crossed by a medial and a discal very indistinct paler waved line: hind wing paler. Underside pale umber-brown: hind wing greyish-speckled, with an ill-defined discal and a submarginal grey-speckled-bordered brown fascia, and a discocellular lunule. Body, palpi, and legs dark brown; tarsi with pale bands.

Expanse 1¼ inch.

Hab. In coll. Dr. Staudinger and F. Moore.

DESCRIPTIONS

OF

INDIAN LEPIDOPTERA HETEROCERA

FROM THE

COLLECTION OF THE LATE MR. W. S. ATKINSON.

BY

FREDERIC MOORE, F.Z.S. ETC.,

LATE ASSISTANT CURATOR, INDIA MUSEUM, LONDON.

Family PYRALIDÆ.

Genus LOCASTRA, *Walker.*

LOCASTRA LATIVITTA, n. sp.   (Plate VII. fig. 1.)

Allied to *L. cuproviridalis. Male* smaller: fore wing with an imperfect transverse sub-basal and discal zigzag black line ; the basal and outer areas glossy greenish-ochreous ; the medial area white, clouded with greenish-ochreous and with two black-and-white tufted spots within the cell ; a slight white tuft also at base of the wing : hind wing white, with a dusky brownish-ochreous apical band, which ends in a slender line at the anal angle ; a very small spot on lower median vein close to the band. Thorax, head, and legs above glossy greenish-ochreous ; abdomen white, with blackish dorsal bands ; palpi ochreous-white, with blackish tip ; legs with brown and white bands.

Expanse 1¼ inch.

Hab. Darjiling. In coll. Dr. Staudinger.

PANNUCHIA, n. g.

Allied to *Locastra.* Fore wing with the costal vein having a triangular notched glandular fold beyond the middle, the subcostal branches beneath being swollen and distorted ; discocellular very concave, the radials much recurved downward, the upper radial from middle of discocellular, lower from end of the cell ; two upper medians from beyond end of the cell, lower near the end : hind wing with the costal vein curved upward near the end :

PART III.—*Sept.* 5, 1887.   n

two subcostals from beyond end of the cell; the cell short; discocellular bent very obliquely outward near upper end and extending much beyond upper end of the cell. Labial palpi small, slender, extending to vertex, third joint short, pointed; maxillary palpi short, slender; antennæ very minutely bipectinated, and with a rather long stout basal tuft, which is clothed with long hairy scales, projecting over the thorax.

Type *P. ænescens*.

PANNUCHA BASALIS, n. sp.   (Plate VII. fig. 2.)

Male. Fore wing reddish-ochreous; crossed by an indistinct outwardly-oblique ante-medial blackish sinuous line and a recurved postmedial line; a marginal row of spots, each with an indistinct yellowish border; basal area mostly white; the outer border with some indistinct blackish longitudinal streaks: hind wing cinereous ochreous-brown; cilia ochreous, with black spots. Thorax whitish, reddish in front; palpi and legs reddish; tibiæ and tarsi with blackish bands; tuft at base of antennæ brown; abdomen cinereous-ochreous, black-speckled above, anal tuft reddish.

Expanse $1\frac{2}{10}$ inch.

Hab. Darjiling. In coll. Dr. Staudinger.

PANNUCHA ÆNESCENS, n. sp.

Male and female. Fore wing olivaceous cinereous-ochreous, with an erect antemedial and a double postmedial irregular transverse sinuous black-speckled line; the medial area sparsely black-speckled, including a spot in the cell; a triangular black basal costal patch, and a submarginal apical and a posterior patch; a slender black dentated marginal line: hind wing pale purplish cinereous-brown. Body, palpi, and legs greenish ochreous-yellow; legs with blackish bands.

Expanse, ♂ $1\frac{1}{4}$, ♀ $1\frac{2}{8}$ inch.

Hab. Darjiling. In coll. Dr. Staudinger and F. Moore.

TELIPHASA, n. g.

Fore wing elongated, narrow; exterior margin oblique and slightly convex; costal vein extending two thirds the length; first subcostal at one half before end of the cell, second at one fifth, bifid, fourth bifid, the fifth emitted at one fourth from it beyond the cell; discocellular inwardly oblique, concave; upper radial from end of the cell in a line with subcostal, lower radial and upper median joined together to half beyond the cell; middle median from extended lower angle at one fourth beyond end of the cell, lower median at one fifth beyond end of cell; submedian recurved. Hind wing short, broad; exterior margin very obliquely convex; costal vein extending to apex; subcostal two-branched at a short distance beyond end of cell; first subcostal curved upward and almost touching the costal; cell more than one third the length; discocellular outwardly oblique, very concave anteriorly; radial and upper median joined together to half beyond the cell; middle median from extreme lower angle of cell one fourth beyond its end, lower median at one fifth before the end; two submedians and an internal vein at equal distances apart. Body moderate; thorax stout ; labial palpi long, ascending, densely clothed with squamous hair, second joint short, third joint very long, extending half its length above the vertex and much like a miniature hare's foot; antennæ finely bipectinated to tip, with a short ascending tuft on basal joint ; head with two short tufts reverted over fore part of thorax ; maxillary palpi short, thick.

Type *T. orbiculifer.*

TELIPHASA ORBICULIFER, n. sp.

Fore wing ænescent olivaceous-yellow, crossed by a medial zigzag and a recurved post-medial sinuous black-speckled line ; medial area transversely speckled with a few black scales, a black spot in the middle and a larger oval slightly white-centred spot at end of the cell ; basal area thickly black-speckled, the speckles more concentrated below base of the cell in the female ; outer margin also sparsely black-speckled, the speckles mostly disposed in the male at the apex and posterior angle ; a marginal black dentated line ; cilia alternated with pale black : hind wing pale yellowish-cinereous in male, purplish-cinereous in female, with a slender curved discal purplish-brown line and broad marginal band, both being most distinct in the female ; a marginal pale blackish linear lunular line ; cilia alternated with pale black. Body ænescent-ochreous ; thorax black-speckled ; abdomen with blackish-speckled bands ; legs with blackish and white terminal bands.

Expanse, ♂ 1⅜, ♀ 1⅘ inch.

Hab. Darjiling. In coll. Dr. Staudinger and F. Moore.

TELIPHASA NUBILOSA, n. sp.

Allied to *T. orbiculifer.* Fore wing darker-coloured ; the transverse sinuous black-speckled lines broader and less defined ; the basal and outer area much blacker and denser speckled ; the white-marked spot at end of the cell smaller and lunate : hind wing dark purplish cinereous-brown, paler at the base, and with a dark spot at end of the cell. Thorax black-speckled ; abdomen and legs with very dark black bands.

Expanse, ♂ 1⅜, ♀ 1¾ inch.

Hab. Darjiling. In coll. Dr. Staudinger and F. Moore.

Genus ORTHAGA, *Walker.*

ORTHAGA OBSCURA, n. sp.

Male. Fore wing dull olivaceous-ochreous ; crossed by a curved antemedial and a re-curved postmedial sinuous black-speckled line ; the basal area partly and the medial area entirely, and the upper and lower outer border black-speckled ; marginal spots also black-ish ; a well-defined spot within the cell and a lunule at the end : hind wing pale cinereous-brown. Cilia yellowish-cinereous. Body dull olivaceous-ochreous ; abdomen and tarsi with brownish bands.

Expanse 1⅟₁₀ inch.

Hab. Darjiling. In coll. Dr. Staudinger and F. Moore.

Note. Near to *O. pyratalis.*

Genus TAURICA, *Walker.*

TAURICA SIKKIMA, n. sp.

Male. Ochreous olivaceous-brown ; fore wing crossed by an erect sinuous black-speckled antemedial line and an irregular angulated postmedial line ; the basal area to near the former line densely black-speckled ; the medial area from near the inner to the outer line also thickly black-speckled ; a few speckles also dispersed on the outer border ; a row of black marginal dots and spots on the cilia ; a black-speckled spot at end of the cell : hind wing pale cinereous ochreous-brown, with indistinct ochreous streaks below the apex ; cilia yellowish, with brown spots. Thorax, head, palpi, and fore legs reddish-ochreous ; abdomen cinereous-ochreous, blackish-speckled ; tarsi with black bands. *Female*: fore wing darker olive-brown ; with more uniformly but sparsely black-speckled transverse lines ; marginal spots and cell-spot as in male : hind wing cinereous-brown ; subapical streak paler ; cilia pale pinkish, brown-spotted.

Expanse, ♂ $1\frac{5}{10}$, ♀ $1\frac{7}{10}$ inch.

Hab. Darjiling. In coll. Dr. Staudinger and F. Moore.

SCOPOCERA, n. g.

Male. Fore wing narrow, rather long ; cell half the length ; first subcostal at one third before end of cell, second approximate ; discocellular concave, radial from lower end ; middle median close to end of cell, lower at one third ; a few fine short hairy scales projecting along the subcostal from middle of the cell ; hind wing narrow, exterior margin very oblique, convex ; cell two fifths the length ; two subcostals from end of cell ; discocellular bent acutely in the middle ; radial from lower end ; middle median close to end of cell, lower at one third. Body moderately slender ; labial palpi ascending, very long, slender, second joint extending more than half its length above the vertex, third joint about half length of second, acicular ; maxillary palpi short, tufted with a few fine long hairs ; antennæ submoniliform, compressed, with a very long dense flat curved basal tuft, which projects hindward beyond base of the thorax, the tuft composed of broad flat scales above and longish hairy scales beneath. *Female*: tuft at base of antennæ absent.

Type *S. pyraliata.*

SCOPOCERA PYRALIATA, n. sp.

Fore wing olivaceous-ochreous, crossed by an inwardly-oblique antemedial sinuous whitish line and a postmedial recurved line ; a marginal row of black points ; basal and medial area below the cell, and contiguous to the transverse lines, clouded with dark cupreous-brown ; a black spot at end of the cell : hind wing whitish olivaceous-cinereous, with a recurved pale-bordered brown discal line and marginal lunular spots. Cilia with an inner blackish line and outer spots. Thorax, head, palpi, antennæ, and legs olivaceous-ochreous ; abdomen with brown bands ; legs with brown bands.

Expanse, ♂ ♀ 1 inch.

Hab. Darjiling. In coll. Dr. Staudinger and F. Moore.

SCOPOCERA SINUOSA, n. sp. (Plate VII. fig. 3.)

Female. Fore wing obscure olivaceous-ochreous, crossed by an inwardly-oblique ante-medial and a postmedial indistinct pale-brown whitish-bordered sinuous line, and less defined marginal points; some ochreous-brown hairy scales below base of the submedian, within the cell, and along the lower subcostal, and also across the disc, on the lower submedian and lower median, the two latter with reddish intervening scales: hind wing olivaceous-cinereous, almost white at the base, brownish exteriorly; crossed by a discal pale sinuous line, and less distinct marginal spots; some black hairy scales on middle of the lower subcostal, lower median, and submedian, the median having some intervening reddish scales. Thorax ochreous; abdomen whitish-cinereous, with black dorsal bands; palpi and legs entirely ochreous.

Expanse 1$\frac{1}{16}$ inch.

Hab. Darjiling. In coll. Dr. Staudinger.

SCOPOCERA VARIEGATA, n. sp. (Plate VII. fig. 4.)

Female. Fore wing with the basal third dark cinereous-brown, the medial area pinkish-ochreous, and the outer border dull purplish-red; the basal bordered by a single inwardly-oblique dusky sinuous line, and the outer area by a pale double-bordered ochreous narrow curved band, the medial area being crossed by an ochreous fascia; some white-tipped hairy scales on the upper and lower subcostal and at base of lower median: hind wing dull cinereous-yellow, crossed by a slender waved discal brown line and an ochreous outer narrow fascia; some ochreous-red hairy scales below the cell, forming an indistinct short discal fascia. Thorax dark cinereous-brown, abdomen paler: palpi and legs entirely ochreous.

Expanse 1$\frac{3}{8}$ inch.

Hab. Darjiling. In coll. Dr. Staudinger.

SCOPOCERA MINOR, n. sp.

Male and female. Fore wing aenescent-ochreous, with a very few black scales dispersed about the medial area, some black streaks along the costa, a small black spot near base of cell, a spot below it, a transverse streak at its end, a speckled streak on posterior margin opposite the latter; an irregular recurved transverse discal narrow speckled black band, and a marginal row of points; these markings are most prominent in the female: hind wing pale aenescent-cinereous, brownish externally, with a slender indistinct pale waved discal line, and blackish marginal points. Palpi and front of head blackish; tuft of antennae in male black-speckled; fore legs with blackish bands.

Expanse $\frac{8}{10}$ inch.

Hab. Darjiling. In coll. Dr. Staudinger and F. Moore.

SARAMA, n. g.

Allied to *Scopocera.* Venation similar. Cells longer and the lower median nearer the end. Labial palpi long, stout, erect, laxly squamous, second joint extending half its length above the vertex, third joint pointed and slightly tufted near the tip; maxillary palpi with a slender tuft of fine long hairs reaching to nearly end of the labial; legs stouter, shorter; tibiæ laxly squamous, slightly hairy beneath; spurs shorter; antennæ very finely serrated and biciliated to the tip, with a long clavate tuft covered with broad lax scales at the side, and with hairy scales along the top, at the tip, and beneath.

SABAMA ATKINSONII, n. sp.

Whitish. Fore wing clouded with ochreous, crossed by an ill-defined antemedial and a postmedial black sinuous line; a black patch at the base, above and below the middle, and also on the disc, the latter patch traversed by a series of white points: hind wing with a pale cinereous-brown discal sinuous line, apical band, and marginal lunular line. Cilia with cinereous-black spots. Abdomen with black lateral bands; thorax in front, tuft of antennæ in front, and base of palpi pale ochreous; legs with blackish bands.

Expanse 1 inch.

Hab. Darjiling. In coll. Dr. Staudinger and F. Moore.

BANEPA, n. g.

Fore wing elongated, almost rectangular; exterior margin slightly excavated at each end, convex in the middle; cell more than half the length; first subcostal at one eighth, second from end of cell, quadrifid; discocellular deeply concave, radials from close to each end; middle median about one sixth, lower at half before end of cell: hind wing broad, exterior margin convex in middle; cell about two fifths the length; subcostals from end of cell, upper partly joined to costal; discocellular deeply concave, radial from close to lower end; middle median near end of cell, lower at one half. Body stout; labial palpi long, projecting obliquely upward, very laxly squamous in front, second joint reaching two thirds beyond the head, third joint about half length of second, slender, cylindrical; maxillary palpi large, reaching to level of second joint of labial, broadly tufted at tip; antennæ in male bipectinated, the branches ciliated and shaft squamous; antennæ setaceous in female; legs stout, fore and middle tibiæ laxly squamous; spurs very long, stout, nearly equal.

Allied to *Aglossa*.

BANEPA ATKINSONII, n. sp.

Fore wing dark ochreous-brown, glossed with cinereous; crossed by a waved slender ochreous antemedial line and a postmedial denticulated line; a blackish spot with outer ochreous-speckled border at end of cell; some black costal ringlet-spots and streaks before the apex; cilia ochreous, brown-spotted, and with an inner brown line: hind wing pale cinereous-ochreous, minutely brown-scaled, the scales denser across the disk and there forming a waved fascia; cilia with brown inner line. Thorax, palpi, and legs dark brown; legs speckled with ochreous; abdomen pale cinereous-ochreous, tip brown.

Expanse, ♂ 1⅜, ♀ 1⅞ inch.

Hab. Darjiling. In coll. Dr. Staudinger and F. Moore.

RODABA, n. g.

Fore wing rather broad; apex very acute, falcate; exterior margin convexly-angular in the middle; cell more than half the length; first subcostal about one third before end of cell, second close to end, third from the end, trifid; discocellular concave, radials from each end; middle median close to end of cell, lower at nearly one third: hind wing rather long, apex convex, exterior margin angular in the middle; cell nearly half the length; subcostals from end of cell; discocellular bent in middle, radial from lower end; middle median close to end, lower at two fifths before the end. Body moderately slender, abdomen not extending beyond hind wings; labial palpi ascending, clothed with long lax hairy scales, second joint extending three fourths its length above the vertex, third joint short; maxillary palpi slender, reaching the vertex; antennæ slender, setaceous; legs slender, smooth, spurs moderately slender.

RODABA ANGULIPENNIS, n. sp.

Fore wing dark purple-red, crossed by a curved antemedial and a straight broad postmedial pale streaked cinereous fascia; the costal edge also cinereous; a faint dusky spot at end of cell: hind wing brownish-cinereous. Cilia cinereous, edged with white. Thorax, head, and palpi purple-red; abdomen brownish-cinereous, with ochreous-brown dorsal bands; legs brown.

Expanse 1 inch.

Hab. Darjiling. In coll. Dr. Staudinger and F. Moore.

Genus STEMMATOPHORA, *Guén.*

STEMMATOPHORA RUDIS, n. sp.

Purplish reddish-brown: fore wing crossed by a slightly-curved pale yellowish-bordered brown antemedial line, a straight postmedial line, and a slender marginal line; a small blackish spot at end of cell, and some yellow points on middle of the costal edge: hind wing paler, crossed by paler similar lines, the discal line being slightly recurved. Legs ochreous beneath.

Expanse $1\frac{3}{10}$ inch.

Hab. Darjiling. In coll. Dr. Staudinger and F. Moore.

Genus PYRALIS, *Linn.*

PYRALIS ASSAMICA, n. sp. (Plate VII. fig. 5.)

Male. Pinkish ferruginous-brown: fore wing crossed by an outwardly-curved antemedial whitish line and a nearly straight postmedial line, the broad medial area being of a paler tint than the outer and basal area; a blackish lunule at end of the cell: hind wing with two similar whitish lines, both being bent upward above the anal angle. Thorax and apex of abdomen brighter ferruginous-brown; abdomen with pale, almost whitish segmental bands.

Expanse, ♂ $1\frac{3}{8}$ inch.

Hab. Assam (September). In coll. Dr. Staudinger.

PYRALIS ANGULIFASCIA, n. sp.

Fore wing very dark violet-brown, crossed by a broad medial pale olivaceous-ochreous band, the inner border of which is slightly bent outward in the middle, and the outer border bent inward to below end of the cell; the band is slightly brown-speckled and has some brown streaks along the costal edge; a slight darkish spot at end of cell; a marginal row of blackish points: hind wing cinereous-brown, palest at the base, with a brown marginal line. Cilia cinereous-white, with a brown inner line. Fore tibiæ with ochreous-white bands; tegulæ with long lax pale cinereous-brown spatular hairy scales.

Expanse $\frac{6}{10}$ inch.

Hab. Darjiling. In coll. Dr. Staudinger and F. Moore.

PYRALIS PALLIVITTATA, n. sp.

Dark violaceous-brown: fore wing crossed by an outwardly-curved antemedial pale-bordered line and a slightly-waved postmedial line; the medial area being pale violaceous brownish-ochreous; a slender black short streak at end of cell and some short white-bordered black streaks on middle of the costal edge; a slender pale marginal line: hind wing paler, crossed by two pale-bordered similar lines, the medial area being also palest; a pale slender marginal line.

Expanse $\frac{8}{10}$ inch.

Hab. Darjiling (*Atkinson*); N.W. Himalaya. In coll. Dr. Staudinger and F. Moore.

Genus DOTHTHA, *Walker*.

DOTHTHA SIMILATA, n. sp.

Near to *D. suffusalis.* Olivescent purple-brown: fore wing with the transverse antemedial pale line regularly curved outward, the submarginal line disposed at twice the distance from the margin; costal spots distinct: hind wing with a well-defined antemedial and discal duplex sinuous blackish line: these lines being most defined in the female. Cilia deep yellow, with purple inner line.

Expanse, ♂ $\frac{7}{10}$, ♀ $\frac{8}{10}$ inch.

Hab. Darjiling. In coll. Dr. Staudinger and F. Moore.

Genus ENDOTRICHA, *Zeller*.

ENDOTRICHA LORICATA, n. sp.

Dark purplish cinereous-brown: fore wing crossed by an inwardly-oblique waved slender white antemedial line, and an indistinct brown duplex submarginal line; the medial area somewhat purplish-red, with a small brown spot at end of the cell; some white-bordered black marks along the costal edge: hind wing with a paler medial band, which is bordered on each side by a waved whitish-bordered blackish line. Thorax, head, palpi, and fore legs olivescent purplish-brown; legs with pale bands.

Expanse $\frac{7}{10}$ inch.

Hab. Calcutta. In coll. Dr. Staudinger and F. Moore.

Family ENNYCHIIDÆ.

Genus PORPHYRITIS, *Hübner*.

PORPHYRITIS SIKKIMA, n. sp.

Allied to *P. puniceulis*, of Europe. Fore wing dark purplish-brown, with a red-ochreous lower basal triangular patch, a transverse discal irregularly-recurved constricted band, and a very small spot at upper end of the cell: hind wing dark cuprescent-brown, with an ochreous-yellow short medial transverse irregularly-clavate broad band. Cilia alternately edged with ochreous-yellow. Body, head, and palpi above purplish-brown; thorax with a few reddish-ochreous scales; legs brownish above; base of palpi, body, and legs beneath pale ochreous.

Expanse $\frac{5}{10}$ inch.

Hab. Darjiling. In coll. Dr. Staudinger.

Genus SYLLYTHRA, *Hübner*.

SYLLYTHRA DICOLOR, n. sp.

Fore wing deep purple: hind wing cinereous-white. Thorax, front of head, and palpi above deep purple; base of palpi, pectus, body above and beneath, and legs cinereous-white.

Expanse, ♂ $\frac{6}{10}$, ♀ $\frac{9}{10}$ inch.

Hab. Assam (*Atkinson*); Calcutta (*Rothney*). In coll. Dr. Staudinger and F. Moore.

Differs from *S. imbutalis*, which has a more regularly triangular form of fore wing, the hind wing being yellowish, with some indistinct marginal brown dots, the palpi ochreous, and the legs red.

Family ASOPIIDÆ.

Genus ÆDIODES, *Guén.*

ÆDIODES AUSTRUSALIS, n. sp.

Male and female. Violaceous-brown. Both wings with similarly-disposed markings, as in *Æ. inscitalis*; these are indistinctly defined, and are of a pale violaceous tint on the fore wing, and are mostly covered with brownish scales on the hind wing, whereas in *Æ. inscitalis* the markings are prominently semidiaphanous and of a pale yellow colour. Cilia also darker.

Expanse $\frac{8}{10}$ inch.

Hab. Calcutta. In coll. Dr. Staudinger and F. Moore.

Genus SAMEA, *Gu'n.*

SAMEA QUINQUIGERA, n. sp. (Plate VII. fig. 14.)

Cinereous-brown, iridescent: fore wing with some yellow streaks bordering a transverse antemedial brown line; a yellow quadrate spot in middle of the cell, two small

C

spots below end of the cell and a large transverse broad spot beyond the cell, contiguous to which is a small spot at the outer upper end : hind wing with the basal two-thirds yellow, crossed by an irregular discal brown line ; a dentate brown-lined mark at end of the cell, and a transverse streak below it. Base of abdomen yellowish ; base of palpi, pectus, and fore legs beneath whitish ; fore femora and tibiæ brown. Cilia cinereous.

Expanse 1⅜ inch.

Hab. Darjiling. In coll. Dr. Staudinger.

Genus MABRA, *Moore.*

MABRA FLAVOFIMBRIATA, n. sp.

Purplish cinereous-ochreous. Both wings with the exterior borders and cilia pale ochreous : fore wing with an indistinct recurved discal transverse brown line, and a lunule at end of cell : hind wing with a very slight brown discal line and marginal lunular line. Base of palpi, body beneath, and legs white ; fore femora and tibiæ pale ochreous.

Expanse ⁴⁄₈ to ⁵⁄₈ inch.

Hab. Calcutta. In coll. Dr. Staudinger and F. Moore.

Differs from *M. eryxalis* (*Asopia eryxalis*, Walker)—which also occurs at Calcutta—in both wings being of a purplish cinereous-ochreous from the base to near the outer border, whereas in *M. eryxalis* the discal area only is of that colour, and also has a slender submarginal brown line.

Genus HEDYLEPTA, *Lederer.*

HEDYLEPTA CONTUBERNALIS, n. sp.

Allied to *H. abruptalis.* Of a paler ochreous-yellow ; transverse lines disposed in the same manner, but much less defined : fore wing with the outer line sinuous, the inward curve not reaching the lower end of cell, and its lower end straight ; there is also a small spot in middle of the cell, as well as the one at its end, a cluster of brown scales between end of the cell and the discal line, and a very slight brownish submarginal fascia : hind wing with the inner line not extended to the cell-mark, but extending towards the disk and having a cluster of brown scales at its end beyond the cell, a brown submarginal fascia beyond the outer line. Fore legs not banded with black ; abdomen with white segmental bands.

Expanse ⁶⁄₈ inch.

Hab. Parisnath Hill, Bengal (Sept.). In coll. Dr. Staudinger and F. Moore.

HEDYLEPTA GEMELLA, n. sp.

Yellowish-ochreous ; brownish along the costa to exterior borders ; marginal line lunular, slender, brown. Both wings crossed by an extremely indistinct brownish irregular discal line, which is curved inwards to below end of the cell : fore wing with a small black spot in middle of the cell, and two lunules at its end : hind wing with two similar but very indistinctly-defined lunules at end of the cell, these being more defined in the female.

Expanse ⁵⁄₈ inch.

Hab. Calcutta. In coll. Dr. Staudinger and F. Moore.

Genus AGROTERA, *Schrank.*

AGROTERA FERRUGINATA, n. sp.

Pale ferruginous-brown : fore wing crossed by an outwardly-oblique angulated ante-medial slender brown line, and a curved discal sinuous line ; outer border clouded with darker brown : hind wing pale ferruginous-yellow, with the lower basal area thickly speckled with dark ferruginous-brown scales and some long ferruginous hairs depending from the median vein ; apical angle brown-speckled, crossed by a dark-brown discal somewhat irregularly-undulated line and a marginal line. Cilia edged with white. Abdomen with ferruginous-brown bands ; base of palpi, pectus, and fore legs white.

Expanse $\frac{9}{10}$ inch.

Hab. Darjiling. In coll. Dr. Staudinger and F. Moore.

PATANIA, n. g.

Fore wing long ; costa slightly arched towards end, apex pointed, exterior margin slightly oblique and convex ; cell half the length ; first subcostal at one sixth, second and third very close together near end of cell, third bifid ; discocellular outwardly oblique, radials from near ends ; middle median close to end of cell, lower at one third : hind wing rather narrow, exterior margin very oblique ; cell one third the length ; two subcostals from end of cell ; discocellular concave, radial from lower end ; two upper medians from end of cell, lower at one-third. Body rather stout, tegulæ rather long and lax ; labial palpi obliquely ascending, reaching the vertex, compactly squamous, third joint minute, conical ; maxillary palpi slender, pointed ; antennæ setaceous ; legs long. smooth ; hind tibiæ with a basal tuft beneath, spurs rather long, unequal.

Type *P. concatenalis (Botys concatenalis,* Walk.).

PATANIA SEMIVIALIS, n. sp. (Plate VII. fig. 6.)

Larger than *P. concatenalis.* Violaceous-brown : cilia cinereous-brown : fore wing with the ochreous-yellow medial costal band somewhat broader and longer across the disk, its lower inner end pointed, and it is crossed by a dark-brown costal vein ; the yellow spot within the cell is absent. Base of palpi and legs ochreous ; tip of femora and band on fore tibiæ brown.

Expanse $1\frac{3}{8}$ inch.

Hab. Darjiling. In coll. Dr. Staudinger.

Family HYDROCAMPIDÆ.

Genus CATACLYSTA, *Hübner.*

CATACLYSTA OCHRIPICTA, n. sp.

Allied to *C. hamalis,* Snellen. Of a much deeper and brighter ochreous-yellow colour, with similarly-disposed sericeous bands, which are pale yellow, not white as in *C. hamalis* ; the outer transverse bands and the discal band on fore wing only having a dark border, and the latter band only on its inner side ; the outer band on both wings is also broader, and that on the hind wing shorter ; there are three white-centred spots, the inner one being incipient.

Expanse, ♂ $\frac{8}{16}$, ♀ 1 inch.

Hab. Cherra Punji (October). In coll. Dr. Staudinger and F. Moore.

Genus CYMORIZA, *Guén.*

CYMORIZA LINEALIS, n. sp.

Ochreous-brown : fore wing crossed by a slender white erect waved subbasal and an antemedial line, an angulated medial and discal line, and a submarginal denticulated line ; a small white streak above end of the cell and a spot below the cell : hind wing crossed by similar waved lines converging to the anal angle, each with blackish-speckled borders. Body with whitish bands ; legs whitish ; fore tibiæ brownish. Cilia white.

Expanse $\frac{7}{16}$ inch.

Hab. Calcutta ; Assam. In coll. Dr. Staudinger and F. Moore.

CYMORIZA INEXTRICATA, n. sp. (Plate VII. fig. 7.)

Male. Umber-brown : fore wing crossed by a slender white acutely-angulated antemedial line, an inwardly-curved medial line, an irregular zigzag discal line, which is curved inward to lower end of the cell, is then acutely bent outward and again inward on the submedian vein, followed by a submarginal angular line with short white streaks extending from it below the apex to the discal line ; the medial area bordering the line is whitish, and the basal and discal interspaces finely black-speckled : a similar speckled dentate spot at end of the cell : hind wing paler brown, crossed by a broad medial whitish band, which is bordered by a slender brown line, the outer line being waved, beyond which is a very slender submarginal brown-bordered whitish line. Cilia whitish. Thorax and abdomen with slender white bands ; palpi black, bordered by white ; legs whitish ; fore tibiæ brown ; tip of middle and hind tibiæ also brown.

Expanse $1\frac{1}{16}$ inch.

Hab. Khasia Hills. In coll. Dr. Staudinger.

CYMORIZA RIVULARIS, n. sp. (Plate VII. fig. 8.)

Umber-brown. Fore wing crossed by a slender white waved subbasal line, a medial angulated line, and a discal irregular sinuous line, the latter interrupted by curving inward beneath the cell towards the medial line, followed by a submarginal series of dentate spots : a white spot at end of the cell, a similar spot below it, and some streaks before the upper end of the discal line : hind wing with a broad medial white band, which is bordered by a broad brown line, the outer line waved, the inner short ; a marginal series of white dentate spots ; cilia alternated with white and with a brown inner line. Abdomen with slender white bands ; base of palpi white ; legs whitish ; fore tibiæ brown.

Expanse $1\frac{1}{16}$ inch.

Hab. Darjiling. In coll. Dr. Staudinger.

CYMORIZA MARGINALIS, n. sp.

Dark brown. Both wings crossed by an antemedial inwardly-oblique waved white line, a short costal band before the apex, and a slender submarginal line, which latter joins a broader short band above the angle, followed by a reddish-ochreous outer band and a marginal black-pointed line; a white spot with blackish border at end of each cell. Cilia white, with inner blackish line. Palpi and legs brownish-white; tibiæ with a brownish band.

Expanse $\frac{8}{10}$ inch.

Hab. Darjiling. In coll. Dr. Staudinger.

Genus HYDROCAMPA, *Latr.*

HYDROCAMPA BENGALENSIS, n. sp.

Pure white. Fore wing with three subbasal transverse curved series of three fuliginous brown spots, a darker spot at end of the cell, one above it, followed by a curved angulated discal band, a submarginal and a marginal curved band, the latter interrupted: hind wing with a subbasal brown spot, an antemedial and a discal partly confluent band enclosing a blackish cell-spot, beyond which is a recurved angulated submarginal and a marginal band. A brown spot on the tegulæ, and bands on abdomen; fore femora and tibiæ brown.

Expanse $\frac{9}{10}$ inch.

Hab. Calcutta. In coll. Dr. Staudinger.

PRAMADEA, n. g.

Fore wing long; costa arched near end; apex pointed, exterior margin slightly oblique, convex hindward; cell fully half the length; first subcostal at one eighth before end, second and third close to end, approximate, third bifid; discocellular outwardly oblique, concave: upper radial near the middle, lower close to end; middle median close to end of cell, lower about one fifth: hind wing rather long, apex slightly angular, exterior margin very oblique, convex, anal angle truncate; cell one third the length; subcostals from end of cell; discocellular deeply concave, radial close to lower end; middle median close to end, lower about one fourth. Body moderately slender; labial palpi ascending, reaching the vertex, squamous, second joint very broad, truncate, third joint minute, conical; maxillary palpi slender, squamous at tip; antennæ slender, setaceous; legs long, slender, smooth; spurs slender, inner very long, outer very short.

PRAMADEA DENTICULATA, n. sp.

Pale cinereous olivaceous-brown: fore wing crossed by an antemedial outwardly-curved black-speckled bordered ochreous-white sinuous line, a recurved discal similar denticulated line, a less distinct submarginal line, and a marginal row of points; a small oval spot in middle of cell, and a lunule at its end; the black-speckled outer border of the discal line mostly extending to the submarginal line, and that of the inner line broad and distinct at its lower end: hind wing darker than fore wing, with a discal zigzag ochreous-white sinuous line, and a less distinct submarginal line; a marginal row of black points. Abdomen with black apical bands; palpi above black; band on fore tibiæ black; base of palpi and pectus white.

Expanse 1⅛ to 1¾ inch.

Hab. Cherra Punji (October); Khasia Hills. In coll. Dr. Staudinger and F. Moore.

Family SICULIDÆ.

Genus RHODONEURA, *Guén.*

RHODONEURA NEVINA, n. sp.

Male. White : fore wings with numerous short delicate brown strigæ, which are disposed transversely from the base and along the costal border to near the apex; two transverse series of short darker-brown streaks on lower part of the disk, of which the outer one between the middle and lower median is oblique, followed by two series of small and less distinct strigæ, which are disposed between the veins; a dot also near the apex: hind wing with several transverse equidistant rows of short brown strigæ. Abdomen with short dorsal brown bands; fore legs and all the tarsi with brown bands.

Expanse 1 inch.

Hab. Darjiling. In coll. Dr. Staudinger and F. Moore.

Allied to *R. tetraonalis.*

Genus PHARAMBARA, *Walker.*

PHARAMBARA RETICULATA, n. sp.

Male and female. Reddish-brown : both wings with numerous very indistinct and slender short waved black transverse strigæ; fore wing also crossed by, apparently, ten more distinct, mostly equidistant, continuous lines; the inner lines being erect and slightly waved, the discal lines irregular and angulated inward to the costa, and the outer lines waved: hind wing with apparently six similar transverse lines, which are wavy and curved. In some specimens the outer lines are bifid towards the lower end.

Expanse 1½ inch.

Hab. Darjiling. In coll. Dr. Staudinger and F. Moore.

PHARAMBARA ALTERNATA, n. sp.

Pale violaceous yellowish-ochreous, hind wing palest. Both wings crossed by numerous waved slender brown strigæ, which are mostly continuous, and some form more or less irregular transverse lines with ochreous-brown alternate interspaces; on the fore wing there are two inner, an interrupted discal, and a thrice irregularly interrupted outer brown interspace, and on the hind wing a medial and discal lower brown interspace. Thorax, head, palpi, and legs brownish-ochreous.

Expanse, ♀ 1⅜₀ inch.

Hab. Darjiling. In coll. Dr. Staudinger and F. Moore.

PHARAMBARA INTIMALIS, n. sp.

Yellowish-ochreous. Wings almost covered with broad transverse mostly-confluent ochreous-brown strigæ : fore wing with a white lunule and a lower black spot at end of cell ; a blackish curved streak below the apex : hind wing with a black medial transverse band. Fore wing angular in middle of exterior margin : hind wing convexly angular in middle of exterior margin.

Expanse $\frac{6}{12}$ to $\frac{7}{12}$ inch.

Hab. Calcutta. In coll. Dr. Staudinger and F. Moore.

Much like *Microsca striatalis*, Swinhoe. Differs in the angulated form of both wings, and also in the broader strigæ.

PHARAMBARA HAMIFERA, n. sp.

Silky lilacine ochreous-white : fore wing speckled and clouded with ochreous-brown ; with a darker-brown curved discal fascia, a shorter antemedial fascia, and a streak before posterior angle ; the costal and lower discal interspaces whitish ; some black dots between the median branches, and a white-bordered hook-shaped mark below the apex : hind wing transversely brown-speckled ; with a medial and discal lower dark ochreous-brown band. Thorax, band on abdomen, palpi, and fore legs ochreous-brown ; tip of palpi and bands on fore legs white.

Expanse 1 inch.

Hab. Assam ; Calcutta. In coll. Dr. Staudinger and F. Moore.

Nearest to *P. pallida* (*Microsca pallida*, Butler) from Japan.

PHARAMBARA EMBLICALIS, n. sp.

Female. Purplish-ferruginous : fore wing with the outer border broadly yellow, crossed by indistinct ferruginous streaks ; some confluent yellow strigæ at end and below the cell : hind wing with some indistinct yellow strigæ near the base, and the middle of outer border also yellowish.

Expanse, ♀ $\frac{7}{10}$ inch.

Hab. Calcutta (June). In coll. Dr. Staudinger.

Genus MICROSCA, *Butler.*

MICROSCA FASCIATA, n. sp. (Plate VII. fig. 22.)

Lilacine-ochreous, with ochreous speckles ; a broad red fascia extending from the apex of fore wing and decreasing hindward to a dark purple narrow band across the hind wing : the lower edges of the band bordered by silvery scales, some of which are also disposed on basal area of the hind wing ; at the costal end of the band is a white streak, a slender streak below the apex, and the posterior angle also white. On the fore wing is a black mark at end of the cell and some spots beyond the end ; some black spots also on base of hind wing. Body ochreous ; fore tibiæ and the tarsi with white bands.

Expanse $1\frac{2}{10}$ inch.

Hab. Darjiling. In coll. Dr. Staudinger.
Allied to *M. trifasciata* (*Botys trifasciata*, Moore).

MICROSCA LOBULATA, n. sp. (Plate VII. fig. 12.)

Pale silky ochreous-yellow: fore wing with the basal two-thirds clouded with dark
ferruginous, ending in a darker transverse discal decreasing fascia with a lobate lower end;
an outwardly-oblique subapical ferruginous line, which joins the fascia on the costa, and
is thence continued to the apex, below which is an angular line above the posterior angle;
discal interspaces traversed by indistinct ferruginous strigæ: hind wing with a dark ferruginous subbasal band; the basal and outer area traversed by ferruginous strigæ. Body,
palpi, and fore legs dark purplish-ferruginous; middle and hind legs paler; front of thorax,
the costal border, and abdomen above suffused with chalybeous-grey in some lights.
Expanse 1 inch.

Hab. Darjiling. In coll. Dr. Staudinger.

Genus MOROVA, Walker.

MOROVA ANGULALIS, n. sp.

Male. Purplish ochreous-red. Wings speckled with minute chalybeous scales; crossed
by short brown strigæ; some white streaks along the costal edge of fore wing. Cilia pure
white, alternated with red in the middle.
Expanse $\frac{7}{16}$ inch.

Hab. Calcutta (*Atkinson*); Rangoon (*Watt*). In coll. Dr. Staudinger and F. Moore.

CAMADENA, n. g.

Fore wing long, rather broad; apex produced to a point, exterior margin oblique,
uneven, convexly angular in the middle; cell three fifths the length; first subcostal more
than half before end of cell, second close to end, third and fifth from the end, third bifid;
discocellular concave in the middle, bent near each end, radials from the angles; middle
median at one sixth, lower at half before end of cell: hind wing long, apex produced,
pointed; exterior margin very oblique, uneven, hardly convex in the middle; cell nearly
half the length; two subcostals from end of cell, upper free from costal; discocellular bent
near lower end, radial from the angle; middle median at one sixth, lower at more than half
before end. Body not extending beyond hind wings; labial palpi curved upward, reaching
to vertex, laxly squamous, third joint short; maxillary palpi not visible; antennæ setaceous;
legs very long, slender; spurs long, slender, unequal.

CAMADENA VESPERTILIONIS, n. sp. (Plate VII. fig. 13.)

Pale ochreous-yellow. Both wings numerously covered with transverse brownish-
ochreous strigæ: fore wing crossed by a slight ochreous-brown inwardly-oblique subbasal
and a similar medial fascia, a small patch below the apex, and a short lower discal fascia:
hind wing with a similar-coloured medial fascia and a narrow submarginal fascia. Thorax,

head, palpi, and fore legs brownish-ochreous; abdomen, middle and hind legs paler; fore tibiæ and tarsi with white bands.

Expanse $1\frac{6}{10}$ inch.

Hab. Darjiling (July). In coll. Dr. Staudinger.

Family SPILOMELIDÆ.

Genus HARITALA, *Moore.*

HARITALA AURORALIS, n. sp. (Plate VII. fig. 17.)

Straw-yellow. Fore wing crossed by seven reddish-ochreous bands, the first three being outwardly oblique, the fourth erect, the others inwardly oblique, the fifth and sixth being united at their lower end: hind wing with a broad lower basal reddish-ochreous band, which is bordered outwardly by a slender discal line, this line being indented and touching a yellow lunule at end of the cell; a paler ochreous contiguous discal band traversed by a slender yellow indented submarginal line. Cilia with a slender inner ochreous band. Thorax and abdomen with reddish-ochreous bands; second joint of palpi and fore legs ochreous; middle and hind legs paler; tarsi whitish.

Expanse $\frac{8}{10}$ inch.

Hab. Cherra Punji (October). In coll. Dr. Staudinger.

HARITALA RECURRENS, n. sp. (Plate VII. fig. 11.)

Male. Pale yellow. Fore wing crossed by four deep ochreous bands, the first and second band oblique and straight, third triangularly dilated at the costal end and looped outward beyond the cell, fourth band curving below and touching end of the loop, its lower end being broad and containing a blackish-scaled spot; a basal ochreous and two black costal spots, each being equidistant: hind wing with a subbasal, medial, and a discal deep ochreous band, the first short, the other two contiguous at their lower end. Both wings also with a slender brown marginal line and inner cilial ochreous band. Thorax, abdomen, and fore legs with ochreous bands; palpi at tip also ochreous.

Expanse 1 inch.

Hab. Darjiling. In coll. Dr. Staudinger.

Genus POLYTHLIPTA, *Lederer.*

POLYTHLIPTA DISTORTA, n. sp. (Plate VII. fig. 25.)

Nearest to the Javan *P. cerealis*, Lederer (Felder, Nov. Voy. pl. 135. f. 34), and to *P. vagalis*, Walker. Differs in the markings being of a pale fuliginous-brown colour. On the fore wing the irregular subcostal band is continued to the base, as in *P. vagalis*, the white cell-spots being confluent, the portion below end of the cell irregularly zigzag across the veins; the transverse angular discal band is sinuous on both sides. On the hind wing

p

the discoidal mark is slender, the angulated discal band and the marginal markings being well separated.

Expanse 1₁⁴₀ inch.

Hab. Darjiling. In coll. Dr. Staudinger.

POLYTHLIPTA PERAGRATA, n. sp. (Plate VII. fig. 15.)

Male. Larger than *P. vagalis*; with darker brown interspaces: fore wing with the white spots larger and not black-bordered, those within and below the cell quadrate, the lower discal spot not excavated below end of the cell, the upper discal spot convex on its outer edge, and the marginal spots broader: hind wing with a short broad uniform streak at end of the cell, the discal angular band more towards the middle, the two outer white portions large and well defined. Body with dark brown lateral bands; legs white; palpi and fore tibiæ tipped with brown.

Expanse 1₁⁶₀ inch.

Hab. Darjiling. In coll. Dr. Staudinger.

Family MARGARONIIDÆ.

Genus GLYPHODES, *Guén.*

GLYPHODES CHILKA, n. sp. (Plate VII. fig. 9.)

Male. Pale violaceous ochreous-brown: fore wing with a very small bluish-bordered semidiaphanous white spot near end of the cell, a quadrate spot in the middle, and a large constricted oval spot at the end of the cell, the latter partly encircled by a slender pale line; an obliquely-triangular white spot situated partly beneath the middle cell-spot, in front of which is a short slender outwardly-curved pale line; some blackish speckles near base of the posterior margin: hind wing with the basal two thirds semidiaphanous ochreous-white, the outer margin broadly pale ochreous-brown, bordered by a discal narrow blackish band which is angled on the lower median; a slight blackish marginal lunular line. Sides of thorax, second joint of palpi, and band on fore legs black.

Expanse 1½ inch.

Hab. Darjiling. In coll. Dr. Staudinger.

Genus SYNCLERA, *Lederer.*

SYNCLERA TIBIALIS, n. sp.

Straw-yellow: fore wing crossed by five equidistant olive-brown bands; from upper end of the third another band extends to the lower end of the fourth, the lower end of the third extending along the margin and joining the outer band: hind wing with three similar bands, the outer band marginal and joined to the lower end of the middle band. Thorax with a brown band down the middle; femora and tibiæ in front and abdomen beneath with black bands; hind tibiæ tufted above with long black hairs.

Expanse 1 inch.

Hab. Darjiling. In coll. Dr. Staudinger.
Nearest allied to *S. gastralis.*

Genus MARGARONIA, *Hübner.*

MARGARONIA FRATERNA, n. sp.

Closely allied to *M. celsalis.* Fore wing with similarly disposed markings, which, with the costal border, are all vinous-brown—not yellow as in *M. celsalis*; the mark at end of the cell being also about twice the width and marked by a broad inner line; the apex of the wing has a dark brown patch, on which is a geminated white spot: hind wing with the white cell-spot less defined, and the indistinct pale-brown discal zigzag line disposed nearer the margin. Palpi and band on fore legs dark brown.

Expanse 1 inch.

Hab. Cherra Punji. In coll. Dr. Staudinger.

PITAMA, n. g.

Fore wing rather long; cell more than half the length; first subcostal at one fourth before end of cell, second and third close to end, third bifid, fifth from end of cell; disco-cellular outwardly oblique, slightly concave, radials from upper end and near lower end; middle median close to end of cell, lower about one third: hind wing broad, triangular; apex convex; cell one third the length; two subcostals from end, upper partly joined to the costal; discocellular outwardly oblique, deeply concave, radial from close to lower end, middle median close to end of cell, lower at one fourth. Body moderately stout; labial palpi obliquely ascending, pointed in front, projecting in the form of a rostrum, laterally broad, laxly squamous; maxillary palpi squamous, truncate at the tip, reaching to level of labial; antennæ simple; legs long, slender, smooth; fore tibiæ thickened, middle tibiæ laxly squamous above; spurs very long and slender, unequal.

PITAMA LATIVITTA, n. sp. (Plate VII. fig. 21.)

Olivescent-white. Both wings with a very broad outer iridescent purplish-brown band: fore wing with the costal border also brown, extending its width from the cell to the margin, its inner border being black-speckled; some black speckles within the cell contiguous to the costal border: hind wing with the band of the same width throughout, but having its inner border evanescent at the anal end. Thorax and abdomen olivescent-white; palpi, side of thorax, tip of abdomen, fore femora, and tibia brown; base of palpi and legs white.

Expanse 1¾ inch.

Hab. Darjiling. In coll. Dr. Staudinger.

RHAGOBA, n. g.

Wings ample. Fore wing elongated; cell half the length; first subcostal at one third before end of cell, second and third contiguous, close to end of the cell, third bifid; disco-cellular outwardly oblique, slightly concave, radials from near the end; middle median

D 2

close to end of cell, lower about one third: hind wing short, broad, exterior margin convex; cell short, less than one third the length, broad; subcostals from end of cell; discocellular slightly concave, radial from close to end; two upper medians from end of cell, lower about one fourth. Body very stout; abdomen rather short; labial palpi obliquely ascending, stout, broad, squamous, third joint minute, obtuse; maxillary palpi slender, compact at the tip; antennæ setaceous, minutely ciliated in male; legs rather stout, long, fore and middle tibiæ laxly squamous; spurs slender, inner very long, outer short.

Type *R. octomaculata* (*Filodes octomaculata*, Moore).

RHAGOBA BIMACULATA, n. sp.

Dark sepia-brown. Base of wings with steel-blue reflections: fore wing with a small oblique hyaline white spot at end of the cell. Thorax and abdomen with steel-blue reflections; palpi entirely brown; a band on fore tibiæ and all the tarsi white.

Expanse $1\frac{1}{4}$ inch.

Hab. Darjiling. In coll. Dr. Staudinger.

Genus FILODES, *Guén.*

FILODES PATRUELIS, n. sp.

Differs from *F. fulvidorsalis* and *F. mirificalis* in its somewhat larger size. Wings much paler in colour, being of a pale violaceous-brown; both sexes have the base of the fore wing less fulvous than in *F. fulvidorsalis,* and the black spots distinct. Both wings with an oblique transverse discal blackish fascia.

Expanse, ♂ $1\frac{3}{10}$, ♀ $1\frac{6}{10}$ inch.

Hab. Calcutta; Cherra Punji. In coll. Dr. Staudinger and F. Moore.

CHAREMA, n. g.

Male. Fore wing long, narrow; apex acute, exterior margin very oblique; first subcostal at two thirds before end of cell, second and third approximated, second close to end of cell, third bifid, fifth from end of cell; discocellular slightly concave, radials from near the ends; middle median close to end, lower at nearly one third: hind wing short, triangular; cell one third the length; subcostals from end of cell, upper slightly touching the costal; discocellular concave, radial from lower end; two upper medians from end of cell, lower about one fourth before the end. Body moderately stout; abdomen extending one third beyond hind wings; thorax with very long lax hairy divergent tegulæ, which extend to nearly half the abdomen; labial palpi ascending to a little higher than the vertex, stout, very broad, compactly squamous, third joint broadly conical; maxillary palpi slender; antennæ slender, setaceous, minutely biciliated; legs long, squamous, spurs slender, unequal.

Type *C. noctescens.*

CHAREMA NOCTESCENS, n. sp.

Olivescent umber-brown. Wings of a uniform tint throughout: fore wing with an indistinct blackish transverse antemedial outwardly-curved line, and an irregular postmedial line, which is curved inward below end of the cell ; a small spot in middle of the cell and a lunule at its end: hind wing with traces of an irregular discal darker line and a cell-spot. Cilia brown. Base of palpi, body beneath, femora, and tibiæ pale ferruginous; tarsi whitish. Expanse $1\frac{6}{10}$ inch.

Hab. Darjiling. In coll. Dr. Staudinger and F. Moore.

CHAREMA ALBOCILIATA, n. sp.

Male. Pale cinereous vinous-brown. Cilia edged with white: fore wing crossed by a slender black sinuous antemedial line and an irregular recurved postmedial line ; a black dot in middle of the cell and a lunule at the end: hind wing with an irregular recurved discal sinuous line and a lunule at end of cell. Base of palpi and legs whitish ; fore tibiæ with a brown band.

Expanse $1\frac{2}{10}$ inch.

Hab. Calcutta. In coll. Dr. Staudinger and F. Moore.

CHAREMA IMBECILIS, n. sp. (Plate VII. fig. 23.)

Male. Pale ochreous-brown : fore wing crossed by an extremely-indistinct brown ante-medial sinuous line, and a recurved discal line : hind wing with a similar indistinct discal line ; a marginal row of brown points. Cilia ochreous. Thorax, head, and tip of palpi brownish-ochreous; legs pale ochreous-brown ; base of palpi, bands on fore tibiæ, middle tibiæ beneath, and tarsi white.

Expanse $1\frac{3}{10}$ inch.

Hab. Darjiling. In coll. Dr. Staudinger.

Allied to *C. vinacealis.* Distinguished from it in colour, and in the discal sinuous line on both wings curving inward below the cell.

Family HAPALIIDÆ.

CHOBERA, n. g.

Male. Fore wing long, very narrow, apex convex, exterior margin very oblique ; cell very long, nearly two thirds the length ; first subcostal at one fifth, second and third approximate, close to end of cell, third bifid ; discocellular inwardly oblique, concave, radials at nearly equal distances apart and from the ends ; middle median close to end of cell, lower about one third: hind wing short, triangular, apex convex, exterior margin very oblique, concave in the middle, anal angle lobular ; cell one third the length ; subcostals from the end, upper joined to costal to two thirds its length ; discocellular deeply concave, radial from lower end ; middle median close to end of cell, lower about one third. Body slender, abdomen extending half beyond hind wings ; thorax with long lax hairy scales ; labial palpi obliquely ascending, reaching level of vertex, broad, pointed and rostriform in front, laxly squamous beneath ; maxillary palpi slender, reaching above the vertex ; antennæ slender, setaceous ; legs slender, long, smooth ; spurs long, slender, unequal.

CHOBERA PALLIDA, n. sp.

Male. Very pale brownish-ochreous. Cilia whitish. Thorax, head, palpi above, and bands on fore legs brighter ochreous ; base of palpi, pectus, and legs white.

Expanse $1\frac{2}{16}$ inch.

Hab. Calcutta. In coll. Dr. Staudinger and F. Moore.

Genus CIRCOBOTYS, *Butler.*

CIRCOBOTYS LIMBATA, n. sp.   (Plate VII. fig. 24.)

Male. Pale purplish brownish-ochreous. Both wings with a narrow ochreous-yellow marginal band, which is narrowest at the posterior end ; cilia also yellow : fore wing with the costal border edged with white, and with very faint traces of a transverse discal sinuous line. Tip of abdomen, front of head, palpi above, and fore legs ochreous ; base of palpi, pectus, band on fore legs, the middle and hind legs white.

Expanse $1\frac{3}{8}$ inch.

Hab. Darjiling. In coll. Dr. Staudinger.

Genus CONOGETHES, *Meyrick.*

CONOGETHES ALBOFLAVALIS, n. sp.

Male. Creamy white : both wings with the exterior border and cilia ochreous-yellow : fore wing with black spots, one being at the base, two on the costa near the base, a short streak from the costa about one third from the apex, a small spot at the apex, one at end of the cell, three below the cell, five on the posterior margin, and one on the middle of the exterior border ; on the middle of the disk is a trace of a short slender black line. Thorax and abdomen with black spots ; legs and tarsi with black bands.

Expanse $\frac{8}{16}$ inch.

Hab. Darjiling. In coll. Dr. Staudinger and F. Moore.

Genus BOTYODES, *Guén.*

BOTYODES INCONSPICUA, n. sp.

Pale cinereous œnescent-brown : fore wing with two short transverse whitish streaks within the cell, an irregular-shaped pointed discal spot beyond, below which is a narrower spot ; a small oval spot before the apex : hind wing with a broad medial transverse uneven-bordered whitish band, on which is an irregular-shaped cell-spot. Cilia along anal angle whitish. Fore and middle femora and tibiæ tipped with cinereous-black.

Expanse $1\frac{3}{16}$ inch.

Hab. Darjiling. In coll. Dr. Staudinger and F. Moore.

BOTYODES FRATERNA, n. sp. (Plate VII. fig. 16.)

Male. Allied to *B. costalis.* Much smaller: fore wing with the ochreous costal border narrower and duller-coloured, the cell-spots very small, indistinct, and slender, the transverse fascia obsolescent; in addition to the pale yellow discal spot there is a yellow triangular spot on middle of the posterior margin: hind wing with the basal area pale yellow, and a very small blackish cell-spot. Abdomen with a broad pale-yellow basal band; base of palpi and legs whitish; fore tibiæ brown.

Expanse 1 inch.

Hab. Darjiling. In coll. Dr. Staudinger.

BOTYODES COSTALIS, n. sp.

Male. Near to *B. scinisalis.* Smaller. Darker pale glossy cuprescent-cinereous; fore wing with the middle of the costal area only ochreous; the two blackish cell-spots more prominent, with traces of an antemedial and postmedial darker fascia; a small yellow constricted spot on middle of the disk: hind wing palest at the base, with traces of a large darker cell-spot and medial transverse fascia. Base of palpi and tarsi whitish.

Expanse 1¼ inch.

Hab. Darjiling. In coll. Dr. Staudinger and F. Moore.

BOTYODES LEOPARDALIS, n. sp. (Plate VII. fig. 26.)

Male. Straw-yellow: fore wing with some iridescent-brown basal spots, a large transverse subbasal spot, contiguous to which is a transverse curved antemedial line, and beyond a broad irregular-shaped medial yellow-spotted band, the upper part of which is composed of the large ordinary end cell-spot; beyond is a discal macular irregular line and marginal row of spots, the intermediate apical area being also brown; a small spot in middle of the cell: hind wing with a small brown cell-spot, a transverse medial and discal zigzag macular line, and a slender marginal line. A black dorsal band near tip of abdomen; femora and tibiæ with black bands.

Expanse 1⁵⁄₁₆ inch.

Hab. Darjiling. In coll. Dr. Staudinger.

Genus HAPALIA, *Hübner.*

HAPALIA NIGRESCENS, n. sp.

Dark violet-brown; hind wing paler brown. Cilia of both wings ochreous-yellow: fore wing with the costal edge ochreous-yellow; crossed by a very indistinct blackish antemedial and a waved postmedial diffused line; a blackish spot at end of the cell. Base of palpi, pectus, fore tarsi, middle tibiæ and tarsus above pure white; fore femora and tibiæ, middle and hind legs above ochreous; fore tibiæ with blackish band.

Expanse 1¹⁄₁₆ inch.

Hab. Darjiling. In coll. Dr. Staudinger and F. Moore.

HAPALIA KASMIRICA, n. sp. (Plate VII. fig. 28.)

Cinereous-brown: fore wing with an ochreous-yellow zigzag subbasal band commencing from below the costal vein, and a sinuous recurved discal band, the latter terminating in a large lower patch; a marginal series of short longitudinal yellow streaks; medial and basal area speckled with yellow scales: hind wing paler; with a broad medial discal ill-defined ochreous-yellow fascia, and some submarginal yellow speckles. Body speckled with yellow scales; abdomen with slender white bands; base of palpi, body beneath, and legs white; fore tibiæ and tarsi above cinerous-brown.

Expanse $1\frac{1}{8}$ inch.

Hab. Kashmir. In coll. Dr. Staudinger.

Allied to *H. lupulinata* (*Botys silacealis*, Hübn.).

HAPALIA BAMBUSALIS, n. sp.

Ochreous-yellow: fore wing crossed by a very indistinct outwardly-oblique antemedial and an inwardly-oblique postmedial sinuous brownish-ochreous line, the latter bent inward to below end of the cell, followed by a similar submarginal fascia: hind wing with an indistinct straight medial transverse ochreous line and a marginal sinuous fascia. Body beneath and legs white; second and third joint of palpi, band on fore tibiæ, and streak along middle tibiæ ochreous; base of palpi white.

Expanse, ♂ $1\frac{4}{10}$, ♀ $1\frac{7}{10}$ inch.

Hab. Darjiling. In coll. Dr. Staudinger and F. Moore.

Allied to *H. zealis*, Guén. (*Botys zealis*).

HAPALIA ROBUSTA, n. sp. (Plate VII. fig. 27.)

Yellow. Fore wing suffused with ochreous-brown at the tip; crossed by a brown antemedial zigzag line, with an acute outward point below the cell, an irregular recurved narrow sinuous postmedial line, the outer margin of the wing being also brown, except the yellow lunules at both ends of the band; a brown spot in middle of the cell and a quadrate spot at its end; veins across the disk also mostly brown: hind wing with a large brown dentate slender discal band, and a broader marginal band. Body, tip of palpi, and fore tibiæ brown; base of abdomen yellow; base of palpi and legs white.

Expanse $1\frac{4}{10}$ inch.

Hab. Darjiling. In coll. Dr. Staudinger.

HAPALIA OBLITA, n. sp.

Cinereous ochreous-brown: fore wing with some ill-defined transverse waved pale-yellow streaks below the cell, a small spot in middle of the cell, a quadrate spot at its end, and a larger quadrate spot beyond the cell; from the latter some pale streaks extend to the apex: hind wing with the base of costa, a broad tapering medial sinuous-bordered band, and a narrow denticulated discal band, pale yellow. Cilia pale yellow. Base of palpi white; fore legs brownish; middle and hind legs white.

Expanse $1\frac{3}{10}$ inch.

Hab. Darjiling. In coll. Dr. Staudinger and F. Moore.

HAPALIA INDISTANS, n. sp.

Cinereous brownish-ochreous: fore wing crossed by a yellowish-bordered brown sinuous antemedial and a recurved postmedial line ; a yellowish quadrate spot at end of the cell : hind wing with a broadly yellowish-bordered brown sinuous recurved discal line, and a straight streak at end of the cell. Base of palpi and legs beneath whitish.

Expanse $1\frac{1}{2}$ inch.

Hab. Darjiling. In coll. Dr. Staudinger and F. Moore.

Also occurs at Dalhousie, N.W. Himalayas.

HAPALIA DORSIVITTATA, n. sp. (Plate VII. fig. 18.)

Dull straw-yellow : fore wing with the costal border and exterior margin iridescent ochreous-brown ; crossed by a slender brown antemedial sinuous line, and a recurved postmedial line ; a brown dot in middle of the cell, a spot at the end, and a marginal row of minute dots : hind wing with an iridescent ochreous-brown medial fascia and outer band, the latter bordered by a marginal row of brown dots. Front of thorax, tip of palpi, and abdomen dark ochreous-brown ; abdomen with pure white segmental bands ; base of palpi and pectus white ; fore and middle legs above brown ; bands on fore legs, middle tibiæ, and tarsi beneath white, hind legs yellowish.

Expanse $1\frac{1}{8}$ inch.

Hab. Darjiling. In coll. Dr. Staudinger.

HAPALIA FLAVOFASCIATA, n. sp. (Plate VII. fig. 19.)

Purpurescent ochreous-brown. Both wings with a transverse postmedial straight yellowish band ; exterior margins yellowish along the edge, with a row of dark brown points. Cilia yellowish. Abdomen with yellowish segmental bands ; base of palpi, bands on fore legs, and middle legs beneath pure white.

Expanse $1\frac{9}{10}$ inch.

Hab. Darjiling. In coll. Dr. Staudinger.

Near to *H. dorsivittata.*

HAPALIA FASCIATA, n. sp. (Plate VII. fig. 20.)

Purpurescent ochreous-brown. Both wings with a transverse discal yellow band, which on the fore wing is waved on both its sides, and on the hind wing tapering to its lower end ; a yellowish line along the outer margin. Cilia ochreous-yellow. Abdomen with slender pale-yellow segmental bands ; base of palpi, bands on fore legs, and middle legs beneath white ; hind legs yellowish.

Expanse $1\frac{2}{3}$ inch.

Hab. Darjiling. In coll. Dr. Staudinger.

Genus EBULEA, *Guén.*

EBULEA DICHROMA, n. sp.

Fore wing yellowish-ochreous : median vein dusky towards end of the cell ; cilia edged

E

with white: hind wing white. Body yellowish-ochreous; base of palpi and pectus white;
legs whitish; fore tibiæ brownish above.

Expanse 1⅗ inch.

Hab. Darjiling. In coll. Dr. Staudinger and F. Moore.

EBULEA OBLIQUATA, n. sp.

Pale yellow, base of hind wing whitish: fore wing with the costal base and the lower
outer border brownish-ochreous; an indistinct transverse antemedial and an irregular
recurved postmedial slender ochreous-brown sinuous line, the upper end of the latter ending
obliquely inward on the costa, and below which is an oblique brown streak extending from
near middle of the costa to below middle of the posterior margin: hind wing with an irregular recurved discal ochreous-brown sinuous line. Both wings with a brown marginal line
and inner cilial line. Abdomen whitish at the base, ochreous at the tip, with a white-bordered black anal band; edges of frontal tuft, base of palpi, body beneath, and legs whitish.

Expanse 1 inch.

Hab. Calcutta. In coll. Dr. Staudinger.

EBULEA BAMBUCIVORA, n. sp.

Fore wing very pale brownish-ochreous, with a faint dusky cinereous longitudinal shade
from end of the cell; costal edge whitish: hind wing ochreous-yellow, with a cinereous
apical shade; cilia white. Thorax, head, and palpi above brownish-ochreous; edges of
the vertex, base of palpi, and legs white; abdomen yellow.

Expanse 1⅒ inch.

Hab. Calcutta. In coll. Dr. Staudinger and F. Moore.

"Larva pale green; turning red before pupating. Lives in rolled-up leaves of
Bamboo: April. Moths emerged May 8th and 13th." (*Atkinson.*)

Genus PIONEA, *Guén.*

PIONEA NOBILIS, n. sp. (Plate VII. fig. 29.)

Light yellow, with a sulphur tinge: fore wing with the base suffused with purplish-
brown; two small blackish spots in middle of the cell and a brown-bordered dentate
mark at the end; two small marks also below the cell; a transverse outer discal purplish-
brown decreasing fascia, which is dilated and outwardly-diffused towards the apex: hind
wing with a very indistinct minute brown spot at end of cell, a recurved discal line, and
slender submarginal fascia. Thorax, tip of palpi, maxillary palpi, and bands on fore legs
purplish-brown; abdomen above and front of head ochreous-brown; base of palpi, pectus,
and abdomen beneath white.

Expanse 1⅒ inch.

Hab. Darjiling. In coll. Dr. Staudinger.

Genus UDEA, *Guén.*

UDEA RENALIS, n. sp.

Near to *U. hypatidalis* and *U. ferrugalis*. Larger: fore wing reddish-ochreous, the outer border purple-brown ; crossed by an antemedial black-dotted line, and a postmedial denticulated line ; a large oval spot in middle of the cell and a large reniform spot at the end ; a dot also below the cell ; cilia purplish-brown : hind wing pale yellowish, with a minute blackish upper and lower spot at end of cell, a very indistinct recurved discal sinuous line, and a slender marginal purplish-brown band ; cilia yellowish, with brown inner line. Body, palpi above, and bands on fore legs purplish-brown ; base of palpi, abdomen beneath, and legs whitish.

Expanse $\frac{9}{10}$ to 1 inch.

Hab. Darjiling. In coll. Dr. Staudinger.

Family SCOPARIIDÆ.

PARBATTIA, n. g.

Female. Fore wing elongated, rather narrow ; exterior margin very oblique ; cell more than half the length ; first subcostal nearly one third before end of cell, second and third contiguous, close to end of cell, third bifid ; discocellular deeply concave, radials from close to each end ; middle median near end of cell, lower at one third : hind wing rather narrow, triangular, exterior margin very oblique ; cell less than one third the length at its upper end and nearly half at lower end ; subcostals from end of cell, upper joined in its middle to the costal ; discocellular outwardly oblique, very concave, radial from lower end ; middle median close to end of cell, lower about one third. Body stout ; labial palpi porrected, first and second joints broad, laxly squamous, third joint cylindrical ; maxillary palpi short, squamous ; antennæ setaceous ; legs rather stout, squamous ; spurs long, nearly equal in length.

PARBATTIA VIALIS, n. sp. (Plate VII. fig. 30.)

Female. Fore wing pale vinous-brown, crossed by a pale yellowish erect narrow antemedial band, and a recurved oblique postmedial band ; veins at the base and the medians below end of the cell yellow-streaked ; a small yellow streak within the cell, a quadrate spot at the end, and a small streak beyond the cell ; a marginal row of yellow points : hind wing pale yellowish, with a brown macular recurved discal line, and broad marginal pale brown band ; cilia yellowish, with brown inner line. Body brown ; tegulæ and bands on abdomen white ; all the tibiæ and tarsi with white bands.

Expanse $1\frac{4}{10}$ inch.

Hab. Darjiling. In coll. Dr. Staudinger.

Family CRAMBIDÆ.

Genus SCHŒNOBIUS, *Dup.*

SCHŒNOBIUS BRUNNESCENS, n. sp.

Fore wing pale brownish-ochreous, silky ; with a minute black spot at lower end of the cell and one below its middle ; very faint traces of an oblique macular discal line, and a marginal row of points : hind wing whitish, apex slightly suffused with ochreous, and with

E 2

a more or less obsolescent brown macular line. Thorax, head, palpi, and fore legs brownish-ochreous; abdomen and hind legs whitish.

Expanse $\frac{9}{10}$ to 1 inch.

Hab. Calcutta. In coll. Dr. Staudinger and F. Moore.

Genus CRAMBUS, *Fabr.*

CRAMBUS AURIVITTATUS, n. sp.

Fore wing pure white, glossy; with a longitudinal golden-yellow band extending below the cell from the base to middle of the exterior margin, the upper edge of which along the median vein being most sharply defined; extreme apex of wing also slightly tipped with golden-yellow; a slender marginal line and an interciliary line golden-brown; four black dots on the marginal line at end of the band: hind wing cinereous-white. Thorax, head, and palpi white; tegulæ and fore legs brownish-ochreous; abdomen brownish-white.

Expanse 1 inch.

Hab. Darjiling. In coll. Dr. Staudinger.

Genus EROMENE, *Hübner.*

EROMENE TRIPUNCTATA, n. sp.

Fore wing pale brownish-ochreous; apex and exterior border reddish-ochreous; a black recurved line extending below the apex and partly along the costa, enclosing a white subapical spot; three black spots on middle of exterior margin; the marginal line also black; cilia cupreous-brown: hind wing brownish-white. Body, palpi, and legs pale brownish-ochreous.

Expanse $\frac{7}{10}$ inch.

Hab. Darjiling. In coll. Dr. Staudinger and F. Moore.

Genus CIRRHOCHRISTA, *Lederer.*

CIRRHOCHRISTA BRYOZALIS. (Plate VII. fig. 10, ♀.)

Margaronia bryozalis, Walker, Catal. Lep. Het. B. M. xix. p. 976 (1859), ♂.

Cirrhochrista ætherialis, Lederer, Pyral. Wien. Ent. Monat. 1863, p. 440, pl. 17. fig. 9, ♂.

Female. Pure white. Much larger than the male: fore wing with the costal and marginal brownish-ochreous dentate band broader, an oblique transverse streak extending from the subbasal point to the posterior margin, the preapical point extending to that below the apex, and the lower end of the marginal band broadly dilated, forming a quadrate patch at the posterior angle: hind wing with a much broader medial marginal oblique ochreous-brown streak. A dorsal brownish-ochreous band on thorax and abdomen of both sexes; labial and maxillary palpi of male white above and beneath, the sides being brownish-ochreous; pectus and legs white; fore tibiæ and tip of fore and middle femora brownish-ochreous; antennæ ochreous.

Expanse, ♂ $\frac{9}{10}$, ♀ $1\frac{1}{10}$ inch.

Hab. Darjiling. In coll. Dr. Staudinger and F. Moore.

Occurs also in China, Borneo, and E. Australia.

CIRRHOCHRISTA ACCIUSALIS.

Margaronia acciusalis, Walker, Catal. Lep. Het. B. M. xix. p. 977 (1859), ♂.

Pure white. *Female* much larger than the male: fore wing with the narrow brownish-ochreous costal band, the marginal band, and the transverse antemedial and postmedial slender indistinct line and cell-streak as in male: hind wing also with the two similar transverse lines as in male. Body white; sides of front, palpi, and fore femora and tibiæ, and bands on the tarsi brownish-ochreous.

Expanse, ♂ $1\frac{3}{10}$, ♀ $1\frac{4}{10}$ inch.

Hab. Darjiling. In coll. Dr. Staudinger and F. Moore.

Genus ESCHATA, *Walker*.

ESCHATA ARGENTATA, n. sp.

Much larger than *E. gelida*, Walker. Fore wing metallic shining silvery-white, crossed by a postmedial very slender golden-yellow irregular line, which curves outward towards the apex and then descends in a recurved wavy manner, followed by a similar-coloured almost erect wavy submarginal line; some minute black scales between the two lines; one or two small marginal spots above anal angle and two short slender streaks at the apex, the inner one curved; cilia at the apex and posterior angle golden yellow: hind wing less shining silvery-white. Thorax metallic white; abdomen above dull white, the two basal segments ochreous; head, palpi, and legs dull white; fore tarsi with ochreous bands.

Expanse $1\frac{9}{10}$ inch.

Hab. Darjiling. In coll. Dr. Staudinger.

ESCHATA CONSPURCATA, n. sp.

Male. Glossy silky-white: fore wing crossed by a very indistinct golden-yellow slender recurved postmedial and a submarginal line; some marginal small black lunular spots above the angle and one at the apex: hind wing fuliginous-brown to near the borders, with a distinct marginal black lunular line. Abdomen dusky-brown, base ochreous, tip white; labial and maxillary palpi with lateral black streak; fore legs and all the tarsi with ochreous bands.

Female. Fore wing with the yellow transverse lines obsolescent: hind wing slightly fuliginous-brown only at the base; tarsal and abdominal bands bright ochreous.

Expanse, ♂ $1\frac{8}{8}$, ♀ $2\frac{2}{8}$ inches.

Hab. Darjiling. In coll. Dr. Staudinger and F. Moore.

Family ENNOMIDÆ.

Genus CROCALIS, *Triet.*

CROCALIS SIMILARIA, n. sp.

Near to *C. lentiginosaria*. Smaller in expanse. Both wings of a darker brownish-ochreous: fore wing sparsely speckled with black scales and dusky strigæ, the latter being clustered across the middle; a white-centred discal spot and transverse blackish line similar, the latter with more prominent white points: hind wing differs in having a white-centred black spot at end of cell, and a blackish transverse slightly-waved discal line.

Expanse 1⅝ inch.

Hab. Darjiling. In coll. Dr. Staudinger and F. Moore.

Genus SELENIA, *Hübn.*

SELENIA DENTILINEATA, n. sp.

Olive-yellow. Wings with indistinct darker slender strigæ, crossed by a slender lilacine-white sinuous inwardly-oblique antemedial line and a postmedial line, the latter extending to near apex, before which it is bent acutely inward to the costa; the interior border of both lines darker olive-yellow, and the exterior line clouded with ochreous-red; some white marginal dentate markings; cilia reddish-ochreous, edged with white.

Expanse 1¾ inch.

Hab. Darjiling. In coll. Dr. Staudinger and F. Moore.

Allied to *S. calcearia* (*Hyperythra calcearia*, Walker, Catal. xx. p. 132).

Genus PERICALLIA, *Steph.*

PERICALLIA OLIVESCENS, n. sp.

Olive-brown. Wings with a few scattered minute black scales. Fore wing crossed by an indistinct waved postmedial brown line, before which is a semidiaphanous white spot situated between the base of middle and lower median vein: hind wing crossed by a similar indistinct brown waved line, before which is a large brown-speckled semidiaphanous white spot at end of the cell, and some smaller spots below it; two discal similar white spots, and a yellow lunule at anal angle. Cilia ochreous-brown, edged with white.

Expanse 1¼ inch.

Hab. Darjiling. In coll. Dr. Staudinger.

PERICALLIA SIKKIMA, n. sp.

Male. Dusky reddish greyish-brown. Wings more or less reddish basally, and with ill-defined reddish submarginal dentate patches; also crossed by numerous indistinct slender brown strigæ. Fore wing crossed by a blackish zigzag antemedial line, and an oblique postmedial straight duplex line, the latter curved inward to the costa on the fore wing, the angle being bordered within by a pale yellowish-speckled patch; costal edge streaked with white; a blackish spot at upper end of the cell: hind wing crossed by a medial black duplex line; a yellowish-speckled mark at end of the cell, above and below which is a transparent white spot. Collar, head, palpi, and legs dark purplish-grey.

Expanse 1½ inch.

Hab. Darjiling. In coll. Dr. Staudinger and F. Moore.

Genus ENDROPIA, *Guén.*

ENDROPIA ALDIFRONS, n. sp.

Male. Dark ferruginous-brown. Wings with numerous short transverse cinereous strigæ. Fore wing crossed by a cinereous-bordered blackish outwardly-curved antemedial line, and an angulated sinuous postmedial line, followed by a less-defined cinereous-bordered dentated submarginal line : hind wing with a cinereous-bordered blackish discal angulated sinuous line, and a less-defined dentated submarginal line. Cilia edged with cinereous. Underside ferruginous-red ; strigæ black ; outer transverse line prominent. Head white above ; vertex, palpi above, and fore legs ferruginous-brown ; palpi beneath and legs ferruginous-red.

Expanse 1⅜ inch.

Hab. Darjiling. In coll. Dr. Staudinger.

ENDROPIA ANTICLEATA, n. sp.

Female. Cinereous umber-brown. Fore wing crossed by a dark-brown sinuous subbasal line, a duplex antemedial and medial line, and a single postmedial line, the two latter lines indistinctly defined posteriorly, the outer line formed by vein-points below the subcostal vein, the lower veins being also contiguously lined with dark-brown and white points : hind wing paler cinereous-brown. Edge of collar, basal joint of antennæ, and terminal bands on legs whitish.

Expanse 1½ inch.

Hab. Darjiling. In coll. Dr. Staudinger.

Genus EPIONE, *Dup.*

EPIONE OBLIQUILINEA, n. sp.

Male. Ochreous-yellow. Wings with ochreous-brown strigæ, and an oblique ochreous-brown narrow band extending from apex to below middle of the abdominal margin ; the inner border of the band angulated, the outer border bounded by a slender straight cinereous-white line, beyond which the broad outer margin is clouded with cinereous-brown. Cell-spots ochreous-brown, with white centre. Fore wing also with an ill-defined sinuous antemedial ochreous-brown line. Collar, speckles on the abdomen and legs, and palpi above ochreous-brown.

Expanse 1 1/10 inch.

Hab. Darjiling. In coll. Dr. Staudinger and F. Moore.

EPIONE ADUSTATA, n. sp. (Plate VIII. fig. 20.)

Male. Wings reddish-ochreous, with cinereous-brown transverse strigæ, which are most numerous and confluent on the basal area, and across the outer area form a broad submarginal band ; the inner border of the latter is recurved and the outer border angulated ; the

medial area bordering the band and the outer margin being bounded by yellow lunules.
Cell-spots black. Body cinereous-brown; head, palpi, and legs ochreous-brown speckled.
Expanse 1 inch.

Hab. Khasia Hills. In coll. Dr. Staudinger.

Genus RUMIA, *Dup.*

RUMIA TRIDENTIFERA, n. sp.

Lemon-yellow. Fore wing marked with minute cinereous-brown speckles, which are
most numerous on the male; a large broad tridentate red spot at end of the cell, a sinuous
cluster of speckles at the apex, a subbasal and a basal costal spot, two small spots on
the posterior margin, and a row of spots on the exterior margin, all these spots being
more or less blotched with dark cinereous-brown; some red speckles also along the costal
border, and a curved discal row of brown vein-points: hind wing paler yellow, with a
cinereous-brown dentate spot at end of the cell, and an indistinct recurved discal denticulated line. Sides of thorax, front of head, palpi above, fore tibiæ, tip of femora and
tibiæ, and tarsi red; body and legs yellow.

Expanse $1\frac{7}{10}$ inch.

Hab. Darjiling. In coll. Dr. Staudinger and F. Moore.

Nearest to *R. mimulina*, Butler.

Genus CAUSTOLOMA, *Lederer.*

CAUSTOLOMA ACUTIPENNIS, n. sp.   (Plate VIII. fig. 7.)

Pale yellow. Both wings with some slender brownish-ochreous strigæ: fore wing
produced at the apex; with an ochreous-brown medial inwardly-oblique transverse irregular-shaped fascia, which ends posteriorly on the submedian vein and is dilated along
the costa to near the apex, and contains a pearly-white lunule at end of the cell, beyond
which is a submarginal recurved series of points: hind wing produced below the apex;
with an ochreous-brown quadrate cell-spot containing a pearly-white dot; beyond which
is a similar submarginal series of points. Front, palpi, and bands on fore legs ochreous-
brown.

Expanse $1\frac{1}{10}$ inch.

Hab. Darjiling. In coll. Dr. Staudinger.

PEETULA, n. g.

Allied to *Opisthographis* (*Rumia*, Dup.). Fore wing more regularly triangular; second
subcostal quadrifid; discocellular acutely bent in the middle; the middle median at one
seventh before end of the cell: hind wing convex exteriorly, the margin even. Antennæ
broadly bipectinated in male; palpi short, not extending beyond the front, broad, very
hirsute, apex minute, very obtuse.

PEETULA EXANTHEMATA, n. sp.

Dull ochreous-yellow. Wings with a few scattered cinereous speckles. Fore wing with a purplish-red basal patch, a small medial costal patch, and a large angulated apical patch, the latter clouded with cinereous-brown ; a small cinereous-brown ringlet at end of the cell ; some red speckles on the costal border, and three submarginal red patches, of which the upper and lower are the largest : hind wing with a cinereous-brown clouded red patch at the apex, and another on the middle of the abdominal margin ; also some smaller red submarginal patches. Thorax, bands on abdomen, head, palpi, pectus, and fore legs above purplish-red ; antennæ brown, its basal joint and base of the shaft being white.

Expanse 2 inches.

Hab. Darjiling. In coll. Dr. Staudinger and F. Moore.

Genus CORYMICA, *Walker*.

CORYMICA CAUSTOLOMARIA, n. sp.

Male. Dull yellow. Wings with numerous short brownish-ochreous transverse strigæ. Fore wing with a semidiaphanous vesicle at base of the submedian ; crossed by an excurved oblique slight brownish-ochreous antemedial band, a prominent medial band, an apical patch, and a patch before the posterior angle ; from the upper patch a curved row of points extends to the posterior margin ; hind wing crossed by a brownish-ochreous medial band, a less-defined narrow submarginal band, and a marginal band. A dark brown dot at end of each cell ; cilia dark brown, edged with cinereous; costal edge of fore wing black-and-white speckled. Collar, palpi, and speckles on the legs brownish-ochreous.

Expanse $1\frac{1}{10}$ inch.

Hab. Darjiling. In coll. Dr. Staudinger.

Family OXYDIIDÆ.

Genus MARCALA, *Walker*.

MARCALA FLAVITUSATA, n. sp. (Plate VIII. fig. 6.)

Male. Purplish-ochreous. Wings with a few indistinct darker strigæ. Fore wing with the basal third purplish-ochreous, bordered by an inwardly-oblique purple-brown line ; the medial area paler and suffused with sulphur-yellow towards the apex ; the posterior angle with a darker purple-brown patch ; a cinereous-centred spot at end of the cell, and one at the apex : hind wing with the costal border yellowish ; crossed by a slender pale purplish-brown medial line. Underside ochreous-yellow : fore wing sulphur-yellow towards the apex, with reddish basal blotches, a patch at posterior angle, and a spot at the apex : hind wing with several small scattered red spots. Collar, base of abdomen, palpi, fore and middle tibiæ and tarsi purplish-ochreous ; body and legs ochreous-yellow.

Expanse $1\frac{4}{10}$ inch.

Hab. Darjiling. In coll. Dr. Staudinger.

F

MARCALA OBLIQUARIA, n. sp.

Pale purplish ochreous-brown. Wings crossed by a broad pale ochreous oblique medial band, which has a dark purple-brown bordering line, the outer line being recurved to the costa at its upper end; basal and outer area sparsely traversed by indistinct short darker brown strigæ; a pale-centred brown spot at end of cell of the fore wing. Thorax, front of head, palpi, tip of abdomen, and fore legs pale purplish-brown; base of abdomen ochreous. Expanse $1\frac{2}{10}$ inch.

Hab. Darjiling. In coll. Dr. Staudinger and F. Moore.

MARCALA IRRORATA, n. sp.

Reddish-ochreous. Wings uniformly speckled with short brown strigæ: fore wing crossed by an oblique antemedial and a postmedial red-brown line; the inner line dilated at the costal end, the outer line bent acutely inward at the apical end; a small white-centred lunate brown spot at end of cell: hind wing pale ochreous along costal border, and crossed by a lower discal brown line. Front of head, tip of palpi, fore legs and tarsi brown. Expanse, ♀ $1\frac{6}{10}$ inch.

Hab. Darjiling. In coll. Dr. Staudinger.

Genus OXYDIA, *Guén.*

OXYDIA VULPINARIA, n. sp.

Female. Ferruginous-brown. Both wings with the outer border broadly darker brown; the discal area densely covered and the basal area sparsely speckled with greyish-white scales; crossed by a slender greyish-white scaled oblique antemedial line, and a marginal line, that on the fore wing bent inward to the costa. Underside reddish-ochreous, with scattered cinereous-brown speckles.

Expanse $2\frac{3}{4}$ inches.

Hab. Darjiling. In coll. Dr. Staudinger.

Family ŒNOCHROMIIDÆ.

Genus DECETIA, *Walker.*

DECETIA PALLIDA, n. sp. (Plate VIII. fig. 1.)

Male. Very pale brownish-ochreous. Wings very indistinctly marked with numerous slender brown strigæ; crossed by an oblique ochreous-brown line extending from the apex to above middle of the abdominal margin. Fore wing also with the discal area clouded with darker brown strigæ, a pale-centred spot at end of the cell, and an ill-defined submarginal sinuous line; hind wing with an impressed pale semidiaphanous streak beyond end of the cell, beneath the radial veinlet. Cilia ochreous-brown. Branches of antennæ blackish. Underside ochreous, with the brown line across both wings.

Expanse $1\frac{1}{2}$ inch.

Hab. Calcutta. In coll. Dr. Staudinger.

Genus NOREIA, *Walker.*

NOREIA FLAVA, n. sp. (Plate VIII. fig. 2.)

Female. Bright yellow. Wings crossed by a discal reddish diffused line, which is obsolescent at the costal end. Fore wing also with a cinereous-speckled reddish oval spot at end of the cell, and a streak below the apex. Underside paler, with the discal band broader; spots also broader. Side of palpi, antennæ, and fore legs above brownish.

Expanse 1½ inch.

Hab. Darjiling. In coll. Dr. Staudinger.

NOREIA CERVINARIA, n. sp.

Reddish cinereous-brown. Wings with numerous short dark-brown strigæ. Fore wing crossed by an inwardly oblique antemedial and a postmedial reddish-brown bordered blackish line, the former being bent inward at the costal end; a black spot at end of the cell : hind wing crossed by a discal reddish-bordered black line. Underside paler than above ; the inner line on fore wing absent; the outer line less defined.

Expanse, ♂ 1$\frac{5}{10}$, ♀ 1$\frac{7}{10}$ inch.

Hab. Darjiling. In coll. Dr. Staudinger and F. Moore.

Genus AUZEA, *Walker.*

AUZEA RETICULATA, n. sp. (Plate VIII. fig. 3.)

Female. Brownish-ochreous. Both wings crossed by very slender, more or less continuous brown strigæ. Fore wing with an oblique yellow broad fascia extending from the apex to middle of the posterior margin, where it is nearly obsolescent : hind wing with the exterior margin broadly yellow. Underside marked as above. Body, head, palpi, and legs brownish-ochreous ; fore tibiæ and tarsi blackish.

Expanse 1$\frac{3}{10}$ inch.

Hab. Darjiling. In coll. Dr. Staudinger.

Genus CIMICODES, *Guén.*

CIMICODES FLAVA, n. sp. (Plate VIII. fig. 5.)

Female. Ochreous-yellow. Wings crossed by a few very indistinct slender brown strigæ, an olive-brown oblique line extending from apex of fore wing to middle of the abdominal margin, the apical end being white-speckled ; a short outwardly-oblique olive-brown costal streak above end of the cell ; base of the costal edge brown-speckled. Collar, head, and palpi above brown ; palpi beneath and legs yellow.

Expanse 1$\frac{8}{10}$ inch.

Hab. Cherra Punji. In coll. Dr. Staudinger.

CIMICODES SANGUIFLUA, n. sp. (Plate VIII. fig. 4.)

Female. Wings pale ochreous-red, crossed by numerous short delicate pink strigæ; a pale yellow oblique band extending from apex of fore wing to the middle of abdominal

F 2

margin, the band being broadly dilated to the costa and traversed its entire length by a pale
olive-green line with some white speckles at its apical end. Cilia pale olive-green. Costal
border of hind wing yellowish. Body yellow; thorax and base of abdomen with a few pink
speckles; collar, head, palpi above, and antennæ brown; palpi beneath and legs yellow;
fore and middle tibiæ and tarsi above pinkish.

Expanse 1 ⁶⁄₁₀ inch.

Hab. Cherra Punji (*October*). In coll. Dr. Staudinger.

Genus SARCINODES, *Guén.*

SARCINODES LILACINA, n. sp.

Female. Pinkish-cinereous; crossed by a slender yellow line, which extends from apex
of fore wing obliquely across the hind wing to middle of abdominal margin, faint traces of a
slender brown inner line, and a submarginal series of whitish points; costal edge of fore
wing and cilia red. Body pinkish-cinereous; front of thorax, palpi, fore legs, and tarsi pale
brown.

Expanse 2⅔ inches.

Hab. Cherra Punji (*October*). In coll. Dr. Staudinger.

Family AMPHIDASIDÆ.

Genus AMPHIDASIS, *Treit.*

AMPHIDASIS REGALIS, n. sp.

Whitish. Wings with numerous short transverse dark brown strigæ; crossed by a
prominent thick black undulated antemedial line, and an irregular angulated postmedial
line; inner line broadly bordered within and outer line irregularly without by dark ochreous-
brown; medial area also more or less clouded with pale ochreous-brown, and crossed by
more or less defined zigzag series of conjoined strigæ, which are most apparent in the female
and on the underside. Abdomen brown-speckled and with white dorsal spots and basal
black band; hind part of thorax ochreous-brown; front of head black; legs black-speckled;
tarsi with black bands.

Expanse, ♂ 2½, ♀ 4 inches.

Hab. Darjiling. In coll. Dr. Staudinger and F. Moore.

Genus CUSIALA, *Moore.*

CUSIALA BOARMIOIDES, n. sp.

Male. Pale brownish-white, speckled with brown: fore wing crossed by a dark brown
wavy antemedial line and an irregular sinuous postmedial line, the former bordered inwardly
and the latter outwardly by a sinuous fascia formed by the clustered brown speckles, beyond
which is a similar submarginal fascia and marginal row of black points; a slender black
lunular streak at end of cell: hind wing with similar lunular cell-streak, postmedial sinuous
line, which is angulated beyond the cell, its bordering and submarginal sinuous brown

fascia; marginal points black. Body brown-speckled; thorax in front, head, and legs pale brownish; palpi dark brown.

Female whiter; the brown speckles more prominent; the transverse lines, marginal spots, and cell-streak black.

Expanse, ♂ 2⅛, ♀ 2¾ inches.

Hab. Darjiling. In coll. Dr. Staudinger and F. Moore.

Genus CHORODNA, *Walker.*

CHORODNA ADUMBRATA, n. sp.

Allied to *C. vulpinaria*. *Male.* Differs from the same sex of that species in its smaller size : fore wing with the two discal contiguous ill-defined denticulated lines curved outward to the posterior margin, are more obsolescent anteriorly, and is traversed by a broad blackish fascia from the angle below the apex to near base of posterior margin; the discal spot is about half the size: hind wing with the medial fascia crossing at some distance before the discal spot. Underside with the corresponding differences apparent.

Expanse 2⅜ inches.

Hab. Darjiling. In coll. Dr. Staudinger and F. Moore.

Family BOARMIIDÆ.

Genus MEDASINA, *Moore.*

MEDASINA SIMILIS, n. sp.

Allied to *M. interruptaria*. *Male.* Fore wing longer, exterior margin more oblique; the pale costal band with long dark brown equally-disposed transverse strigæ; the lower dark brown strigose area with sharply-defined upper edge; a short pale medial line, and a continuous pale submarginal line : hind wing bidentate below the apical angle; discal line less sinuous, submarginal line continuous.

Expanse 2⅝ inches.

Hab. Darjiling. In coll. Dr. Staudinger.

MEDASINA DISSIMILIS, n. sp.

Male. Wings pale testaceous, of the same shape as *M. interruptaria* : fore wing slightly but irregularly clouded with darker testaceous-brown, and with regularly-disposed transverse black strigæ, some of which are joined on the costa and there form three or four short broad streaks; discal sinuous black line and submarginal pale line indistinctly defined : hind wing with the short black strigæ from the base to the sinuous discal line, and more slender strigæ along the exterior border, disposed as in *M. interruptaria*.

Expanse 2¾ inches.

Hab. Darjiling. In coll. Dr. Staudinger.

MEDASINA PERSIMILIS, n. sp.

Male. Allied to *M. interruptaria.* Brighter ferruginous: fore wing with the broad pale costal band marked by a few slender dark brown strigæ, disposed only along the edge, and with a pale-bordered lunular oblique medial line, the submarginal pale waved line being somewhat further from the outer margin : hind wing not angulated at the middle of the exterior margin ; medial line much less sinuous and not pale-bordered ; the submarginal pale line continuous.

Expanse 2⅝ inches.

Hab. Darjiling. In coll. Dr. Staudinger and F. Moore.

MEDASINA FRATERCULA, n. sp.

Male. Much smaller than *M. interruptaria,* being nearly one third less in expanse ; the dark brown strigæ uniformly disposed throughout : fore wing with the broad pale costal band marked by short brown strigæ only along the costal edge ; the discal sinuous line diffused, black, sharply defined, more oblique, and disposed nearer the outer margin, the outer border being glaucescent ; a series of submarginal white dots instead of the pale waved line : hind wing not angulated on the outer margin, the discal sinuous line sharply defined, and its outer lower border white-speckled, followed by a submarginal row of white dots.

Expanse 1⅝ inch.

Hab. Darjiling. In coll. Dr. Staudinger and F. Moore.

CALICHA, n. g.

Male. Fore wing rather narrow, elongate, triangular ; exterior margin scalloped ; cell half the length ; first subcostal about one fifth before end of cell, second approximate, trifid, fifth from end of cell ; discocellular bent below the middle, radial from above the angle ; middle median at one fifth, lower at half before end of cell : hind wing broad, exterior margin convex, scalloped ; cell more than half the length ; first subcostal near end of cell ; discocellular bent above middle ; the middle median at one fifth and lower at two fifths before end of cell. Body stout ; thorax clothed with thick compact lax scales ; palpi porrect, short, stout, not extending beyond the front ; antennæ in male broadly bipectinated, the branches ciliated ; in female setaceous ; legs stout, squamous ; spurs stout, rather long.

CALICHA RETRAHENS, n. sp.

Male and female. Dark olive-brown, minutely black-speckled : fore wing crossed by six or seven inwardly-oblique black sinuous diffused lines, each of which is indistinctly bordered by a whitish chalybeous line ; a reddish triangular patch below the costa and on the posterior margin bordering the discal line ; a reddish streak also at posterior angle : hind wing with four or five similar lines with chalybeous border, the discal line also reddish-bordered.

Expanse 2¾ inches.

Hab. Darjiling. In coll. Dr. Staudinger and F. Moore.

Genus MENOPHRA, *Moore.*

MENOPHRA DEFICIENS, n. sp.   (Plate VIII. fig. 23.)

Male. Lilacine ferruginous-brown, paler towards the base ; sparsely speckled with dark brown : fore wing crossed by an antemedial and a postmedial recurved row of small black points, a less-defined marginal row of points, and a medial recurved ferruginous line ; a cluster of blackish speckles at end of cell, and another below the apex : hind wing with a ferruginous medial line, a discal row of pale-tipped black points, and less-defined marginal points towards the apex.

Expanse 1⅝ inch.

Hab. Darjiling.   In coll. Dr. Staudinger.

MENOPHRA TORRIDARIA, n. sp.   (Plate VIII. fig. 27.)

Male. Ferruginous-brown, with numerous short slender brown strigæ : fore wing crossed by a duplex acutely-angular antemedial line, and a curved postmedial similar line ; both lines ferruginous along the centre, the former being acutely bent inward from lower end of cell to near base of posterior margin, and thence extending across base of hind wing, the outer line extending from the apex across the disk ; interbasal and submarginal area tinged with cinereous ; the medial area and a patch on middle of exterior margin pale yellowish-ferruginous ; some white dentate spots below the apex ; a black dot at end of cell, a cluster of speckles near posterior angle, and marginal points : hind wing with the medial area pale yellowish-ferruginous ; a white line extending across the outer border ; a black dot at end of cell and marginal row of points. Thorax, head, palpi, and fore legs dark cinereous-brown ; abdomen pale ferruginous-brown.

Expanse 1⅜ inch.

Hab. Darjiling.   In coll. Dr. Staudinger.

MENOPHRA CONSPICUATA, n. sp.

Pale ochreous-brown. Wings with slender transverse strigæ ; marginal line dark brown. Fore wing crossed by a slightly-curved dark brown denticulated antemedial and a recurved postmedial line, both dilated at the costal end, and the outer line also at its posterior end ; between the two lines is a medial erect line, which is dilated and partly includes a lunule at end of the cell ; beyond is a submarginal broken pale-bordered sinuous fascia : hind wing with a very indistinct similar subbasal and discal line, outer pale-bordered fascia, and cell-spot. Palpi and fore legs with brown bands ; middle and hind legs brown-speckled.

Expanse 1⁷⁄₁₆ inch.

Hab. Darjiling.   In coll. Dr. Staudinger and F. Moore.

MENOPHRA PALLIDARIA, n. sp.

Male and female. Dull pale brownish-ochreous. Wings brownish-white about the middle, with numerous delicate brown strigæ ; marginal line brown, slender. Fore wing crossed by an excurved brown diffused antemedial line, an angulated medial, and a denticulated postmedial line ; the middle line partly including a lunule at end of the cell : hind wing with less-defined slender subbasal and discal denticulated line. Body, middle and hind legs brown-speckled; palpi and fore legs with brown bands.

Expanse $1\frac{6}{10}$ inch.

Hab. Darjiling. In coll. Dr. Staudinger and F. Moore.

MENOPHRA? VIALIS, n. sp. (Plate VIII. fig. 9.)

Male. Yellowish-testaceous; the basal and outer area of both wings traversed by numerous short black strigæ; the medial area obliquely crossed by a broad yellowish band: fore wing also crossed by a black antemedial sinuous line, and a submarginal diffused black sinuous fascia, which is bordered by a slender white outer line: hind wing with a similar but less defined submarginal fascia and white line. Body brown-speckled; abdomen with a basal white band; palpi and bands on legs brown; anal tuft pale testaceous.

Expanse $1\frac{1}{4}$ inch.

Hab. Darjiling. In coll. Dr. Staudinger.

HIRASA, n. g.

Near to *Hemerophila*. Wings shorter; exterior margins less scalloped: fore wing with the first subcostal about one third before end of cell, second and third close to end, second anastomosed to first for a short distance from the base, third trifid; lower median at nearly two fifths before end of cell. Antennæ of male thickish, naked.

Type *H. scripturaria.*

HIRASA SCRIPTURARIA.

Tephrosia scripturaria, Walker, Catal. Lep. Het. B. M. xxxv. p. 1590 (1866).

Hab. Darjiling. In coll. Dr. Staudinger.

HIRASA CONTUBERNALIS, n. sp.

Much darker cinereous-brown than the allied *H. scripturaria,* with the transverse lines thinner and less defined: fore wing with the inner line similar, the outer line recurved and hardly sinuous, and with an inner contiguous darker brown fascia, whereas in *H. scripturaria* it is acutely sinuous towards the upper end; there are no white dentate marks below the apex and the cell-spot is larger: hind wing with the line regularly waved throughout, and with an inner contiguous dark brown fascia, not curved from its lower end and angulated on the subcostal, as in *H. scripturaria.*

Expanse $1\frac{6}{8}$ inch.

Hab. Khasia Hills (*Atkinson*); Shillong (*Austen*). In coll. Dr. Staudinger and F. Moore.

Genus CLEORA.

Cleora, Curtis, Brit. Ent. ii. fol. 88 (1825); Stephens, Catal. Brit. Ins. ii. p. 123 (1829); id. Illust. Haust. iii. p. 181 (1831).

Fore wing elongate, more triangular than in *Alcis* (*A. repandaria*), exterior margin even; first subcostal at one sixth before end of cell, second approximate, both free, third close to end of cell, trifid, sixth from end of cell; discocellular slender, waved, radial from its middle; the middle median close to end of cell, lower at one third : hind wing short, triangular, apex convex, exterior margin almost even; cell more than half the length, first subcostal close to end; discocellular slightly oblique; middle median close to end, lower about one third. Body somewhat stouter and shorter than in *A. repandaria*; palpi shorter and broader, laxly clothed; antennæ broadly bipectinated to near tip, branches slender and finely ciliated; legs squamous, hind tibiæ incrassated, spurs moderately long, slender.

Type *C. cinctaria.*

CLEORA CONIFERA, n. sp.

Female. Yellowish-ochreous, sparsely speckled with purple-brown strigæ : fore wing crossed by a purple-brown waved antemedial and an irregular recurved discal line, a similar coloured streak along middle of the inner line, and an irregular-shaped white-bordered submarginal fascia composed of conical-shaped portions, the middle portion being the largest and the lowest obsolescent; below the apex is also a longitudinal angular streak; a streak at end of cell, one above it, and a marginal row of points : hind wing broadly ochreous-white along costal border; crossed by a short medial and discal line, and a white-bordered similar fascia; a slight spot at end of cell, and a marginal pointed line. Fore legs with brown bands, middle and hind legs brown-speckled.

Expanse $1\frac{3}{10}$ inch.

Hab. Darjiling. In coll. Dr. Staudinger and F. Moore.

CLEORA DUPLEXA, n. sp.

Fore wing yellowish-ochreous, speckled with purplish-brown ; crossed by two partly-confluent curved antemedial and two similar waved discal purplish-brown denticulated bands, followed by a submarginal series of dentate spots, and two short streaks below the apex, a spot at end of the cell, and one above it on the costa : hind wing ochreous-white, the exterior border and cilia pale ochreous; a faint brown spot at end of cell, a curved discal line, some streaks at and above anal angle, and a slender marginal pointed line. Body yellowish-ochreous ; palpi and bands on legs purple-brown.

Expanse $1\frac{3}{10}$ inch.

Hab. Darjiling. In coll. Dr. Staudinger and F. Moore.

CLEORA IDÆOIDES, n. sp.

Cinereous-white, speckled with pale ochreous-brown ; crossed by a slight blackish denticulated pale ochreous-brown bordered antemedial and discal line, and a submarginal similar brown fascia traversed by a sinuous whitish line ; a slight black streak at end of the cells, and a marginal pointed line. Body speckled with brown scales ; palpi and bands on fore legs brown.

Expanse, ♂ ♀, 1 inch.

Hab. Darjiling. In coll. Dr. Staudinger and F. Moore.

CLEORA OCHRIFASCIATA, n. sp.

Male and female. Pale ochreous, very sparsely speckled with minute brown scales, which are more numerous in the female; crossed by three ochreous-yellow fasciæ, which are traversed by the sinuous antemedial irregular-angulated discal and submarginal sinuous black-speckled lines; a small black spot at end of the cell, and marginal row of points; costal border of hind wing cinereous. Body black-speckled, banded with ochreous-yellow; palpi brown; fore legs and tarsi with brown bands.

Expanse, ♂ 1, ♀ 1$\frac{1}{10}$ inch.

Hab. Darjiling. In coll. Dr. Staudinger and F. Moore.

CLEORA IRRORATA, n. sp.

Male and female. Whitish. Numerously covered with minute cinereous-black scales and short strigæ: fore wing crossed by five ochreous-yellow irregular fasciæ; an antemedial and submarginal sinuous black-speckled diffused line, a postmedial and marginal series of black points, and a small mark at end of cell: hind wing with three slight ochreous-yellow fasciæ, the two outer bordering the medial and discal series of black points, which in the female are black-speckled diffuse sinuous lines; a small black mark at end of cell, and marginal points. These markings are more prominent in the female. Body with ochreous bands and black speckles; palpi and bands on legs brown.

Expanse 1$\frac{2}{10}$ inch.

Hab. Darjiling. In coll. Dr. Staudinger and F. Moore.

Genus PSEUDOCOREMIA.

Pseudocoremia, Butler, Proc. Zool. Soc. 1877, p. 394.

Fore wing elongate, triangular, apex obtusely pointed; cell nearly two thirds the length; first subcostal at one sixth, base slightly touching the costal near its end, second contiguous, free, third from close to end, trifid, sixth from end of cell; discocellular concave, radial from middle; the middle median close to end of cell, lower more than one third: hind wing broad, apex and exterior margin convex, even; cell more than half the length; two subcostals from end; discocellular bent in middle, radial from angle; middle median near end, lower at more than one third. Body slender, long; palpi porrect, laxly clothed beneath, second joint extended entire length beyond head, third joint minute; antennæ broadly bipectinated to near end, the branches ciliated; legs long, compactly squamous, spurs long.

Type *P. fragosata.*

PSEUDOCOREMIA VARIEGATA, n. sp.

Male. Ochreous. Wings clouded with brown strigæ and numerously speckled with white scales: fore wing crossed by a black-speckled basal, antemedial, and a submarginal fascia, and an angular discal black-pointed line, each bordered by a slender white-speckled line; a black lunule at end of cell, and a marginal row of points; a reddish-ochreous shade beyond and below the cell: hind wing with the costal border pale ochreous, a discal and submarginal black sinuous line with white-speckled outer line and reddish-ochreous border-ing shade ; a slight black lunule at end of cell and a marginal line. Body speckled with white and brown scales; abdomen with a white basal band; palpi and bands on legs brown. Expanse $1\frac{3}{16}$ inch.

Hab. Darjiling. In coll. Dr. Staudinger and F. Moore.

PSEUDOCOREMIA SEMIALBA, n. sp.

Allied to *P. pannosaria.* Fore wing cinereous-brown, with a few scattered black slender strigæ ; veins also slightly black-speckled ; an ill-defined blackish-speckled basal patch, and a slender streak at end of cell ; a transverse antemedial indistinct black wavy line dilated at each end, a discal sinuous line which mostly show only the points, followed by a submarginal more prominent sinuous white-bordered black fascia, the outer border having a slight testaceous patch below the apex and in the middle : hind wing cinereous-white, with a slight cinereous-brown cell-spot, a row of discal dots, and a broad marginal band traversed with pale streaks. Body cinereous-brown ; tegulæ blackish ; bands on legs and abdomen whitish.

Expanse $1\frac{4}{16}$ inch.

Hab. Darjiling (*Atkinson*); Khasia Hills (*Austen*). In coll. Dr. Staudinger and F. Moore.

PSEUDOCOREMIA IMBECILIS, n. sp.

Male. Fore wing brown, numerously white-speckled ; base blackish-speckled ; crossed by a discal slightly-curved series of black dentate points, which are sinuously bordered on the inner side with white and by a testaceous sinuous line on the outside ; a submarginal white-bordered black sinuous fascia, and a marginal row of points ; cilia alternated with white ; apex and middle of exterior border slightly pale testaceous : hind wing cinereous-white, with a slight brown spot at end of cell, and some brown-speckled strigæ from the anal angle and along the outer margin. Body brown, with whitish bands ; legs with brown bands.

Expanse $1\frac{3}{8}$ inch.

Hab. Darjiling. In coll. Dr. Staudinger.

PSEUDOCOREMIA ALBIFERA, n. sp.

Female. Fore wing pale testaceous, black-speckled, and with a darker transverse ill-defined basal, subbasal, and medial fascia, followed by a denticulated discal line, and an interrupted submarginal white sinuous line ; a large white costal patch between the medial fascia and discal line, and a similar testaceous-white patch disposed obliquely below it from the discal line to the exterior margin ; marginal points black ; cilia alternated with white : hind wing pale brownish-cinereous, with traces of a darker sinuous recurved discal line ; marginal line dark brown. Body brownish-cinereous ; thorax and abdomen speckled with testaceous and black scales ; palpi and bands on legs brown.

Expanse $1\frac{5}{16}$ inch.

Hab. Darjiling. In coll. Dr. Staudinger and F. Moore.

o 2

Genus ALCIS.

Alcis, Curtis, Brit. Ent. iii. fol. 113 (1825); *id.* Guide, p. 157 (1829); Stephens, Catal. B. Ins. ii. p. 124; *id.* Illust. Haust. iii. p. 184 (1831).

Deileptenia (part), Hübn. Verz. p. 316.

Boarmia (part), Treit.

Male. Fore wing elongate, triangular; exterior margin uneven; cell fully half the length; first subcostal at one eighth, second and third contiguous, third trifid, sixth from end of cell; discocellular concave, outwardly oblique, radial from middle; the middle median near end of cell, lower at two fifths: hind wing rather long, apex very convex, exterior margin scalloped; cell more than half; first subcostal near end; discocellular bent in the middle; the middle median near end of cell, lower at one third. Body long, extending beyond hind wings, slender; antennæ long, broadly bipectinated to near tip, the branches finely ciliated; palpi ascending to vertex, clothed with long lax hairy scales in front, third joint very short, thick, obtusely conical; legs long, squamous; fore tarsi very long; hind tibiæ very long, stout, sheathing a basal tuft of fine long hairs beneath, hind tarsi less than half its length; tibial spurs short.

Type *A. repandaria.*

ALCIS VAGANS, n. sp.   (Plate VIII. fig. 16.)

Pale ochreous-brown, partly olivaceous-whitish about the disk; numerously covered with slender brown transverse strigæ: fore wing crossed by an antemedial darker brown duplex zigzag line, and a similar pale-centred recurved denticulated postmedial line, beyond which is a pale zigzag submarginal line, and a marginal row of black points; a pale-centred oval spot at end of cell: hind wing whitish along the costa; an ill-defined darker brown lunular spot at end of cell, discal pale-bordered denticulated line, submarginal white zigzag line, and blackish marginal points. Body brown-speckled; front of head, palpi, and bands on fore legs dark brown. Underside whitish, marked with brown strigæ; a darker brown submarginal band, with whitish marginal patches on the fore wing.

Expanse 2¼ inches.

Hab. Darjiling. In coll. Dr. Staudinger.

ALCIS SIKKIMA, n. sp.

Male. Olivaceous umber-brown. Wings very sparsely flecked with short slender transverse strigæ: fore wing crossed by an inwardly-oblique waved duplex diffused blackish antemedial line, and a duplex postmedial sinuous pale-centred line, beyond which is an incomplete pale-bordered sinuous fascia and marginal lunular line; between the two lines is a recurved line encompassing a large oval spot at end of cell: hind wing with a blackish-speckled inner line, a lunule at end of cell, a whitish lunular-bordered sinuous discal line, incomplete pale-bordered submarginal sinuous fascia, and marginal black lunular line. Thorax with black bands; palpi at the side and legs with black bands.

Expanse 2⅗ inches.

Hab. Darjiling. In coll. Dr. Staudinger.

Alcis vicina, n. sp.

Male. Brownish-ochreous: fore wing sparsely flecked with short slender brown strigæ, which are regularly disposed transversely along the costa and more irregularly on the basal area and the exterior border; crossed by a slender black antemedial sinuous line, and an angulated postmedial line, between which is a medial zigzag line, and beyond a pale-bordered blackish sinuous submarginal fascia, and a marginal lunular line; a black lunule at end of cell: hind wing sparsely speckled with short blackish transverse strigæ; crossed by a straight black antemedial line, recurved discal line, submarginal pale-bordered sinuous fascia, and a prominent marginal line, the two former being obsolescent on the anterior margin; a lunular spot at end of cell. Thorax, and abdomen above, and palpi with blackish-speckled bands; fore legs with darker bands.

Expanse 2 inches.

Hab. Darjiling. In coll. Dr. Staudinger and F. Moore.

DARISA, n. g.

Nearest to typical *Boarmia* (*B. consonaria*). Wings larger, similar in pattern above : fore wing with the costal vein bifid near its end; first subcostal about one sixth, free, second trifid, fifth from end of cell; discocellular concave, radial from middle; the middle median close to end, lower at nearly one half: hind wing longer; cell half the length; subcostal and middle median close to end of cell, lower median at one third. Body long; thorax laxly clothed; abdomen with lateral tufts; palpi obliquely porrect, broad; antennæ broadly bipectinated to near tip, the branches stout and densely ciliated.

Type *D. mucidaria* (*Boarmia mucidaria*, Walk. Suppl. p. 1581).

DARISA MAXIMA, n. sp.

Male and female. Brownish-ochreous. Wings very sparsely speckled with black scales and short strigæ, which are most apparent externally : fore wing crossed by an antemedial and postmedial slender indistinct black sinuous line, beyond which is a greyish black-speckled sinuous interrupted fascia, with whitish outer border; the two former lines and the outer margin bordered by a brighter ochreous-brown black-speckled fascia, and a similar fascia crosses the middle; a slight dusky lunule at end of cell : hind wing thickly black-speckled across the base; a slender black discal sinuous line, an ill-defined submarginal blackish-speckled pale-bordered sinuous fascia, both being outwardly-bordered by brighter ochreous-brown. Side of palpi and bands on fore and middle legs blackish; hind legs speckled.

Expanse, ♂ 2⅝, ♀ 2⅚ inches.

Hab. Darjiling. In coll. Dr. Staudinger and F. Moore.

ASTACUDA, n. g.

Nearest *Catoria.* *Male.* Fore wing more regularly triangular, apex somewhat pointed, exterior margin oblique; first subcostal bifid near its end (single in the female), very slightly touching the costal near the end, third trifid ; discocellular almost straight, slightly oblique ; middle median close to end of cell, lower at half : hind wing broader, cell half the length ; middle median at one eighth, lower at half.　Body stout ; palpi obliquely porrected, narrower, smooth ; antennæ bipectinated to one third the tip, the branches very long, stout, and densely ciliated ; legs stouter.

Type *A. cineracea.*

ASTACUDA CINERACEA, n. sp.

Dark cinereous, with a pale purplish tint ; brown-speckled : fore wing crossed by a slender blackish antemedial and a postmedial pale cinereous-white bordered sinuous line, between which is a similar-shaped speckled fascia, and beyond a submarginal pale-bordered sinuous fascia, and a marginal row of black points ; each line dilated at the costal end ; a blackish lunule at end of cell : hind wing crossed by a medial blackish-speckled fascia, encompassing a lunular cell-mark ; a sinuous discal line, a pale-bordered fascia, and marginal points.　Front of head, palpi, and fore legs above dark brown.

Expanse, ♂ 2¾, ♀ 2¼ inches.

Hab. Darjiling.　In coll. Dr. Staudinger and F. Moore.

ASTACUDA AMPLA, n. sp.

Male. Differs from *A. cineracea* in being of a pale brownish colour ; markings similarly disposed, but diffused and less defined, the pale borders white ; the medial sinuous fascia being partly obsolescent and its upper end placed immediately above and confluent with the cell-spot ; cell-spot on both wings also broadly oval (not lunate).　On the hind wing the discal sinuous line is situated nearer end of the cell.　*Female* more distinctly and thickly brown-speckled on a whiter ground-colour ; the markings more diffused and less defined than in same sex of *A. cineracea,* the pale borderings also whiter.　Underside paler, whiter, with indistinct brown strigæ ; discal line broken and composed of strigæ ; marginal band paler, the band on hind wing being narrow and submarginal.

Expanse, ♂ 2, ♀ 2¾ inches.

Hab. N.E. Bengal (*Grote*) ; Cherra Punji (*Atkinson*).　In coll. Dr. Staudinger and F. Moore.

Genus CATORIA, *Moore.*

CATORIA OLIVESCENS, n. sp.

Differs from *C. sublavaria* (*Boarmia sublavaria,* Guén.) in being of a pale dull olivaceous colour, instead of white ; markings similarly disposed.　On the fore wing the cell-spot is larger, the postmedial series of points less dentate, the submarginal and marginal series of spots quite round.　On the hind wing the cell-spot is also much larger, oval in shape, and has a pale lunular centre, the inner speckled fascia is broader, and the outer series of markings more distinctly rounded.　On the underside the black marginal band is broader, and the black cell-spot on fore wing is lunular and extends entirely across end of the cell.

Expanse 1⅝ inch.

Hab. Darjiling.　In coll. Dr. Staudinger and F. Moore.

Genus CHOGADA, *Moore.*

CHOGADA FRATERNA, n. sp.

Nearest to the Ceylonese *C. alienaria.* Wings longer; mostly pale brownish-ochreous, with numerous short dark brown slender strigæ : fore wing crossed by an inwardly-curved antemedial black sinuous line, and a more distinct white-bordered postmedial line, the inner line bordered and the outer margin traversed by dark ochreous-brown sinuous fasciæ; submarginal line white; a dentated lobate lunule at end of cell : hind wing with a brown-speckled medial fascia encompassing a white lunule at end of cell; a white-bordered black sinuous discal line; outer border ochreous-brown, traversed by white sinuous streaks. Thorax, abdomen, palpi, fore and middle legs with dark brown bands ; hind legs brown speckled.

Expanse 1⅞ inch.

Hab. Darjiling. In coll. Dr. Staudinger and F. Moore.

BURICHURA, n. g.

Allied to *Chogada.* Wings larger, broader, venation similar : fore wing more triangular. Thorax and abdomen more robust. Antennæ with short very fine cilial tufts ; palpi stouter, shorter, third joint very short and obtuse, joints almost connected.

Type *B. imparata.*

BURICHURA IMPARATA.

Boarmia imparata, Walker, Catal. Lep. Het. B. M. xxi. p. 372 (1860).

Hab. Darjiling. In coll. Dr. Staudinger and F. Moore.

AMRAICA, n. g.

Wings ample. *Male.* Fore wing elongate, triangular, exterior margin slightly scalloped ; cell fully half the length; first subcostal near to end of cell, bifid from near the base, third very close to end, trifid, sixth from end of cell ; discocellular inwardly concave, radial from the middle ; the middle median close to end of cell, lower at two fifths: hind wing broad, short, exterior margin convex, scalloped ; cell more than half; subcostals from end ; discocellular outwardly concave, bent below the middle, radial from the angle ; middle median close to end of cell, lower nearly one third. Body very stout ; palpi porrect, very short, thick, not projecting beyond the head, third joint minute, obtusely conical ; antennæ broadly pectinated to one third the tip on outer side only, the branches ciliated, the inner side with very short tufted ciliæ ; legs stout, coarsely squamous ; spurs rather stout.

Female. Wings longer: fore wing less triangular. Antennæ stout, simple.

AMRAICA FORTISSIMA, n. sp.

Pale ochreous-brown : fore wing crossed by a blackish zigzag antemedial line, and a less-defined similar postmedial line, followed by a submarginal whitish zigzag line ; a large dusky oval spot at end of cell ; basal area, an irregular fascia across the middle, a fascia outside the postmedial line, the apex and marginal border darker ochreous-brown : hind wing with a large oval dusky spot at end of cell, a diffused sinuous discal fascia, and outer margin dark ochreous-brown, the latter bordered by the sinuous whitish line. Thorax above dark ochreous-brown; collar whitish; front of head, palpi, and fore legs dark brown.

Expanse, ♂ 2⅝, ♀ 3¾ inches.

Hab. Darjiling. In coll. Dr. Staudinger and F. Moore.

Note.—*Boarmia ponderata*, Feld., pl. 125. f. 18, is an allied species.

LASSABA, n. g.

Fore wing moderately short, broad, triangular; apex obtuse; cell more than half the length; costal bifid near end; first subcostal about one sixth before end of cell, second contiguous, trifid, fifth from end of cell; discocellular outwardly-oblique, concave, radial from the middle; the middle median at one sixth, lower at half: hind wing broad, apex convex, exterior margin dentated; cell more than half the length; discocellular outwardly-oblique, radial from the middle; first subcostal and middle median each very near end of cell, lower median nearly half. Body moderately slender; palpi porrect, projecting one third beyond the head, clothed with very lax hairy scales beneath; antennæ in male very broadly bipectinated to one third the tip, branches ciliated; antennæ in female setaceous; legs long, slender, spurs slender.

LASSABA CONTAMINATA, n. sp.

Male and female. White: fore wing with a few dark brown strigæ disposed mostly along the costa, on the veins, and along the exterior border; crossed by a dark brown-speckled narrow zigzag antemedial band, an incomplete postmedial, and a submarginal band, the space between the two latter being clouded with ferruginous-brown; a black streak at end of cell, which is contiguous to a broader costal streak above it and a paler streak obliquely below it: hind wing sparsely flecked with brown; a blackish dot at end of cell; crossed by a discal row of blackish points, the outer border clouded with ferruginous-brown; marginal lunular slender line, black. Tip of palpi and bands on fore legs dark ferruginous-brown; hind legs black-speckled.

Expanse, ♂ 2⅛, ♀ 2⅗ inches.

Hab. Darjiling (*Atkinson*); Chumbi Valley, Sikkim (*Elwes*). In coll. Dr. Staudinger and F. Moore.

Genus GNOPHOS, *Treit.*

GNOPHOS GRANITARIA, n. sp.

Greyish-brown, palest in the male; numerously speckled with black scales: fore wing crossed by an indistinct black-speckled antemedial duplex sinuous line, a postmedial series of points, a medial zigzag fascia, and a more prominent submarginal grey-bordered sinuous fascia; a black streak at end of the cell: hind wing with a less distinct black-speckled cell-spot, discal points, and pale-bordered submarginal fascia. These markings on both wings are bordered by a pale ochreous shade, which is most distinct across the disk. *Female* darker, more thickly and blacker scaled, the transverse markings, though blacker, are more diffused.

Expanse 1¾ inch.

Hab. Sind Valley, Kaschmir; Chamba. In coll. Dr. Staudinger and F. Moore.

GNOPHOS TEPHROSIARIA, n. sp.

Male. Brownish-white. Wings with minute short brown strigæ; slightly brownish clouded across the disk. Fore wing crossed by an incomplete very ill-defined antemedial, medial, discal, and a submarginal blackish denticulated line, followed by a marginal row of points; each line most apparent and dilated at the costal end; middle line including a lunule at end of the cell: hind wing with less-defined slender similar denticulated lines, and a marginal pointed line. Underside white, with narrow cinereous medial band including the cell-spot, and a broad outer submarginal band.

Expanse 1⅝ inch.

Hab. Darjiling. In coll. Dr. Staudinger.

Genus PINGASIA, *Moore.*

(Syn. HYPOCHROMA, part, *Guén.*)

PINGASIA RUFOFASCIATA, n. sp.

White. Both wings with the basal area very sparsely speckled with minute ochreous scales; crossed by a slightly-recurved postmedial slender black denticulated line, followed by two regularly-formed red-scaled broad sinuated fasciæ; marginal line and cell-streak slender, black: fore wing also with a slender black antemedial line. Underside with a narrow fuliginous-brown submarginal band, that on the fore wing slightly touching the outer margin below the apex; cell-streak also distinct. Body ochreous-speckled; thorax and head reddish in front; a black line across the vertex; palpi above and fore legs above with brown bands.

Expanse 1⅞ inch.

Hab. Darjiling. In coll. Dr. Staudinger and F. Moore.

Larger than *P. ruginaria,* Guén.; differs from it in the postmedial line being situated nearer the outer margin, and not undulated, the fasciæ above being red, and both entire; whereas in *P. ruginaria* the outer fascia is interrupted, and the band on the underside is also about half the width. It is also distinct from *P. irroraturia,* which has a regularly-formed sinuous postmedial line.

PINGASIA APICALIS, n. sp.

Male. Olivaceous-green (fading to ochreous-yellow); numerously speckled with very short black strigæ: fore wing crossed by an indistinct diffused blackish erect wavy antemedial line, and a prominent black sinuous postmedial line; a black lunule at end of the cell, and a marginal row of narrow spots; subbasal interspace paler, the medial area and outer discal area clouded with purplish-brown; a prominent white patch at the apex, and some minute discal points: hind wing crossed by a prominent black sinuous postmedial line, clouded beyond by purplish-brown; a black lunule at end of cell, and narrow marginal spots. Cilia alternated with purplish-brown, and partly white-edged. Body above, palpi, and fore legs brown-speckled.

Expanse 2 inches.

Hab. Darjiling. In coll. Dr. Staudinger.

Nearest to *P. vigens,* Butler (Types Lep. Het. B. M. vi. pl. 116. f. 3).

PINGASA SIMILIS, n. sp.

Cinereous olivaceous-green, fading to cinereous-ochreous or whitish-cinereous; medial and discal area crossed by slender rather long purplish-brown strigæ; marginal line slender, brown: fore wing with the costal border irregularly purplish-brown; crossed by a slender waved brown antemedial line, and a slightly sinuous discal line with prominent points; a dark purple-brown lunule at end of cell, a patch below the apex, and a streak from the posterior angle: hind wing crossed by a brown slender sinuous discal line, with contiguous outer purplish-brown patches.

Expanse $1\frac{3}{8}$ to $1\frac{7}{8}$ inch.

Hab. Darjiling. In coll. Dr. Staudinger and F. Moore.

Nearest to *P. costistrigaria.*

DINDICA, n. g.

Fore wing rather short, narrow, exterior margin slightly oblique and convex, posterior margin long; cell fully half the length; first subcostal about one eighth, second from end of cell, quadrifid, sixth also from the end; discocellular bent before the middle, radial from below the angle; middle median close to end, lower nearly half before end: hind wing short, very broad, somewhat quadrate, apex very convex, exterior margin dentate, abdominal margin long; cell half the length at upper end; two subcostals from end; discocellular outwardly-oblique, bent inwards near upper end, radial from below the angle; two upper medians from end of cell, lower at nearly one third before the end. Body moderately stout, thorax clothed with long lax spatular-tipped woolly hairs: abdomen with lax dorsal tufts of spatular hairs; palpi porrect, broad, compressed, laxly clothed beneath, third joint decumbent, short, thick; antennæ long, in male bipectinated to one third the tip, the branches short and ciliated; legs smooth, long, spurs stout.

Type *D. basiflavata.*

DINDICA BASIFLAVATA.

Hypochroma basiflavata, Moore, Proc. Zool. Soc. 1867, p. 632.

Hab. Darjiling. In coll. Dr. Staudinger and F. Moore.

DINDICA LEOPARDINATA.

Hypochroma leopardinata, Moore, Proc. Zool. Soc. 1867, p. 634.

Hab. Darjiling. In coll. Dr. Staudinger and F. Moore.

Genus PACHYODES, *Guén.*

PACHYODES PICTARIA, n. sp.

Male. Fore wing olivaceous-green, the basal and medial area with darker strigæ; costal border with short purple-black streaks and a broad triangular dentate-bordered streaked patch before the apex; a slight similar patch at the posterior angle; a slender black angular streak at end of cell, and a marginal slender lunular line: hind wing white; outer border partly olivaceous-green; a purplish-black speckled interrupted discal fascia, which is bordered at the anal end by a brighter patch; a black lunular streak at end of cell, and marginal line. Thorax and head olivaceous-green; abdomen cinereous-white; front of head and bands on legs purple-black.

Expanse 2 inches.

Hab. Darjiling. In coll. Dr. Staudinger and F. Moore.

PACHYODES ORNATARIA, n. sp.

Male. Differs from *P. hæmataria* (II.-Schœff. Exot. Schmett. f. 205) on the fore wing in the two basal lines being nearer together and disposed obliquely outward, as well as being more irregular; the purple-brown strigæ more broadly dispersed, the marginal elongated spots longer: hind wing with brown strigæ clustered at the base, cell-streak much shorter; the black medial discal streak and the succeeding points very prominent, that on the lower median and internal vein forming a band to the margin; the longitudinal red subanal streak with black centre; marginal streaks also broader. On the underside the fore wing is thickly speckled along the costal border, and there is no lower basal and discal black spot. On the hind wing, in addition to the three anterior spots, there is a small black discal spot midway between the cell and outer margin; a lengthened subanal longitudinal spot and two superposed anal spots. All the legs are dark brown; the thoracic and abdominal crests are bell-shaped, with black edges.

Expanse 2 to 2½ inches.

Hab. Darjiling (*Atkinson*); Cherra Punji (*Austen*). In coll. Dr. Staudinger and F. Moore.

Family GEOMETRIDÆ.

Genus GEOMETRA, *Linn.*

GEOMETRA TUMIDILINEA, n. sp.

Pale glaucescent-green. Both wings crossed by a cinereous purplish-brown discal sinuated line, which is outwardly swollen in the middle of its length and at the lower end; marginal line and a slender streak at end of the cell also purplish-brown. Cilia pale purple with whitish inner line. Fore wing also with an erect slender antemedial line, and a few speckles along the costal border. Abdomen, front, and fore legs purplish-brown.

Expanse 1½ inch.

Hab. Darjiling. In coll. Dr. Staudinger and F. Moore.

GEOMETRA BISERIATA, n. sp.

Male. Glaucescent-green. Fore wing with an antemedial series of three small purplish-brown speckled vein-spots, and a recurved postmedial series of similar spots, the penultimate upper and the lowest being the largest, and the latter white-speckled; some minute speckles also along the costal edge, and a slender spot at end of the cell: hind wing with a purple-brown and white-speckled cell-spot, one also on middle of the abdominal margin; some minute speckles scattered about the discal area, and a row of marginal points; cilia cinereous-white. Front, fore legs above, and antennæ purplish-brown.

Expanse 1 8⁄10 inch.

Hab. Darjiling. In coll. Dr. Staudinger.

Genus THALASSODES, *Guén.*

THALASSODES LUNIFERA, n. sp.

Olive-green. Wings angulated in the middle of the exterior margin. Fore wing crossed by a darker curved subbasal and a recurved postmedial sinuous line; a small white-centred brown spot at end of the cell; a cinereous-speckled purple-brown patch and adjacent cilia at the posterior angle, and some subapical speckles above it: hind wing with a darker recurved medial sinuous line, the upper end of which is brown-speckled; a brown spot, centred with a white lunule, at end of the cell; some marginal brown speckles below the apical angle.

Expanse $1\frac{7}{10}$ inch.

Hab. Cherra Punji. In coll. Dr. Staudinger.

Genus AGATHIA, *Guén.*

AGATHIA? DIVARICATA, n. sp. (Plate VIII. fig. 15.)

Olive-green. Fore wing crossed by a narrow ochreous-brown outwardly-recurved ante-medial band, which meets a similar almost erect postmedial band, the latter being bordered by two partly confluent outer bands, and followed by a marginal band, all being posteriorly confluent: hind wing with a similar coloured narrow band curving from near base of the abdominal margin to a transverse recurved discal band, beyond which are two partly-confluent lower outer bands, the upper end of the latter merging into a cinereous ochreous-brown apical patch, bordered by a darker marginal band. All the bands traversed by a cinereous-whitish line.

Expanse $1\frac{3}{8}$ inch.

Hab. Cherra Punji. In coll. Dr. Staudinger.

Family EPHYRIDÆ.

Genus ANISODES, *Guén.*

ANISODES LUNULOSA, n. sp. (Plate VIII. fig. 8.)

Ferruginous, sparsely black-speckled: fore wing crossed by a yellowish basal speckled band, the medial area numerously traversed by confluent yellow strigæ, beyond which is an irregular recurved discal series of well-defined yellow lunular spots, of which the upper is apical and large: hind wing with the basal area traversed by confluent yellow strigæ, and crossed by a discal series of well-defined yellow lunules ; a black dot at end of each cell. Tip of palpi and fore legs brown.

Expanse $1\frac{5}{8}$ inch.

Hab. ——? In coll. Dr. Staudinger.

Genus ARGIDAVA, *Walker.*

ARGIDAVA IRRORATA, n. sp.

Male and female. Pale cinereous-ochreous. Fore wing numerously speckled with minute olivescent-brown scales; crossed by a subbasal, antemedial, discal, and a submarginal row of brown points, and a marginal row of linear spots; cell-mark also linear: hind wing with very faint traces of a discal and a submarginal olivescent-brown fascia. Legs speckled with olivescent-brown.

Expanse, ♂ $1\frac{2}{10}$, ♀ $1\frac{4}{10}$ inch.

Hab. Darjiling. In coll. Dr. Staudinger and F. Moore.

Family IDÆIDÆ (ACIDALIDÆ, *Auct.*).

BARDANES, n. g.

Male. Fore wing triangular, broad; with an elongated lappet or fold on the upperside, extending from the base of the costa to near the apex, its end being spatulate and very coarsely scaled, the interior surface of both the lappet and the costa being smooth, nacreous, and enclosing a compact series of hairy scales the whole length between the costal and subcostal veins; cell short, only two fifths the length; costal vein forming the anterior edge of the fold; first subcostal at one fifth before end of cell, quadrifid, fifth close to the end and joined to first for a short distance above its base, sixth also from end of cell; discocellular slightly oblique, radial from its middle; the middle median close to end of cell, lower at one third: hind wing short, broad; exterior margin very convex; cell very short, one fourth the length; two subcostals at one fifth beyond end of cell; discocellular very slender, erect, radial from its middle; two upper medians at one fifth beyond the cell, lower close to the end. Body short, rather stout; palpi porrect, slender, smooth, extending a little beyond the front; antennæ biciliated; legs rather stout, long; middle spurs and two pairs on hind tibiæ long and slender.

BARDANES PLICATA, n. sp. (Plate VIII. fig. 22.)

Male. Ochreous-yellow. Both wings crossed by ill-defined ochreous-red subbasal sinuous narrow bands, and five outer but more prominent bands, and marginal pointed line; the outer bands on the fore wing and the costal fold being of a cinereous purplish-ochreous. Collar, front, and palpi, and bands on abdomen ochreous-red; legs above cinereous-ochreous.

Expanse 1 inch.

Hab. Darjiling. In coll. Dr. Staudinger.

Genus HYRIA, *Steph.*

HYRIA UNDULOSATA, n. sp.

Red. Fore wing with the costal border and a broad erect medial band cinereous-red and slightly black-scaled; the band with undulated outer border; basal area crossed by yellow wavy lines, and the outer area by undulated lines; the median and submedian veins with several blackish points; marginal points black: hind wing with the basal half cine-rous-red, the border angulated; outer area with yellow wavy lines; the veins with black points; marginal points black.

Expanse 1 inch.

Hab. Darjiling. In coll. Dr. Staudinger.

Nearest allied to *H. bicolorata.*

RUNECA, n. g.

Male. Fore wing triangular, apex pointed; cell half the length; first subcostal at two fifths before end of cell, second at one fourth, quadrifid, slightly touching the first near its base, sixth from end of cell; discocellular recurved, radial from its middle; the middle median very close to end of cell, lower at two fifths: hind wing broad, rather quadrate; cell fully half the length; two subcostals from end of cell; discocellular recurved, radial from its middle; the middle median very close to end of cell, lower at nearly half. Body slender; abdomen with lateral tufts; antennæ finely bipectinated to tip; palpi slender, porrect, extending a little beyond the head; fore and middle legs smooth, tarsi long, middle spurs slender; hind tibiæ long, flattened, densely fringed along the outer edge, spurs not visible, tarsi short.

RUNECA FERRILINEATA, n. sp. (Plate VIII. fig. 13.)

Ochreous-white. Wings with a few minute scattered brown scales. Fore wing with the costal and outer border tinged with ferruginous-red; crossed by an antemedial and post-medial outwardly-curved brown-scaled ferruginous-red lines, which are obsolescent in their middle, beyond which is a slender submarginal denticulated line, and a prominent black straight marginal line; a minute black dot at end of the cell: hind wing with a brown-scaled ferruginous diffused medial line, two less-defined denticulated submarginal lines, and prominent black marginal line; cell-spot black. Cilia ferruginous-red. Collar and bands on abdomen ferruginous-red; front, palpi, fore and middle legs brownish-ochreous; palpi black-tipped.

Expanse $\frac{8}{10}$ inch.

Hab. Darjiling. In coll. Dr. Staudinger.

Has much the aspect of a *Drapetodes.*

Genus IDÆA, *Treit.*

IDÆA UNDULATARIA, n. sp.

Male and female. Cinereous-white. Wings speckled with minute brown scales. Fore wing crossed by an inwardly-oblique antemedial, medial, discal, and two submarginal undulated cinereous-ochreous narrow fasciæ, which are more thickly brown-scaled; cell-spot and marginal points black: hind wing with a medial, discal, and two submarginal similar undulated fasciæ, cell-spot, and marginal points. Body brown-speckled; front, fore femora, and tibiæ dark brown.

Expanse, ♂ $\frac{8}{10}$, ♀ $\frac{9}{10}$ inch.

Hab. Darjiling. In coll. Dr. Staudinger and F. Moore.

IDÆA ALBOMACULATA, n. sp.

Male and female. Olivescent cinereous-ochreous. Both wings with a darker inwardly-oblique discal fascia, and a submarginal series of prominent white spots; a black point at end of each cell. Front, palpi, and fore legs above dark brown.

Expanse 1 to $1\frac{1}{10}$ inch.

Hab. Darjiling. In coll. Dr. Staudinger and F. Moore.

IDÆA? ALBOSIGNATA, n. sp.

Cinereous-white. Fore wing numerously speckled with minute ochreous-brown scales, excepting upon a small oval space at lower end of the cell, which mostly form several darker brown-speckled contiguous sinuous diffused lines, and a marginal lunular line; the outer lines having intervening ochreous borders, and the veins with white and blackish points: hind wing less brown-speckled; crossed by two discal and two submarginal brown-speckled ochreous diffused lines, with blackish vein-points, followed by a distinct marginal lunular line. Cilia with inner brown line. Thorax, head, tip of palpi, and bands on fore legs brown; middle and hind legs brown-speckled; abdomen whitish.

Expanse 1 inch.

Hab. Darjiling. In coll. Dr. Staudinger and F. Moore.

Genus CRASPEDIA, *Hübn.*

CRASPEDIA KASHMIRENSIS, n. sp.

Pure white. Both wings with a very few minute black scales scattered about the basal area; crossed by a medial ill-defined cinereous-ochreous diffused sinuous line, and a slender black sinuous discal line, which is outwardly bordered by a darker cinereous-ochreous macular fascia, beyond which is a similar lunular fascia, and a blackish slender marginal pointed line; cell-spot minute, distinct, black: fore wing also with an antemedial sinuous less-defined cinereous-ochreous line. Front dark brown; fore and middle legs pale brown.

Expanse $\frac{9}{10}$ to 1 inch.

Hab. Kashmir. In coll. Dr. Staudinger and F. Moore.

Nearest allied to the European *C. contignaria.*

CRASPEDIA STIGMATA, n. sp.

Pale ochreous. Wings with a few minute scattered black scales. Fore wing with a transverse excurved wavy indistinct brownish antemedial line, a recurved sinuous medial and discal line, followed by two less-defined sinuous submarginal fasciæ and a marginal row of dots; the inner line with minute black vein-points, the discal and outer line with a subapical and lower intervening prominent ochreous-red patch, the sinuous edges of which are deep black; a minute blackish cell-spot: hind wing with a medial and discal sinuous indistinct line, outer submarginal fasciæ, black marginal dots, and a cell-spot below the inner line. Front, palpi above, and fore legs above dark purplish-brown; palpi beneath and pectus white.

Expanse, ♂ ♀ $1\frac{2}{10}$ inch.

Hab. N.W. India; Solun. In coll. Dr. Staudinger and F. Moore.

CRASPEDIA PERSIMILIS, n. sp.

Nearest to *C. similaria* ('Aid,' pl. 151. fig. 2), from Darjiling. Antennæ of male more broadly and stronger bipectinated. Wings violaceous-ochreous, numerously covered with minute black scales. Fore wing with an indistinct transverse antemedial series of black points, a less-defined cell-streak, and a recurved medial fascia, a more prominent recurved discal series of black points, followed by two ill-defined submarginal sinuous speckled fasciæ, which are most visible posteriorly, and a marginal row of points: hind wing with a similar indistinct cell-streak, medial fascia, discal points, submarginal fasciæ, and marginal pointed line. Underside clearer violaceous-ochreous, with darker cell-streak, medial and discal sinuous line. Front, palpi, and fore legs dark purplish-brown.

Expanse 1½ inch.

Hab. N.W. India: Chumba, Dharmsala, Umballa, Solun. In coll. Dr. Staudinger and F. Moore.

Genus LUXIARIA, *Walker.*

LUXIARIA INTENSATA, n. sp.

Male. Wings ochreous-yellow; almost covered with transverse purple-brown strigæ, and crossed by a darker subbasal, medial, and a broader sinuous discal fascia, the latter bordered by an inner series of darker points. Front, palpi, fore and middle legs above purple-brown.

Expanse 1¼ inch.

Hab. Darjiling. In coll. Dr. Staudinger.

LUXIARIA FASCIOSA, n. sp.

Male and female. Yellowish-ochreous. Wings numerously flecked with slender transverse ochreous-brown strigæ. Fore wing with a clouded purplish ochreous-brown recurved narrow antemedial and medial, and a broader postmedial and submarginal fascia, the latter broken below the apex, and inwardly bordered by a row of darker brown points: hind wing with a narrow medial and broad postmedial and submarginal similar clouded fascia, with inner row of darker brown points. Front, palpi, and fore legs above ochreous-brown.

Expanse 1⅘ inch.

Hab. Darjiling. In coll. Dr. Staudinger and F. Moore.

LUXIARIA OBLIQUATA, n. sp.

Male and female. Whitish-ochreous. Wings sparsely speckled with minute ochreous-brown scales. Fore wing crossed by two inwardly-oblique diffused brownish-ochreous antemedial lines, two medial partly confluent lines, and two similar submarginal lines; the inner medial line with blackish vein-points: hind wing with a single basal, medial, and double discal and submarginal similar diffused lines; the inner discal line being edged with blackish points. Front, tip of palpi, and legs above brown.

Expanse 1⅘ inch.

Hab. Darjiling; Cherra Punji. In coll. Dr. Staudinger and F. Moore.

Genus TIMANDRA, *Dup.*

TIMANDRA RESPONSARIA, n. sp.

Smaller than *T. convectaria*. Fore wing less triangular. Pale brownish-ochreous. Wings numerously speckled with darker minute strigæ; the oblique transverse fascia extending from the apex across both wings less linear, brownish, diffused in male, indistinct in female; the outer slender line on fore wing more wavy, that on the hind wing being obsolescent in the female; cell-spot on fore wing indistinct.

Expanse 1 inch.

Hab. Cherra Punji (October). In coll. Dr. Staudinger and F. Moore.

TIMANDRA ALBIFRONTATA, n. sp.

Female. Purplish cinereous-brown. Both wings crossed by an inwardly-oblique diffused purple-brown postmedial line; cilia purple-brown, edged with white. Fore wing also with a similar parallel antemedial line, some darker costal speckles, and a dot at end of the cell. Vertex above and base of antennæ white; abdomen above purple-brown.

Expanse $\frac{7}{8}$ inch.

Hab. Darjiling. In coll. Dr. Staudinger.

Genus TRYGODES, *Guén.*

TRYGODES FERRIFERA. (Plate VIII. fig. 17.)

Female. White. Fore wing crossed by a broad basal, antemedial, and a postmedial sinuous ferruginous-red speckled band, and a narrow marginal band: hind wing with traces of a similar medial, and a more-defined broad discal, and narrow marginal band. Front and fore legs above brown; body ferruginous-speckled.

Expanse $1\frac{1}{10}$ inch.

Hab. Darjiling. In coll. Dr. Staudinger.

EMODESA, n. g.

Male. Fore wing rather short and broad; costa fringed, much arched at the base, apex very acute, exterior margin convex in the middle; cell disposed towards the middle, more than half the length; costal and subcostal widely separated, both straight from the base to near end; first subcostal at fully two fifths before end of cell, trifid, fourth from end of cell and slightly touching second close to its base, sixth also from end of cell; discocellular outwardly-oblique, radial from angle very close to lower end; middle median at one fourth, lower at three fifths: hind wing short, broad, quadrate, exterior margin obtusely angular in the middle; cell more than half the length, disposed along the middle; first subcostal at two thirds before end of cell, joined to costal from near its base to near its end; disco-cellular outwardly-oblique, radial from angle close to lower end; middle median at one fourth, lower at half. Cilia long. Body slender, very short; palpi slender, porrect, hirsute beneath; front smooth, flat; antennæ pectinated on one side only, the branches broad to one third the tip; legs rather short, smooth; spurs slender, hind tibiæ with one pair of spurs only.

Allied to *Auzata.*

I

EMODESA SINUOSA, n. sp. (Plate VIII. fig. 18.)

White. Fore wing with the costal fringe cinereous; crossed by three subbasal, two discal, and two submarginal cinereous-speckled sinuous lines, the discal lines diffused across the middle, and the outer submarginal line lunular; beyond which is a blackish-pointed marginal line: hind wing crossed by a similar subbasal, two diffused discal, and two submarginal lines, and marginal points. Cilia alternated with cinereous. Abdomen with a broad cinereous subbasal band; branches of antennæ brownish.

Expanse $\frac{9}{10}$ inch.

Hab. Khasia Hills (October). In coll. Dr. Staudinger.

AGNIBESA, n. g.

Allied to *Somatina.* Fore wing more regularly tringular; apex not subfalcate; exterior margin more oblique, posterior margin shorter; cell nearly half the length; first subcostal at one sixth before end of cell, quadrifid, fifth from end of cell, bifid, slightly touching the first halfway between its base and second, the sixth being emitted below at halfway beyond end of cell and its juncture with the first; discocellular outwardly recurved, radial from above its middle; two upper medians from end of cell, lower at nearly one third: hind wing short, exterior margin very convex, sinuous: cell half the length; two subcostals on a footstalk one third beyond the cell; discocellular outwardly oblique, radial from above the middle; two upper medians from end of cell, lower nearly one third. Body slender; palpi porrect, slender; antennæ setaceous; legs long, slender; hind tibiæ slender, spurs moderately long.

Type *A. pictaria.*

AGNIBESA PICTARIA.

Somatina pictaria, Moore, Proc. Zool. Soc. 1867, p. 645.

Hab. Darjiling.

AGNIBESA RECURVILINEATA, n. sp.

Cinereous-white. Fore wing crossed by a much-recurved basal, subbasal, and medial ochreous-brown line, two discal angulated diffused lines, and a less-defined sinuous submarginal line; the medial and discal lines are mostly bright ochreous and partly confluent at their upper ends; a small brown spot at end of the cell; upper marginal line lunular, brown: hind wing with less-defined similar wavy diffused lines, and cell-spot. Front, palpi, and fore legs above ochreous.

Expanse $1\frac{1}{10}$ inch.

Hab. Darjiling. In coll. Dr. Staudinger and F. Moore.

Genus RAMBARA, *Moore.*

(ZANCLOPTERYX, part, *Gven.*)

RAMBARA DENTIFERA, n. sp.

Larger than *R. saponaria.* White. Both wings with a recurved discal series of brown dentate marks, the three upper on the fore wing being black, the third large and prominent;

marginal black spots distinct ; a brown-edged tridentate lunular mark at end of each cell.
Tip of palpi and fore tibiæ cinereous-brown.

Expanse $1\frac{2}{10}$ inch.

Hab. Darjiling. In coll. Dr. Staudinger and F. Moore.

Family MICRONIIDÆ.

Genus ACROPTERIS.

Acropteris, Hübner, Geyer, Zuträge, iv. p. 36, fig. 761 (1832).
Micronia (part), Guen., Walker.

Male. Fore wing triangular, costa slightly arched, apex acute, exterior margin straight ;
cell one third the length, fusiform ; first subcostal at one third before the end, second at
one sixth, trifid towards the end ; fifth from the end, bifid near its base ; discocellular
slender, bent inward near upper end, the lower end much lengthened and extending inward
to one fourth from base of the cell ; radial from angle near upper end ; median vein very
short, inflated, the three branches recurved and starting from the end, the second forked
near its base ; submedian much recurved from the base, with a slender short lower branch ;
slender folds between the veins prominent. Cell in female of normal shape ; the median
branches also in their normal positions : hind wing short, quadrate, anterior margin lobate
at the base ; exterior margin obtusely angular in the middle ; cell two thirds the length ;
first subcostal near end of cell ; discocellular extremely slender, scarcely perceptible, bent
near upper end, radial from the angle ; middle median very close to end, lower at one half.
Body slender, shorter than the hind wings ; palpi very slender, obliquely porrect, second
joint short, third joint very long, linear ; antennæ flattened, with short, stout, broad, barely
separated sinuations ; legs rather short, stout ; middle tibiæ clothed with long fine silky
hairs above, spurs long ; hind tibiæ short, inflated, hairy above, with two pairs of spurs, the
outer short, tarsi short, thick.

Type *A. grammearia,* Hübn.

ACROPTERIS STRIATARIA.

Phal. Geom. striataria, Linn. Syst. Nat. ii. 1, p. 859 (1767) ; Clerck, Icones, pl. 55. f. 4, ♂ .
Phalæna striataria, Fabr. Ent. Syst. iii. 2, p. 131.
Micronia striataria, Guen. Phal. ii. p. 28 ; Walker, Catal. Lep. Het. B. M. xxiii. p. 819.

Hab. Darjiling.

Genus PSEUDOMICRONIA.

Pseudomicronia, Moore, Lep. Ceylon, iii. p. 461.

PSEUDOMICRONIA CÆLATA, n. sp.

Micronia caudata, Walker, Catal. Lep. Het. B. M. xxiii. p. 817 (*nec* Fabricius).

White. Fore wing crossed by five pale cinereous-brown slender strigose bands, of
which the three outer are duplex, and narrower at the costal end than in *P. fraterna* :

I 2

hind wing with four transverse pale cinereous-brown fasciæ, and a submarginal strigose fascia, which is duplex anteriorly; marginal slender line and two caudal spots black.

Expanse 2 inches.

Hab. Darjiling (*Atkinson*); Bombay (*Swinhoe*); S. India (*Mulhouse*). In coll. British Museum, Dr. Staudinger, and F. Moore.

MICRONIDIA, n. g.

Male. Fore wing elongated, rather narrow; apex convex; cell nearly half the length; first subcostal nearly one eighth before end of cell, bifid near end, first slightly touching the costal near the end, third from angle close to end of cell, trifid, sixth from end of cell; discocellular deeply concave, outwardly-oblique, radial from above its middle; the middle median from angle close to end of cell, lower at one third. Hind wing short, broad, somewhat quadrate, exterior margin very convex below the middle; cell half the length; first subcostal close to end; discocellular deeply concave, outwardly-oblique; no radial; middle median from angle close to end of cell, lower at one third. Body slender, short; palpi small, porrect, apex very acute; antennæ setaceous, flattened at the base; legs squamous; hind tibiæ incrassated, with two pairs of short slender spurs.

Type *M. simpliciata.*

MICRONIDIA SIMPLICIATA.

Micronia simpliciata, Moore, P. Z. S. 1867, p. 646.

Hab. Darjiling (*Atkinson*).

DITRIGONA, n. g.

Male. Fore wing triangular; cell more than half the length; first subcostal at one fifth before end of cell, second at one sixth, bifid close to end, fourth from end of cell, bifid near the base, the fourth being bent and slightly touching second at nearly half its length; discocellular bent inward near upper end and outward near lower end, radial from lower angle; middle median at one fifth, lower at half. Hind wing short, triangular, much prolonged hindward, and with a small spatular angle; cell two fifths the length at upper end, and half at lower end; first subcostal at half before end of cell; discocellular outwardly oblique, bent near lower end, radial from the angle; middle median at one fifth, lower at two fifths. Body short; palpi very small, slender, ascending, closely applied, apex pointed; antennæ bipectinated to one third the tip, the branches short, thickish, and slightly ciliated; legs slender, smoothly squamous; middle tibiæ with one pair, and hind tibiæ with two pairs of slender spurs.

DITRIGONA TRIANGULARIA.

Urapteryx triangularia, Moore, P. Z. S. 1867, p. 612.

Hab. Darjiling.

URAPTEROIDES, n. g.

Wings of similar form and pattern to *Urapteryx.* Fore wing broader and less triangular than in *Strophidia* (*S. fasciata*, Cram.); cell nearly one third the length, very broad and triangular; first subcostal at one fourth before end of cell, second and third at equal distances from first and end of cell, third bifid at half its length, fifth and sixth on a footstalk at nearly half beyond end of the cell; discocellular slender, outwardly-oblique, very long, radial from its middle; two upper medians on a footstalk at one fifth beyond the cell, lower close to end; submedian with a short slender lower branch: hind wing more prolonged and narrow hindward than in *Strophidia*; exterior margin more oblique, concave above the caudal angle; cell less than one third the length; first subcostal close to end; discocellular extremely slender, outwardly-oblique, bent above the middle, radial from the angle; two upper medians on a footstalk at about one third beyond the cell, lower very close to the end. Body rather stout, anal tuft conchiform; palpi obliquely porrect, reaching a little beyond the front, laxly squamous beneath, apex short, pointed; antennæ flattened, setaceous; legs rather stout; hind tibiæ not thickened, with two pairs of long stout spurs.

Type *U. astheniata.*

URAPTEROIDES ASTHENIATA.

Micronia astheniata, Guén. Phal. ii. p. 24 (1857); Walker, Catal. Lep. Het. B. M. xxiii. p. 821 (1861); Westwood, Trans. Zool. Soc. 1879, p. 514.

Hab. Borneo (*Wallace*); Darjiling (*Atkinson & Farr*).

Family EROSIIDÆ.

Genus DIRADES, *Walker.*

DIRADES RUPTARIA, n. sp.

White. Fore wing with some slender ochreous-brown transverse strigæ on the costa, which are most distinct from the base, and a patch of strigæ below the apex. Both wings with three broken transverse pale ochreous strigose bands, a marginal line, and interciliary line. Palpi, sides of abdomen, and legs above with ochreous-brown bands.

Expanse $\frac{7}{10}$ inch.

Hab. Calcutta. In coll. Dr. Staudinger and F. Moore.

Allied to *D. conchiferata.*

DIRADES RETICULATA, n. sp.

Pale umber-brown. Wings thickly covered with dark brown and whitish transverse strigæ disposed between the veins, the veins being pale brown; fore wing with an erect whitish antemedial line, which curves inward to the costa and meets a similar postmedial line, the inner border of the entire line being blackish; a whitish-bordered blackish marginal streak below the apex: hind wing with a whitish angular subbasal and a recurved discal line, both with blackish border. Cilia edged with dark brown. Body speckled with dark brown; front, palpi, and bands on fore legs dark brown.

Expanse $\frac{9}{10}$ to 1 inch.

Hab. Darjiling. In coll. Dr. Staudinger and F. Moore.

DIRADES MULTISTRIGARIA, n. sp.

Fuliginous-black. Wings numerously covered with slender whitish transversely-disposed strigæ. Fore wing with a curved marginal series of black points, of which the two upper are large, the lower minute: hind wing with three medially-disposed marginal dentate black spots. Scales on body whitish-edged. Shaft of antennæ white-speckled; front, palpi, and legs fuliginous-black.

Expanse ₁⁰₀ inch.

Hab. Darjiling. In coll. Dr. Staudinger and F. Moore.

HASTINA, n. g.

Wings short. Fore wing broad, triangular; apex falcate, exterior margin convex and biangulated in the middle; cell nearly half the length; first subcostal at one third, quadrifid, fifth from end of cell and slightly touching the first at halfway between its base and second, sixth from below the fifth halfway between its base and juncture with the first; discocellular slender, radial from its middle; the middle median very close to end of cell, lower at two fifths: hind wing short, broad, exterior margin deeply scalloped; cell nearly half the length; subcostal joined to costal from near base to near end of cell; two subcostals at one third beyond the cell; discocellular bent near upper end, radial from the angle; middle median close to end of cell, lower at one third. Body short, rather stout; front smooth; palpi small, slender, decumbent, apex pointed; antennæ slender, simple; legs rather short, smooth; hind tibiæ with two pairs of slender spurs.

HASTINA CÆRULEOLINEATA, n. sp.

Cuprescent olivaceous-brown. Fore wing crossed by five slender smalt-blue scaled subbasal lines, each being bent inward to the costa, and a postmedial angulated line, beyond which are some scattered smalt-blue speckles: hind wing with less-defined similar smalt-blue subbasal lines, discal line, and outer speckles. Thorax and abdomen with blue-scaled bands; shaft of antennæ and a line between their base also blue.

Expanse ₁⁰₀ inch.

Hab. Darjiling. In coll. Dr. Staudinger and F. Moore.

Family CABERIDÆ.

Genus STEGANIA, *Guén.*

STEGANIA LATIFASCIATA, n. sp.

Ochreous. Both wings crossed by dark cinereous-brown strigæ, and a broad medial angular-bordered band, which is marked with ochreous strigæ on the fore wing, and on the hind wing is whitish across its middle and also extends below the apex; a black spot at end of the cell. Abdomen cinereous-brown; palpi and fore legs brownish.

Expanse 1₁⁰₀ inch.

Hab. Darjiling. In coll. Dr. Staudinger and F. Moore.

STEGANIA PURPURASCENS, n. sp. (Plate VIII. fig. 19.)

Dark purplish cinereous-brown. Wings crossed by a very indistinct brown antemedial, medial, and two postmedial lines, the two inner lines curved, the two outer lines zigzag; marginal line slender, brown: fore wing with a pale ochreous-yellow apical spot and two larger spots before the apex, each being traversed by short ochreous-brown strigæ; some very short ochreous strigæ also along the costal edge. Front of head, palpi, and tarsal band pale ochreous.

Expanse 1 inch.

Hab. Darjiling. In coll. Dr. Staudinger.

Genus CORYCIA, *Dup.*

CORYCIA CINERASCENS, n. sp.

Whitish-cinereous; hind wings palest: fore wing very indistinctly marked with short transverse brownish-cinereous strigæ, and a spot at end of cell. Tip of abdomen, palpi, and fore legs brownish-cinereous.

Expanse 1⅜ inch.

♀[*Hab.* Darjiling. In coll. Dr. Staudinger and F. Moore.

CORYCIA ALBA, n. sp.

Glossy-white. Both wings crossed by a faintly-defined brownish-cinereous medial line, and a postmedial speckled fascia: fore wing also with a similar antemedial fascia, and a small spot at end of the cell. Front of head, tip of palpi, and bands on fore legs brown.

Expanse 1¾ inch.

Hab. Darjiling. In coll. Dr. Staudinger and F. Moore.

Family MACARIIDÆ.

Genus MACARIA, *Curtis.*

MACARIA DELETARIA, n. sp. (Plate VIII. fig. 14.)

Male. Allied to *M. indistincta.* Pale violaceous cinereous-brown. Fore wing crossed by an extremely indistinct antemedial, medial, and a discal inwardly-oblique yellowish sinuous fascia; costal border with cinereous-brown strigæ; a speckled pale ochreous patch at the apex: hind wing with a similar indistinct medial and discal sinuous fascia. Cilia ochreous-brown. Underside paler, sparsely speckled with slender brown strigæ. Head and palpi ochreous; legs cinereous.

Expanse 1⁴⁄₁₀ inch.

Hab. Darjiling. In coll. Dr. Staudinger.

MACARIA INDISTINCTA, n. sp.

Male and female. Pale olivaceous cinereous-brown. Fore wing with the costal edge violaceous-brown; crossed by an extremely indistinct antemedial, medial, discal, and a submarginal outwardly-oblique yellowish sinuous fascia, the two outer being violaceous-brown bordered; a violaceous-brown lunule at end of the cell: hind wing with similar indistinct yellowish fasciæ, the two outer with slight brown points. Cilia violaceous-brown. Underside uniform pale violaceous cinereous-brown, without markings.

Expanse $1\frac{2}{10}$ inch.

Hab. Khasia Hills. In coll. Dr. Staudinger and F. Moore.

Somewhat allied to *M. liturata*, of Europe.

Genus GONODELA, *Boisd.*

GONODELA PLACIDA, n. sp.

Pale cinereous brownish-ochreous. Wings with numerous indistinct minute brown strigæ; crossed by an indistinct slender brown inwardly-oblique antemedial and medial line, and a more prominent postmedial slightly-duplex line, each bent inward before the costa, the angle of the outer line being nearly obliterated; outer line bordered by a broad clouded brownish-ochreous irregular fascia. Cell-spot minute, black. Body, palpi, and legs brown-speckled.

Expanse $1\frac{1}{10}$ inch.

Hab. Calcutta (*Atkinson*); Bombay (*Leith*). In coll. Dr. Staudinger and F. Moore.

GONODELA HORRIDARIA, n. sp.

Male. Brownish-cinereous. Wings with numerous prominent brown minute strigæ: fore wing crossed by an ill-defined ochreous-brown waved antemedial and medial diffused line, and a postmedial ochreous-bordered blackish duplex line, curved inward to the costa, the outer line dilated at both ends and with a black spot between the upper and middle median veins: hind wing with an ochreous-brown waved medial line and an ochreous-bordered blackish duplex discal diffused line, which is outwardly bordered by ochreous and black strigose patches and a central discal black spot. Cell-spot and marginal line black. *Female* with all the lines narrower, less defined, and the discal patches mostly obsolescent.

Expanse $1\frac{2}{10}$ inch.

Hab. Darjiling. In coll. Dr. Staudinger and F. Moore.

GONODELA KHASIANA, n. sp.

Male. Violaceous cinereous-brown. Wings minutely brown-speckled: fore wing crossed by an inwardly-oblique waved diffused brown antemedial and medial line, and a prominent postmedial duplex brown line, which is acutely bent inward before the costa, and is externally clouded with dark brown; interspace between the base of median veins and of the submedian, and also a patch below the apex, white: hind wing with a brown waved medial diffused line, and duplex discal line, with outer clouded border and a darker discal spot; a whitish patch at base of median veins. Cell-spot black. Head, palpi, and legs black-speckled.

Expanse $1\frac{3}{10}$ inch.

Hab. Khasia Hills. In coll. Dr. Staudinger and F. Moore.

Genus CARIGE, *Walker.*

CARIGE LUNULINEATA, n. sp. (Plate VIII. fig. 26.)

Cinereous-white. Wings with numerous minute speckles, which are cinereous-brown on the basal area and blackish on the outer area; crossed by a discal recurved black-speckled duplex lunular yellow-centred band, beyond which are some upper and lower submarginal and marginal yellow-bordered black-speckled lunular patches; marginal line yellow; cilia black, with yellow vein-points; cell-spot indistinct. Underside as above. Body, palpi, and legs black-speckled.

Expanse $1\frac{4}{10}$ inch.

Hab. Darjiling. In coll. Dr. Staudinger.

Genus EVARZIA, *Walker.*

EVARZIA TRILINEARIA, n. sp.

Male. Cinereous brownish-ochreous. Wings minutely brown-speckled: fore wing crossed by an inwardly-oblique dark-brown antemedial, medial, and a duplex postmedial line, each bent inward before the costal end, the outer postmedial being more or less obsolescent: hind wing crossed by a dark brown medial and a duplex discal line; a more or less defined blackish spot on the disk. Cell-spot small, blackish; marginal line brown. Body, palpi, and legs brown-speckled.

Expanse $1\frac{3}{10}$ inch.

Hab. Darjiling. In coll. Dr. Staudinger and F. Moore.

Genus ZEHEBA, *Moore.*

ZEHEBA AUREATA, n. sp.

Pale golden-yellow. Fore wing crossed by a very faint brown antemedial and a medial line, and a distinct discal excurved line, the outer margin being purplish-ochreous, bordered by a blackish lunular marginal line; costal end of the inner lines dark brown: hind wing with a faint medial line and curved discal line, with outer purplish-ochreous band; cell-spot minute, brown. Cilia purplish-brown. Some black streaks on costal edge of fore wing. Front of head, tip of palpi, and fore legs ochreous-brown.

Expanse $1\frac{3}{10}$ inch.

Hab. Darjiling. In coll. Dr. Staudinger and F. Moore.

Genus ZARMIGETHUSA, *Walker.*

ZARMIGETHUSA BIANGULIFERA, n. sp.

Male and female. Cinereous-white. Wings thickly speckled with cinereous-brown transverse strigæ, which are obsolescent on the anterior border of the hind wing. Fore wing crossed by an ill-defined outwardly-angulated antemedial and a postmedial reddish-ochreous speckled band; some blackish-cinereous speckles below the apex, marginal points, and a clouded spot at end of the cell: hind wing with a less-defined lower discal reddish-speckled band, blackish-cinereous marginal points, and a clouded cell-spot. Body, palpi, and legs cinereous-brown speckled.

Expanse 1 2/10 inch.

Hab. Darjiling. In coll. Dr. Staudinger and F. Moore.

ZARMIGETHUSA MINOR, n. sp.

Smaller than *Z. extersaria.* Ochreous-white. Wings minutely brown-speckled; cell-point blackish. Fore wing crossed by a slight brown-speckled postmedial line, which is bent inward before the costa, followed by a similar submarginal straight line, between which and the apex the space is clouded with cinereous-brown: hind wing crossed by a medial, discal, and a submarginal slight brown-speckled line, the discal line curved upward to end of the medial line; apex clouded with brown speckles. Body, middle and hind legs brown-speckled; palpi and fore legs cinereous-brown.

Expanse 1 8/10 inch.

Hab. Calcutta. In coll. Dr. Staudinger.

Family FIDONIIDÆ.

Genus PHYLETIS, *Guén.*

PHYLETIS KHASIANA, n. sp.

Male. Ochreous-yellow, palest on the hind wing; costal border and cilia purplish-red. Fore wing with an oblique straight discal purple-red band: hind wing with traces of a few red speckles from middle of the abdominal margin. Underside reddish-speckled; a band on fore wing and two discal sinuous speckled lines on hind wing. *Female* brighter yellow, with the red also brighter. Front, palpi, and legs above purple-red.

Expanse 1 2/10 inch.

Hab. Khasia Hills (October); Cherra Punji (October). In coll. Dr. Staudinger and F. Moore.

PHYLETIS CINERASCENS, n. sp.

Male. Fore wing pale cinereous; crossed by a darker almost straight discal diffused line, and a slightly sinuous pale submarginal line: hind wing cinereous-white, with two indistinctly defined cinereous slightly sinuous discal diffused lines; marginal line cinereous-brown. Thorax cinereous; front and tip of palpi brown; base of palpi and pectus whitish; fore legs above cinereous-brown; abdomen and legs pale cinereous; antennæ brown, base of shaft cinereous.

Expanse 1 4/10 inch.

Hab. Pir Panjal, Kashmir. In coll. Dr. Staudinger.

PHYLETIS SIMILATA, n. sp.

Female. Brownish-ochreous. Both wings crossed by an inwardly-oblique slightly-recurved slender brown antemedial line, a more oblique prominent postmedial line with a diffused outer border, and a slender wavy submarginal line; a small brown spot at end of each cell. Front, palpi, and legs brighter ochreous.

Expanse $1\frac{1}{10}$ inch.

Hab. Khasia Hills; Cherra Punji. In coll. Dr. Staudinger and F. Moore.

Nearest to *P. brunnescens.*

JANARDA, n. g.

Male. Fore wing elongated, triangular, apex acuminate, subfalcate, exterior margin very oblique; cell fully half the length; first subcostal at one fifth, second quadrifid, slightly touching the first above its base; sixth from end of cell; discocellular outwardly-oblique, radial from its middle; the middle median very close to end of cell, lower at two fifths: hind wing short, apex convex, exterior margin undulated; cell half the length; two subcostals at one third beyond the cell; discocellular outwardly-oblique, radial from the middle; the middle median close to end, lower at nearly half. Body short; palpi short, porrect, slender, apex pointed; antennæ serrated and ciliated; fore and middle legs long, very slender, middle tibiæ with a pair of slender spurs; hind legs short, tibiæ slightly flattened towards the end, tufted along the outer edge, spurs obsolete, tarsi short.

JANARDA ACUMINATA, n. sp.

Male. Pale violaceous cinereous-brown. Both wings crossed by an indistinct inwardly-oblique recurved subbasal slender brown line, and a wavy submarginal line; the basal and marginal area slightly darker clouded; cell-point indistinct. *Female* with the transverse lines very indistinctly defined, and the marginal borders darker cinereous-brown. Front and fore legs above dark cinereous-brown.

Expanse, $\male$ $1\frac{0}{10}$, $\female$ $1\frac{8}{10}$ inch.

Hab. Darjiling. In coll. Dr. Staudinger and F. Moore.

GAMORUNA, n. g.

Fore wing triangular; costa arched, apex subfalcate, very acute, posterior angle rounded; cell half the length; first subcostal at one sixth, joined to costal for a short distance near the base, bifid; second close to end of cell, trifid; sixth from the end of cell; discocellular concave, radial from the middle; the middle median close to end of cell, lower at one half: hind wing very broad, exterior margin convex; cell nearly half the length; first subcostal close to end; discocellular outwardly oblique, radial from above the middle; the middle median close to end of cell, lower at two fifths. Body slender, short; palpi longer than the breadth of the head, third joint linear, slender, as long as the second; antennæ minutely pectinated in male; legs long, slender, spurs slender, apical pair of hind tibiæ short.

GAMORUNA PALPARIA.

Panagra palparia, Walker, Catal. Lep. Het. B. M. xxiii. p. 988 (1861).

Hab. Darjiling. In coll. Dr. Staudinger and F. Moore.

Family P A N T H E R I D Æ (Zerenidæ, *auct.*).

Genus METABRAXAS, *Butler.*

METABRAXAS FALCIPENNIS, n. sp.   (Plate VIII. fig. 29.)

Male. White. Fore wing with dark cinereous strigæ along the costal border, a broken excurved subbasal band, a large spot from above lower end of the cell, some smaller transverse discal spots, and a broad broken submarginal band which is medially connected by speckles; a streak also below the apex; some ochreous-yellow bordered black spots at base of the wing: hind wing with some cinereous speckles along the costa and abdominal margin, and a medial, discal, and submarginal broken band, each composed of an upper and lower portion; a slender broken marginal line. Body ochreous-yellow, black-spotted; antennæ, front, and bands on fore and middle legs above cinereous-brown.

Expanse 1½ inch.

Hab. Khasia Hills (October). In coll. Dr. Staudinger.

CULCULA, n. g.

Fore wing elongate, triangular; cell fully half the length; first subcostal at one fifth before end of cell, bifid near its base, third from the end, trifid; discocellular inwardly oblique, bent outward below the middle, radial from above the middle; the middle median near end of cell, lower at two fifths: hind wing short, broad; exterior margin convex; cell more than half the length; subcostal convex at the base, first branch before end of cell; discocellular bent in the middle, radial from angle; the middle median near end of cell, lower at one third. Body stout; abdomen thickly tufted at apex in female; palpi short, squamous; antennæ in male biserrated, ciliated; legs thick, squamous, spurs short, stout.

CULCULA EXANTHEMATA, n. sp.

White. Fore wing with a large ferruginous basal blotch, an interrupted transverse discal cinereous-bordered dark ferruginous macular band, a large cinereous blotch at end of cell, some small blotches below the cell and others scattered along the outer margin and apex: hind wing with a cinereous-bordered ferruginous macular discal broken band, the upper portions being composed of small spots; a large cinereous blotch at end of cell, and a series of small submarginal spots. Tip of tegulæ, band on base of abdomen, apical tuft, middle of thorax and front of head, and pectus ferruginous; palpi, fore legs, and tarsi brown.

Expanse, ♂ 2⅗, ♀ 3¾ inches.

Hab. Darjiling. In coll. Dr. Staudinger and F. Moore.

THALERIDIA, n. g.

Male. Wings semitransparent. Fore wing short, triangular; cell more than half the length; first subcostal at one fourth before end of cell, second at one eighth, trifid close to the end, fifth from end and touching the second at one fifth its length and then bent outward, sixth from below the fifth at halfway between its base and juncture with second; discocellular bent near upper and lower end, the radial from lower angle; middle median about one fifth, lower at fully two fifths: hind wing short, broad, quadrate; cell long, more than half the length at lower end, disposed along the middle; costal and subcostal much arched at the base; first subcostal at half before end of cell; discocellular extremely oblique, slightly concave, bent near lower end, radial from the angle; middle median near end of cell, lower about one third. Body short; front smooth; palpi very small, slender, squamous; proboscis moderate; antennæ broadly pectinated on one side (the inner) only; legs slender, long, spurs slender, hind tibiæ with two pair.

THALERIDIA PRUINOSA, n. sp.

Wings semitransparent, cinereous-white; thickly covered with raised white scales; crossed by an antemedial, postmedial, and two submarginal pale cinereous sinuous lines, the medial line being nearly straight towards its upper end. Body and legs white; antennæ ochreous.

Expanse, ♂ 1 $\frac{3}{10}$ inch.

Hab. Darjiling. In coll. Dr. Staudinger.

Family LARENTIIDÆ.

Genus LARENTIA, *Dup.*

LARENTIA OCHREATA, n. sp.

Fore wing with the basal area brownish-ochreous; crossed by some incomplete basal and subbasal blackish sinuous lines, a moderately broad scalloped-edged angular ochreous-brown medial band, the discal border of which is whitish and the outer area beyond yellow, traversed by parallel indistinct brownish-ochreous sinuous lines, with dark brown costal ends, the outer also with double subapical points; marginal line black-spotted; area of medial band traversed by diffused blackish ringlet lines: hind wing brownish-cinereous, with paler submarginal sinuous line; marginal line brown. Thorax, head, and palpi yellowish-ochreous; bands on legs dark brown; abdomen brownish-cinereous.

Expanse, ♀ 1 $\frac{4}{10}$ inch.

Hab. Darjiling. In coll. Dr. Staudinger.

Allied to *L. variegata*, Moore.

Genus EMMELESIA, *Steph.*

EMMELESIA PICTARIA, n. sp.

White. Fore wing crossed by five inwardly-oblique equidistant undulated pale yellow bands, of which the anterior ends are mostly brownish-ochreous, and the outer bands also slightly ochreous at their posterior ends; some ochreous speckles also between the costal interspaces, and a submarginal sinuous speckled line; cilia yellow, with the apical end ochreous, cell-spot black : hind wing crossed by four pale yellow sinuous bands ; cell-spot black ; cilia yellowish. Band on fore femora and tibiæ ochreous-brown.

Expanse $\frac{9}{10}$ inch.

Hab. Darjiling. In coll. Dr. Staudinger and F. Moore.

Genus EUPITHECIA, *Curtis.*

EUPITHECIA LINEOSA, n. sp.

Cinereous. Fore wing with the basal half brown-scaled, forming a triangular basal band ; the outer half traversed by several oblique parallel brown-scaled diffused lines : hind wing brown-scaled, indistinctly forming transverse diffused lines. Body with dorsal brown bands.

Expanse $\frac{8}{10}$ inch.

Hab. Darjiling. In coll. Dr. Staudinger.

EUPITHECIA USTATA, n. sp.

Fore wing brownish-ochreous ; the basal half mostly covered with obliquely-disposed white confluent strigæ ; exterior border with a white-speckled sinuous submarginal line : hind wing white, with several indistinctly-defined ochreous-speckled wavy lines ; marginal line ochreous. Body white ; front, palpi, and legs ochreous.

Expanse $\frac{8}{10}$ inch.

Hab. Darjiling. In coll. Dr. Staudinger.

EUPITHECIA DECORATA, n. sp.

Male and female. Fore wing ochreous-yellow ; crossed by a purplish-cinereous basal and medial wavy-bordered band, a broad outer fascia traversed by a submarginal row of prominent white dentate spots ; the subbasal interspace also traversed by purplish-cinereous diffused lines ; the edges of the bands dark-speckled ; the middle band with slight parallel inner whitish line ; cell-spot distinct, blackish ; cilia with triangular black spots : hind wing yellowish-cinereous, with an indistinct purplish-cinereous discal and a marginal sinuous fascia ; cell-spot distinct ; cilia with slight purplish-cinereous spots. Thorax, head, palpi, and legs beneath yellow ; palpi and legs above brown-speckled.

Expanse $\frac{7}{10}$ inch.

Hab. Darjiling. In coll. Dr. Staudinger and F. Moore.

EUPITHECIA INCURVATA, n. sp.

Fore wing brownish-ochreous ; crossed by a subbasal and a broad medial angular band, the inner borders of which are darker brown and have a parallel inner line, and the outer borders are edged by a whitish line, followed by a dark brown sinuous fascia traversed by a submarginal whitish line, the upper end of which is curved inward before the apex ; a brown fascia also between the subbasal and medial band : hind wing brownish-cinereous. Body brownish-ochreous, with whitish dorsal bands ; palpi and legs brown-speckled.

Expanse $\cdot\frac{8}{10}$ inch.

Hab. Khasia Hills. In coll. Dr. Staudinger.

EUPITHECIA? GRISEIPENNIS, n. sp. (Plate VIII. fig. 28.)

Female. Fore wing dark lilacine-grey, with an inwardly-oblique angular pale-edged dark greyish-brown basal band, and a parallel medial band, which is narrowest across its middle, is traversed by a pale grey line, and has a black outwardly-oblique spot at end of the cell; a paler greyish-brown submarginal fascia, traversed by a recurved pale line: hind wing with two recurved medial slight grey-brown fasciæ, and a short streak above anal angle. Body dark grey; bands on thorax and abdomen, palpi, and fore legs grey-brown.

Expanse 1$\frac{1}{16}$ inch.

Hab. Cherra Punji (October). In coll. Dr. Staudinger.

ARDONIS, n. g.

Male. Fore wing rather short, broad. apex obtusely pointed; exterior margin erect anteriorly, slightly oblique below the middle median, posterior margin rather short; cell half the length; first subcostal at one fourth, bifid, the first being partly joined to the costal; third trifid, also partly joined to the second; sixth from end of cell; discocellular oblique, radial from its middle; the middle median close to end, lower at two fifths: hind wing very short, somewhat quadrate; costal margin straight, apex convex, exterior margin convex in the middle, anal angle somewhat produced; cell broad, upper end one third the length, lower end nearly half; two subcostals on a footstalk one third beyond the cell; disco-cellular very oblique, radial from above its middle; the middle median at one sixth, lower at one third. Body rather slender; antennæ simple; palpi porrect, extending two thirds beyond the head, rather broad and flattened, third joint pointed; legs long, slender, smooth, middle and hind spurs long and slender.

Female. Fore wing with the lower discal area of the underside of the wing smooth, shining, nacreous, and with a glandular oval depressed patch of raised black scales below the lower median vein. Hind wing with the apical area on the upperside similarly nacreous, and with an oval patch of speckled black scales between the base of the subcostals. Body rather stout; palpi more slender.

ARDONIS CHLOROPHILATA.

Eupithecia chlorophilata, Walker, Catal. Lep. Het. B. M. xxvi. p. 1768 (1862), ♀.

Male. Grass-green. Fore wing with four equidistant violet-brown patches along the costa, and one on middle of the exterior margin, a small spot at end of the cell, one below its base, three obliquely below its middle, and a discal series disposed on the veins: hind wing with the apical area nacreous, a glandular patch of black scales between the base of the subcostals; some transverse subbasal, medial, and discal black speckles, and some also on the anal margin. Underside very pale brownish-ochreous, with browner transverse fasciæ: fore wing with a lower discal nacreous space and a depressed oval glandular patch of black scales below the lower median, the patch being also slightly visible from the upperside. Band on tegulæ, and base of abdomen brown.

Expanse 1$\frac{3}{10}$ inch.

Hab. Darjiling. In British Museum and coll. Dr. Staudinger.

Genus COLLIX, *Guén.*

COLLIX FLAVOFASCIATA, n. sp.   (Plate VIII. fig. 25.)

Pale purplish violet-brown. Both wings crossed by a subbasal, antemedial, and a discal wavy-bordered excurved pale yellow band, and a slender submarginal line; from the discal band a streak also extends below the apex, and a patch to middle of the exterior margin; cell-spot lunular, brown.

Expanse $1\frac{4}{16}$ inch.

Hab. Darjiling. In coll. Dr. Staudinger.

Genus REMODES, *Guén.*

REMODES LINEOSA, n. sp.   (Plate VIII. fig. 10.)

Male. Fore wing broadly lanceolate; the posterior margin very short, exterior margin very convex; olivaceous-green; crossed by apparently twelve darker sinuous lines, of which the lower portions of the medial and discal are diffusedly purplish-black, the interspaces slightly cinerescent; marginal spots black: hind wing pale cinereous-ochreous; lobe brown. Thorax, head, and palpi olivaceous-green; fore and middle legs cinereous-brown; abdomen cinereous-ochreous.

Expanse $1\frac{2}{16}$ inch.

Hab. Darjiling. In coll. Dr. Staudinger.

REMODES FASCIATA, n. sp.

Male and female. Fore wing olivaceous-green; crossed by a single subbasal purplish-brown sinuous line, an inwardly-oblique medial line, a discal linear band, interrupted submarginal marks, and broad marginal spots, the latter being larger in the male than in female; cilia cinereous: hind wing pale purplish-cinereous; lobe of the male brownish. Thorax, head, and palpi olivaceous-green; fore and middle legs above cinereous-brown; abdomen cinereous.

Expanse, δ $1\frac{2}{16}$, φ $1\frac{1}{16}$ inch.

Hab. Cherra Punji (October). In coll. Dr. Staudinger and F. Moore.

Nearest to *R. triseriata* of Ceylon.

REMODES INTERRUPTATA, n. sp.

Male and female. Fore wing olivaceous-green, with longitudinally short purple-black costal, subcostal, median, and submedian dentate markings, which form portions of the ordinary transverse subbasal, medial, and submarginal series of sinuous lines; marginal spot prominent, dentate, blackish; cilia cinereous: hind wing pale purplish-cinereous; exterior margin of male cleft and folded. Thorax, head, palpi above, and legs olivaceous-green, abdomen cinereous; palpi beneath and bands on fore tibiæ dark brown.

Expanse, δ $1\frac{4}{16}$, φ $1\frac{3}{16}$ inch.

Hab. Darjiling; Khasia Hills; Cherra Punji. In coll. Dr. Staudinger and F. Moore.

Male. Fore wing olivaceous-green, with traces of darker sinuous lines, of which the medial and discal are brownish at the costal end; and intervening slightly cinerescent lines; a purple-brown marginal band traversed by a sinuous submarginal cinereous line: hind wing pale cinereous, the exterior margin slightly scalloped, but not cleft or lobed as in other species, the basal vesicle, however, being present. Thorax, head, and palpi olivaceous-green; abdomen cinereous; legs above brownish-cinereous.

Expanse 1⅒ inch.

Hab. Cherra Punji (October). In coll. Dr. Staudinger.

BRABIRA, n. g.

Male. Fore wing long, triangular; apex pointed; cell more than half the length; first subcostal at one fifth before end of cell, quadrifid, fifth from the end, curved upward and slightly touching the second close to its base, sixth also from the end; discocellular concave, radial from its middle; the middle median at one fifth, lower at one third, curving downward: hind wing small, short, narrow, anterior margin very convex at the base, exterior margin produced to an acute angle at end of first subcostal, and also at a lesser angle at end of second subcostal, abdominal margin very short and with an elongated vesicular fold; cell short, disposed along the middle; costal partly joined to subcostal, the subcostals from end of cell; lower veins not visible. Cilia very long. Body short, rather stout; antennæ broadly bipectinated to one third the tip, the branches ciliated; front broad, smooth; palpi porrect, projecting half beyond the head, hirsute; proboscis moderate; legs long, smooth, one pair of slender spurs on middle tibiæ, and two pair on hind tibiæ.

Female. Fore wing less triangular; hind wing of normal shape; discocellular outwardly-oblique, radial from its middle; the middle median near end, lower about one third. Palpi longer; more compact; antennæ minutely pubescent in front.

Type *B. Atkinsonii.*

BRABIRA ATKINSONII, n. sp.

Pale brownish-ochreous. *Male*: fore wing crossed by very indistinct purplish-brown inwardly-oblique subbasal, discal, and a submarginal duplex sinuous line, each being most apparent and dilated at the costal end; cell-spot prominent, broad, blackish: hind wing with traces of darker discal sinuous lines. *Female*: both wings with less-defined traces of duplex sinuous lines; cell-spot on fore wing smaller. Body, head, palpi, and legs beneath ochreous; fore legs above and antennæ purplish-brown.

Expanse, ♂ ♀ 1 inch.

Hab. Darjiling. In coll. Dr. Staudinger and F. Moore.

BRABIRA PALLIDA, n. sp. (Plate VIII. fig. 12.)

Male. Pale cinereous ochreous-white. Fore wing with ochreous-brown basal, subbasal, medial, and subapical costal streaks, which represent the ends of obsolescent lunular bands; the lower portion of the two former visible only by slight brown-speckled points on the

L

median and submedian vein, and the latter by more-defined interrupted lunules; slight marginal points and interciliary line ochreous-brown; a black lunule at end of the cell : hind wing with a brown-speckled medial fascia, and slight ochreous submarginal fascia. Antennæ and bands on fore legs ochreous-brown ; basal joint of antennæ white.

Expanse $1\frac{1}{16}$ inch.

Hab. Darjiling. In coll. Dr. Staudinger.

Genus LYGRANOA, *Butler.*

LYGRANOA ERECTILINEATA, n. sp.

Purplish-cinereous. Fore wing crossed by an erect antemedial and a postmedial diffused brown line; both with an outer reddish line; exterior veins and marginal line brown; a brown dot at end of the cell ; cilia yellowish, with purple-brown points: hind wing with traces of a pale brown medial line. Side of palpi and bands on fore legs purplish-brown.

Expanse $\frac{9}{10}$ inch.

Hab. Khasia Hills (October). In coll. Dr. Staudinger.

Genus LOBOPHORA, *Steph.*

LOBOPHORA SIKKIMA, n. sp. (Plate VIII. fig. 30.)

Male. Fore wing very pale pinkish-cinereous ; crossed by a very ill-defined ochreous-brown angulated subbasal line, a recurved discal denticulated line, and submarginal line ; the medial, discal, and marginal interspaces traversed by less-distinct denticulated lines ; all the lines being mostly apparent by darker brown points on the veins ; marginal points and points also along the posterior margin brown ; cell-spot short, slender, blackish : hind wing paler, with a narrow vesicular fold at base of the abdominal margin ; an indistinct ochreous-brown marginal pointed line. Body pinkish-cinereous ; abdomen with ochreous-brown bands ; front, palpi above, and bands on fore and middle legs ochreous-brown.

Expanse $1\frac{4}{10}$ inch.

Hab. Rungchu Valley, 12,000 ft., Sikkim (October). In coll. Dr. Staudinger.

LOBOPHORA DECORATA, n. sp.

Male and female. Cinereous-white. Fore wing crossed by a very slight black-speckled subbasal line, an ill-defined broad angulated and denticulated-bordered medial band formed of very sparsely disposed minute black scales, beyond which is a sinuous submarginal line and a marginal row of linear spots ; the basal and discal interspaces traversed by pale reddish lunular lines, and the submarginal area by oval red spots ; cilia black-speckled : hind wing with a slender brownish marginal line ; male with a narrow vesicular fold at base of abdominal margin. Front, bands on palpi, and on fore legs dark ochreous-brown.

Expanse $1\frac{1}{10}$ to $1\frac{2}{10}$ inch.

Hab. Darjiling. In coll. Dr. Staudinger and F. Moore.

Genus ANTICLEA, *Steph.*

ANTICLEA PLUMBEATA, n. sp.

Female. Fore wing dark lilacine-grey, with an outwardly-angulated basal blackish band, and a broad sinuous-bordered medial clouded band ; the middle band edged by a slender ochreous line, which is slightly tipped with white at the points and is obsolescent in its middle on the outer border, its apical end being also blackish-bordered ; cilia brownish : hind wing cinereous ; cilia brownish. Body dark grey ; head and palpi ochreous, fore legs with dark brown bands.

Expanse 1⅜ inch.

Hab. Darjiling. In coll. Dr. Staudinger and F. Moore.

ANTICLEA SCHISTACEA, n. sp.

Male and female. Fore wing dark purplish-grey ; with an excurved blackish subbasal band, and a transverse erect medial clouded band with slightly angulated outer border, beyond which is an interrupted white slightly sinuous line : hind wing whitish-cinereous. Body dark grey ; front, palpi, and bands on fore legs blackish.

Expanse 1⁹⁄₁₆ inch.

Hab. Darjiling. In coll. Dr. Staudinger and F. Moore.

ANTICLEA LATERITIATA, n. sp.

Allied to *A. cuprearia.* Fore wing darker ochreous-brown ; crossed by a less-defined broader and more obliquely disposed subbasal and antemedial dark brown wavy linear band, and a less irregular-shaped discal band, followed by ill-defined outer discal denticulated lines and a submarginal row of whitish dentate spots with dark brown borders ; marginal line prominent, dark brown : hind wing cinerescent ochreous-yellow, palest at the base, with a short brown marginal line from the anal angle. Body, palpi, and legs ochreous-brown.

Expanse, ♂ 1½, ♀ 1¾ inch.

Hab. Darjiling. In coll. Dr. Staudinger and F. Moore.

Genus PHIBALAPTERYX, *Steph.*

PHIBALAPTERYX PLURILINEATA, n. sp.

Pale cinereous-brown. Fore wing crossed with inwardly-oblique recurved ill-defined ochreous-brown diffused wavy lines, the exterior lines being more dusky brown ; veins with indistinct minute black points ; a cluster of blackish scales across the middle median vein, and a dot at upper end of the cell : hind wing with similar ochreous-brown wavy lines, which are most apparent from the abdominal margin ; marginal line pointed, blackish. Front of head, palpi, and bands on fore legs blackish.

Expanse 1⁹⁄₁₀ inch.

Hab. Cherra Punji (*Atkinson*); Khasia Hills (*Austen*). In coll. Dr. Staudinger and F. Moore.

Allied to the European *P. vitalbata.*

L 2

Genus SCOTOSIA, *Steph.*

SCOTOSIA EXPANSA, n. sp.

Allied to *S. venimaculata.* *Male and female* larger, cinereous ochreous-brown : fore wing crossed by a similarly disposed subbasal, antemedial, and a discal sinuous blackish band, composed of duplex contiguous lines, which are dilated and clouded with dark brown anteriorly ; interspace between the bands and also the outer border traversed by less distinct similar lines, the outer being mostly defined by brown vein-points, followed by a submarginal pale-bordered sinuous fascia : hind wing paler, crossed by very indistinct pale brown wavy lines, which are mostly defined by vein-points, and a more distinct brownish-speckled submarginal fascia ; marginal dentated line brown. Bands on palpi, and fore legs, dark brown.

Expanse 2 to $2\frac{1}{4}$ inches.

Hab. Darjiling *(Atkinson),* Himalaya *(B. Powell).* In coll. Dr. Staudinger and F. Moore.

SCOTOSIA NUBILATA, n. sp.

Reddish-brown. Fore wing with a very broad subbasal and an exterior marginal dark brown, wavy-bordered band ; the former traversed by parallel wavy chalybeous-white lines and the latter inwardly-bordered by broader and more distinct chalybeous-white speckled lines and outer pale brown lines ; the basal and medial area also traversed by brown wavy lines : hind wing reddish-brown, the apical border and adjoining cilia being yellow ; marginal line black. Body, palpi, and legs reddish-brown.

Expanse $1\frac{8}{10}$ inch.

Hab. Darjiling. In coll. Dr. Staudinger.

Allied to *S. miniosata.*

SCOTOSIA SORDIDATA, n. sp.

Male and female. Ochreous-brown : fore wing crossed by a basal and a broad medial darker brown sinuous-bordered band, both of which are edged by a parallel whitish line ; the bands and the subbasal and discal area traversed by parallel black lines, and the outer border by a clouded fascia traversed by a submarginal sinuous whitish-pointed line ; cell-streak and marginal lunular line black : hind wing with a medial, discal, and outer blackish line, the discal line edged by a white line ; submarginal sinuous white points, and a black marginal line. Palpi and bands on fore legs dark brown.

Expanse $1\frac{1}{2}$ inch.

Hab. Darjiling. In coll. Dr. Staudinger and F. Moore.

Genus ARICHANNA, *Moore.*

ARICHANNA ALBOVITTATA, n. sp.

Male. Fore wing brownish-ochreous, with short dark brown strigæ, which mostly form confluent patches between the veins from the base beyond the middle, and also at the apex ; crossed by a prominent broad white discal fascia, a submarginal slender sinuous line, and a small streak near the base: hind wing white, with a faint cinereous spot at end of the cell, and a recurved discal diffuse line.   Cilia brownish-ochreous.   Body brownish-ochreous ; front of thorax, head, palpi, and legs dark brown.

Expanse 2 inches.

Hab. Darjiling.   In coll. Dr. Staudinger and F. Moore.

ARICHANNA FURCIFERA, n. sp.

Male and female. Fore wing olivescent brownish-ochreous, with dark ochreous-brown strigæ, which mostly form compact patches between the upper and lower veins; crossed by a white subbasal line, a broad, partly confluent medial bifid band, and a broken submarginal bifid line: hind wing brownish-cinereous ;   crossed by a brown medial and submarginal slender lunular band.   Thorax, head, palpi, and legs dark ochreous-brown ; abdomen paler.

Expanse, ♂ 1⅜, ♀ 2 inches.

Hab. Darjiling.   In coll. Dr. Staudinger and F. Moore.

Genus EUSTROMA, *Hübn.*

EUSTROMA DENTIFERA, n. sp.

Fore wing umber-brown ; crossed by a slender excurved erect subbasal, antemedial, and a discal white line ; the first line wavy, the second acutely indented below the lower median, and the outer slightly indented on the middle and lower median and submedian ; followed by a submarginal decreasing series of blackish dentate spots with whitish outer borders, and a broken angular marginal band with white inner border; middle area traversed by a central upper and lower pale-bordered blackish ringleted line and a wavy lateral line ; the basal and subbasal interspace also traversed by pale-bordered lines : hind wing brownish-cinereous, with a slender discal and submarginal paler wavy line ; marginal line dark brown ; cilia pale umber-brown.   Thorax, palpi, and bands on legs dark brown ; abdomen brownish-cinereous.

Expanse 1¹¹⁄₁₆ inch.

Hab. Darjiling and Cherra Punji.   In coll. Dr. Staudinger and F. Moore.

Near to *E. obscurata.*

EUSTROMA TRIANGULIFERA, n. sp.

Cinereous ochreous-brown : fore wing with a broad dark brown outwardly-oblique wavy-bordered basal band, a triangular medial costal band, and a narrow angular marginal band ; the two former bordered by white, edged with a brown line, and the latter by a white line only ; the oblique medial interspace and the outer border traversed by a wavy white line ; marginal line dark brown : hind wing crossed by a white discal and submarginal line, both being sinuous posteriorly, beyond which is a narrow dark brown white-edged medial marginal band.   Band on palpi and speckles on the legs dark brown.

Expanse 1⁷⁄₁₆ inch.

Hab. Darjiling.   In coll. Dr. Staudinger and F. Moore.

EUSTROMA SIDERIFERA, n. sp.

Male. Fore wing cinereous violaceous-brown; crossed by a slender sinuous white subbasal, antemedial, and postmedial line; the basal and subbasal interspace also with a less-defined white lunular line, the medial area with indistinct diffused whitish lines; the discal area with slender bordering duplex white line, the upper parts of which are macular, beyond which is a submarginal row of white spots; a blackish lunule at the end of the cell: hind wing brownish-cinereous; cilia brownish. Thorax, head, palpi, and bands on fore legs violet-brown; abdomen cinereous.

Expanse 1¾ inch.

Hab. Darjiling. In coll. Dr. Staudinger.

EUSTROMA DECURRENS, n. sp.

Cidaria oblongata, Walker, Catal. Lep. Het. B. M. xxv. p. 1402 (*nec* Guénée).

Fore wing purplish-brown; crossed by eight outwardly-oblique, narrow, straight white lines, which proceed from the costa and extend to the lower discal area, which is broadly yellowish and clouded with brownish-ochreous at the posterior angle; two short lines also extend upward from the posterior margin between the first and second basal lines, there is also a short line between the second and third, one between the fifth and sixth, and another between the sixth and seventh, the seventh and eighth being crossed in their middle; marginal line black, white-bordered, and thrice broken posteriorly; cilia pale lined: hind wing brownish-cinereous, with two discal waved whitish lines, and a brown marginal line. Thorax, head, palpi, and fore legs above purplish-brown, with whitish bands; abdomen and legs brownish-cinereous.

Expanse 1⅜ inch.

Hab. Nynee Tal. In coll. Dr. Staudinger and F. Moore.

EUSTROMA MUSCICOLOR, n. sp.

Fore wing olivescent-brown; crossed by a broad darker brown basal and medial band, and a less-defined narrow marginal band; edges of the inner band erect, wavy, of the outer somewhat angulated; the bands traversed by parallel blackish lines; the subbasal and discal interspace also traversed by diffused lines: hind wing brownish-cinereous; cilia olivescent-brown. Thorax, head, and palpi olivescent-brown; legs with blackish bands; abdomen brownish-cinereous, speckled with brown.

Expanse 1½ inch.

Hab. Darjiling. In coll. Dr. Staudinger and F. Moore.

EUSTROMA PORPHYRIATA, n. sp.

Male. Fore wing ochreous-red, crossed by four outwardly-oblique straight basal white lines, two zigzag antemedial lines, and two postmedial posteriorly zigzag lines, with two outer parallel similar lines; the medial area also traversed by less-defined parallel white lines: hind wing whitish-cinereous, with some ochreous-red anal lunules, and a paler marginal fascia. Body ochreous-red, streaked with whitish-cinereous; legs brown-speckled. Expanse $1\frac{2}{10}$ inch.

Hab. Darjiling. In coll. Dr. Staudinger.

Genus HARPALYCE, *Steph.*

HARPALYCE KASHMIRICA, n. sp.

Female. Fore wing ochreous-yellow, with an angulated basal, and an erect sinuous-bordered medial transverse ochreous band, followed by a less-defined ochreous discal sinuous fascia and marginal points; the bands paler centred and with brown-speckled edges, the middle band being constricted below the cell: hind wing yellowish-white, with traces of a slight ochreous recurved medial sinuous line. Thorax, head, palpi, and legs ochreous-yellow; fore and middle legs above brown-speckled; abdomen whitish.

Expanse $1\frac{3}{10}$ inch.

Hab. Margan Pass, Kashmir. In coll. Dr. Staudinger.

Genus CIDARIA, *Treit.*

CIDARIA ALBOFASCIATA, n. sp.

Male and female. Fore wing reddish-brown, crossed by a narrow basal darker brown band, and a broad medial white band, followed by a submarginal row of dark-bordered white dentate spots; a white patch at the apex, and a large patch extending from medial band to the exterior margin; the medial band also bordered by brown wavy lines traversed by a discal row of points, and also partly brown-clouded across its middle; cell-spot large and black: hind wing white, with a brown cell-dot and marginal line; cilia pale ochreous. Body reddish-brown; palpi and bands on legs dark brown.

Expanse $\frac{9}{10}$ inch.

Hab. Darjiling. In coll. Dr. Staudinger and F. Moore.

CIDARIA MACULATA, n. sp.

Fore wing dark olivescent-brown; marked with a pure white lower basal spot, a subbasal constricted band, an upper and lower discal spot, an apical spot, and a spot on middle of the exterior margin; also with some intervening discal and submarginal white dots: hind wing white, with an olivescent-brown white-speckled marginal band. Body dark brown; a white band across the thorax and base of abdomen; legs with slight white bands.

Expanse $\frac{9}{10}$ inch.

Hab. Darjiling. In coll. Dr. Staudinger and F. Moore.

CIDARIA AFFINIS, n. sp.

Fore wing cinereous-brown; crossed by a basal and a denticulated-bordered dark dusky-brown medial band, followed by a discal parallel line, and a marginal angulated fascia traversed by a slender sinuous white line; the fascia mostly apparent at the apical end; bands also traversed by blackish lines: hind wing cinereous-white, with an indistinct brown discal pointed line, and marginal line; cilia pale ochreous. Body cinereous-brown; abdomen with reddish bands; palpi and legs with blackish bands.

Expanse 1 inch.

Hab. Darjiling. In coll. Dr. Staudinger and F. Moore.

Allied to *C. seriata*.

CIDARIA SERIATA, n. sp.

Male and female. Fore wing reddish-brown; crossed by a wavy-bordered basal band, and a denticulated medial cinereous band, both edged by a cinereous-white line, and traversed by blackish lines; the latter also with a cell-streak; beyond is a discal parallel black line, and a submarginal series of pure white spots, of which the middle one is large and geminate, and the upper spots with blackish-clouded borders; marginal line black, with white points in the female: hind wing cinereous-white, with indistinct brownish-pointed discal line, marginal line, and ochreous cilia. Body reddish-ochreous; bands on legs blackish.

Expanse 1 inch.

Hab. Darjiling. In coll. Dr. Staudinger and F. Moore.

CIDARIA CURCUMATA, n. sp.

Male and female. Fore wing yellowish-ochreous; crossed by a basal, subbasal, and a broad angular-bordered medial violaceous-brown band, followed by some discal points, and a broken outer fascia, which is traversed by a pale sinuous line; marginal line black; medial band also traversed by parallel diffused blackish lines; cilia brown: hind wing brownish-cinereous, palest apically; with a slender darker sinuous discal and submarginal line; marginal line brown. Thorax, head, and palpi ochreous.

Expanse, δ $1\frac{2}{10}$, φ $1\frac{3}{10}$ inch.

Hab. Darjiling. In coll. Dr. Staudinger and F. Moore.

CIDARIA BICOLOR, n. sp.

Fore wing cinereous-black; crossed by indistinct darker subbasal and discal sinuous lines, the interspaces between the latter being more or less pale, which across the disc form a whitish fascia; this fascia being more apparent on the underside; a similar pale patch also at the apex: hind wing whitish, with the base and a marginal band, as well as the cilia, blackish-cinereous; the band with a white patch at the apical end. Body, palpi, and legs cinereous-black.

Expanse $1\frac{1}{2}$ inch.

Hab. Sikkim, 11,000 feet, August (*Blanford*); Chumbi Valley (*Elwes*). In coll. Dr. Staudinger and F. Moore.

CIDARIA? OBLIQUISIGNA, n. sp.

Whitish. Fore wing very sparsely speckled with minute ochreous-brown scales; crossed by a duplex ochreous-brown speckled curved basal sinuous line, an upper medial and discal duplex lines, which are joined together below the cell and are outwardly biangulated between the upper and middle medians, below which are lower discal irregular curved lines ; the inner discal and lower curved line being almost black, the discal line with a parallel outer ochreous-red line, followed by submarginal diffused red-speckled lines, and a blackish marginal line ; a prominent outwardly-oblique black streak across end of the cell : hind wing with traces of a recurved discal and two submarginal brown-speckled diffused sinuous lines ; marginal line distinct. Front, antennæ, fore and middle legs above ochreous-brown ; abdomen whitish.

Expanse 1 inch.

Hab. Darjiling. In coll. Dr. Staudinger and F. Moore.

Family EUBOLIIDÆ.

Genus ONYCHIA, *Hübner.*

ONYCHIA LATIVITTA, n. sp.

Fore wing pale cinereous ferruginous-brown ; with minute dark brown strigæ ; crossed by a broad medial dark cuprescent-brown band, the inner edge of which is erect and slightly curved, the outer edge angulated beyond the cell ; exterior margin slightly suffused with darker brown ; cell-spot indistinct : hind wing brownish-cinereous, with some darker speckles from the anal angle, and a faint slender discal recurved line is apparent in the male.

Expanse, ♂ $1\frac{4}{10}$, ♀ $1\frac{6}{10}$ inch.

Hab. Himalayas (Dalhousie, Darjiling). In coll. F. Moore and Dr. Staudinger.

Allied to the European *O. mæniaria* (*Eubolia mæniaria*, Dup.).

ONYCHIA VIOLACEA, n. sp. (Plate VIII. fig. 21.)

Fore wing violaceous-brown ; with some very slender indistinct blackish transverse strigæ ; crossed by a straight diffused ochreous-brown antemedial line, and an angulated discal line, between which, and nearest the outer, is a less-defined brown slender recurved fascia ; cell-point black, slender ; marginal line slender, brown : hind wing pale violaceous cinereous-brown, with some indistinct brown strigæ from the abdominal margin, and a more distinct short line above the anal angle ; cilia brown. Body violaceous-brown ; palpi and legs ochreous with brown speckles.

Expanse $1\frac{3}{10}$ inch.

Hab. Darjiling. In coll. Dr. Staudinger.

Allied to *O. grisea* (*Nadagara grisea*, Butler).

Family TORTRICIDÆ.

Genus CERACE, *Walker.*

CERACE PERDICINA, n. sp.

Male. Fore wing with the costal and posterior border black, the intervening middle area from the base to the apex and also along the exterior margin being scarlet ; numerous

M

short transverse yellow streaks disposed along the costal border, and rows of yellow spots between the veins: hind wing reddish-ochreous, with a single submarginal row of black spots, and a marginal row of minute dots. Thorax black, with yellow streaks; head yellow; palpi black above; body reddish-ochreous; legs yellow, with black bands.

Expanse 1¾ inch.

Hab. Darjiling. In coll. Dr. Staudinger.

TOPADESA, n. g.

Male. Fore wing broad; costa much arched at the base, apex acute, produced, exterior margin very slightly oblique, angulated in the middle, posterior margin convex at the base; cell three fifths the length, disposed along middle of the wing; first subcostal at two fifths before end of cell, second close to end, trifid, fifth from the end and joined to third for a short distance close to its base, sixth also from the end; discocellular concave, radial from close to lower end; two upper medians from angles at end of cell, lower at one third before the end: hind wing rather narrow, apex convex, exterior margin very oblique, convex in the middle; costal vein recurved at the base, slightly touching subcostal; cell more than half the length; two subcostals at a short distance beyond the cell; discocellular bent in the middle, radial from near lower end; two upper medians on a footstalk half beyond the cell, lower about one fifth before the end. Body moderately stout, smooth; palpi porrect, very long, extending three fourths beyond the head, second joint curved at the base, third joint fully half of second, pointed; clothed with short compact scales; antennæ setaceous; legs squamous, femora broad, flattened, middle and hind pairs of spurs long, slender, unequal; a prominent foliaceous appendage from base of abdomen beneath hind femora.

TOPADESA SANGUINEA, n. sp.

Fore wing bright red, irregularly clouded with cinereous-brown along the costal and exterior borders, posterior border, and across the middle; a small basal space, the apical and exterior border including the cilia narrowly bright yellow; before the middle is a transverse wavy black line, and beyond a postmedial bright red line; a prominent white spot at end of the cell: hind wing pale cinereous-yellow. Thorax, head, and palpi above cinereous-brown; palpi red along the side, yellow beneath; legs yellowish; abdomen above pale brownish-cinereous.

Expanse 1 inch.

Hab. Darjiling. In coll. Dr. Staudinger and British Museum.

Genus GRAPHOLITHA, *Treit.*

GRAPHOLITHA LOBIFERANA, n. sp.

Male. Fore wing umber-brown, with paler transverse striæ; a brown elongated lobate mark disposed longitudinally on the posterior border from near the middle, before which are some whitish lower subbasal and basal transverse incurved streaks; some short silvery-white outwardly-oblique streaks along the outer half of the costa and two inwardly-oblique similar streaks before the apex; a submarginal series of black points, of which the lower are silvery-white bordered; cilia whitish with brownish edge: hind wing brownish-cinereous, cilia paler. Body umber-brown, abdomen paler, anal tuft pale ochreous; tegulæ with whitish edges; front and palpi above whitish; fore and middle legs, and tarsal bands brown.

Expanse $\frac{8}{12}$ inch.

Hab. Darjiling. In coll. Dr. Staudinger and F. Moore.

Family TINEIDÆ.

Genus DAVENDRA, *Moore.*

DAVENDRA FLAVIBASA, n. sp.

Male. Fore wing with the basal half deep yellow, the outer half brownish-ferruginous, which is brightest in its middle; cilia brown posteriorly, yellow anteriorly: hind wing and abdomen dark brown; cilia yellowish-cinereous. Thorax deep yellow; collar and front dark brown; head and palpi yellowish; palpi with black-speckled bands; legs brown, with slight yellowish bands, paler beneath; antennæ dark brown.

Expanse $\frac{8}{16}$ inch.

Hab. Darjiling. In coll. Dr. Staudinger.

This insect quite agrees with the type of *Davendra* (*D. Mackwoodii*) in its venation and other characters.

Genus TOXALIBA, *Walker.*

TOXALIBA UMBRIPENNIS, n. sp.

Dark cuprescent umber-brown. Extreme costal edge of the fore wing and edges of the cilia cinereous-ochreous. Costal border of the hind wing, abdomen, and legs brownish-cinereous. Head and palpi bright ochreous; antennæ blackish.

Expanse $1\frac{1}{10}$ to $1\frac{2}{10}$ inch.

Hab. Darjiling. In coll. Dr. Staudinger and F. Moore.

Distinguished from *T. reductella* by the broader and shorter wings.

Genus EUPLOCAMUS, *Latr.*

EUPLOCAMUS STRIGOSA, n. sp.

Male. Pale yellowish-white: fore wing with a broad quadrate subbasal, a small antemedial, and a large postmedial costal brown patch; the lower basal, medial, and outer areas crossed by broad zigzag, partly continuous strigæ, which are mostly brown anteriorly, and olivaceous posteriorly; cilia with alternated brown tips: hind wing crossed by indistinct cinereous-brown zigzag strigæ; marginal points brown. Tegulæ in front, and a band across middle of the thorax, base of palpi, and fore legs brown; front, palpi above, tip of abdomen, and hind legs yellowish; antennæ brown.

Expanse $1\frac{3}{12}$ inch.

Hab. Darjiling. In coll. Dr. Staudinger.

Genus ATTEVA, *Walker.*

ATTEVA PULCHELLA, n. sp.

Fore wing dark gilded-yellow, the exterior border being tinged with purple-brown ; a longitudinal series of white spots, of similar size, and mostly disposed as those in *A. niveigutta*: hind wing bright ochreous-yellow, with the outer half and cilia purplish-brown. Thorax and head gilded-yellow, thorax white-spotted, front edged with white; palpi purple-brown, with white bands; antennæ brown, the basal joint being white; legs brown above, yellow beneath; tip of tibiæ and tarsal joints banded with white.

Expanse $1\frac{1}{5}$ inch.

Hab. ——? (*Atkinson*). In coll. Dr. Standinger.

Note. This is nearest allied to the Australian *A. albiguttella* (*Oeta albiguttela*, Zeller, Verh. zool.-bot. Ges. 1873, p. 230), the type of which has been examined.

Genus HYPONOMEUTA, *Zeller.*

HYPONOMEUTA BRUNNESCENS, n. sp.

Pale cinereous-brown: fore wing with five longitudinal series of black spots, terminating on the disc, and the upper row extending along edge of the costa, beyond which is a sub-marginal series of spots disposed somewhat in shape of the figure 8: hind wing and abdomen paler cinereous-brown. Thorax with black spots in front ; palpi and fore legs dark brown.

Expanse $1\frac{2}{10}$ inch.

Hab. Darjiling. In coll. Dr. Staudinger.

Genus ADELA, *Latr.*

ADELA ATKINSONII, n. sp.

Male. Fore wing with the basal area metallic ochreous-green, the apical area and cilia cupreous-red, the division being formed by a postmedial transverse outwardly-oblique narrow yellow band, which is bordered on both sides by a contiguous silvery band; there are also two shorter upper antemedial silvery bands, which are preceded by a subbasal longitudinal costal black-edged yellow streak ; apical area thickly speckled with longitudinally disposed, partly confluent yellow speckles: hind wing pale cuprescent-brown, cilia paler. Thorax metallic cuprescent-green ; front and palpi ochreous; fore and middle legs dark metallic cupreous-brown ; tibiæ beneath and tarsal bands yellow ; antennæ steel-blue at the base.

Expanse $\frac{8}{5}$ inch.

Hab. Darjiling. In coll. Dr. Staudinger.

Family PTEROPHORIDÆ.

Genus ALUCITA, *Steph.*

ALUCITA SIKKIMA, n. sp.

Male. Ochreous; shafts of the plumes of both wings tranversely brown-speckled where they are crossed by the bands; a broad basal, a medial, and an outer purplish-brown transverse band, the basal being composed of three almost confluent narrow bands which occupy more than a third of the wing, the medial band is narrow, the outer band is broken into narrower bands on the plumes of the hind wing. Thorax and abdomen with broad brown-speckled bands; palpi brown-speckled, fore legs above with brown-speckled bands, and hind legs with pale bands; shaft of antennæ also brown-speckled. *Female.* Cinereous-white; the transverse brown speckles on shaft of the plumes, and also the bands darker; bands on body and legs also darker.

Expanse, ♂ 1⅔, ♀ 1⁸⁄₁₂ inch.

Hab. Darjiling. In coll. Dr. Staudinger and F. Moore.

ADDITIONAL SPECIES.

Family PAPILIONIDÆ.

PAZALA, n. g.

Papilio, sect. xx., Felder, Spec. Lep. pp. 301, 316 (1864).

Fore wing short, broad, triangular, glossy; costa slightly arched towards the end, apex broad, exterior margin slightly waved, scarcely oblique; cell broad at the end, discocellulars straight, upper longest; first subcostal short and anastomosed to the costal: hind wing triangular; costa oblique, exterior margin very oblique, sinuous anteriorly, scalloped posteriorly, with a long, very slender tail: cell long, somewhat broad; first subcostal at more than half before end of cell, second concave at the base; discocellular bent inward below the middle. Body short, thickly clothed with silky hairs; front of head and palpi densely hairy; femora pilose beneath; antennæ very short, club short and thick.

Type *P. glycerion.*

PAZALA GLYCERION.

Papilio glycerion, Gray, Lep. Ins. of Nepal, p. 6, pl. 3. f. 2 (1831); Westwood, Arc. Ent. pl. 55. f. 3.

Hab. Sikkim (*Atkinson*), Nepal (*Ramsay*).

DABASA, n. g.

Papilio, sect. xxix. subsect. A, Felder, Spec. Lep. pp. 306, 352 (1864).

Male. Fore wing triangular; costa very much arched; exterior margin concave; posterior margin short; cell very broad: hind wing short, broad; exterior margin sinuous, with a rather long outwardly-curved spatulate tail, the angle at end of lower median prolonged; basal area above, and the abdominal border beneath thickly clothed with fine hairs; cell very long, extending two thirds the length, narrow. Body hairy; antennæ slender, club short, broad, and abruptly curved.

Type *D. gyas.*

DABASA GYAS.

Papilio gyas, Westwood, Arcana Ent. i. pl. 11. f. 1 (1841).

Hab. Sikkim (*Atkinson*).

TAMERA, n. g.

Papilio, sect. lx. subsect. C, Felder, Spec. Lep. pp. 320, 368 (1864).

Male. Fore wing triangular; costa arched, apex convex: hind wing short, broad, exterior margin scalloped, the angle at end of middle median dentate and slightly produced. Antennæ slender, with a gradually lengthened club.

Type *T. castor.*

This is quite distinct from the group (*Charus*) of which *Helenus* is the type, with which it is associated by Prof. Wood-Mason (Journ. As. Soc. Bengal, 1880, p. 149), and, with its allies, *Mahadeva*, *Mehala*, and *Abrisa*, approximate to the *Panope* group (*Chilasa*).

TAMERA CASTOR.

Papilio castor, Westwood, Ann. Nat. Hist. ix. p. 37 (1842); Arcana Ent. ii. p. 129, pl. 80. f. 1, 2 (1845), ♂; Wood-Mason, J. A. S. Beng. 1880, pl. 9. f. 1, 2, ♂ ♀; Westwood, P. Z. S. 1881, p. 179, f. 1-4, ♂ ♀.

Papilio pollux, Westw. Ann. N. H. ix. p. 37 (1842); Arc. Ent. ii. p. 129, pl. 90. f. 1.

Hab. Sillhet (*Atkinson*).

MEANDRUSA, n. g.

Papilio, sect. xxix. subsect. B, Felder, Spec. Lep. pp. 306, 352 (1864).

Male. Fore wing triangular; costa much arched towards the end, apex prolonged, falcate, acuminate; exterior margin concave below the apex, lower angle abrupt; posterior margin very short; lower discocellular very concave: hind wing short, much narrowed hindward; exterior margin oblique, sinuous, with a long narrow curved tail; subanal and anal angle obsolete; base of wing and abdominal fold densely hairy; cell very long and narrow. Body somewhat woolly; antennæ short, with a very short broad club.

Type *M. evan.*

MEANDRUSA EVAN.

Papilio evan, Doubleday, Ann. Nat. Hist. xvi. pp. 235, 304 (1845); Gen. D. Lep. pl. 2. f. 2 (1846); Westwood, Cab. Orient. Ent. pl. 31. f. 1, 1 *a* (1848).

Hab. Sibsagur, Assam (*Atkinson*).

ISAMIOPSIS, n. g.

Papilio, sect. xxxvii. (part), Felder, Spec. Lep. pp. 308, 354 (1864).

Male. Fore wing elongated, broad, triangular; costa arched regularly, apex and posterior angle obtusely pointed, exterior margin oblique and straight; cell long, broad : hind wing very short, triangular, exterior margin slightly scalloped; cell short, not reaching half the length of the wing, narrow; first subcostal at less than half before end of the cell, upper discocellular very oblique, acutely bent inwards below the middle.   Antennæ short, with a gradually-thickened lengthened club.

Type *I. telearchus.*

This is a mimic of the *Isamia splendens.*

ISAMIOPSIS TELEARCHUS.

Papilio telearchus, Hewitson, Trans. Ent. Soc. 1852, p. 22, pl. 6. f. 3.

Hab. Khasia Hills (*Atkinson*).

Family NYCTEOLIDÆ.

Genus TYANA, *Walker.*

TYANA FLATOIDES, n. sp.

Male. Fore wing grass-green ; costal edge and cilia pale ochreous, apical edge red ; a reddish-bordered white spot at base of the wing, and a lower reddish-tipped white streak ; a broad reddish constricted patch ascending from middle of the posterior margin : hind wing and abdomen white.   Thorax green ; collar, front, palpi, and basal joint of antennæ ochreous-white, edged with red ; legs and dorsal bands on the abdomen ochreous-white, fore legs reddish above ; antennæ reddish.

Expanse 1½ inch.

Hab. Cherra Punji (October).   In coll. Dr. Staudinger.

Genus CHIONOMERA, *Butler.*

CHIONOMERA TRIANGULIFERA, n. sp.

Male. Fore wing yellow, with some speckles along base of the costal border and outer third of the wing below the apex, including the cilia, dark purple-red ; a large pure white spot at the base and a smaller spot above middle of the cell, both of which are edged with red speckles ; a series of white spots along the exterior margin, the middle spot being large and triangular : hind wing white.   Thorax white ; head, antennæ, tip of palpi, bands on fore legs above and speckles on middle legs purplish-red ; abdomen, base of palpi, and hind legs ochreous-white.

Expanse 1 inch.

Hab. Cherra Punji (October).   In coll. Dr. Staudinger.

CHIONOMERA SANGUINOLENTA, n. sp.

Male. Fore wing yellow, with a very irregular-bordered dark purplish-red subbasal and a broad outer band including the cilia ; both bands joining posteriorly and blotched with purplish-cinereous, the outer band not extending above the subcostal ; costal border and medial intervening area being red-speckled : hind wing white, with broad cinereous outer band. Thorax, head, palpi, antennæ, and legs yellow: thorax red-speckled; palpi, tip of fore tibia, and tarsi with a red band.

Expanse 1 inch.

Hab. Darjiling. In coll. Dr. Staudinger and F. Moore.

CHANDICA, n. g.

Male. Fore wing broad, short; costa arched at the base, apex very obtuse, exterior margin very slightly oblique, posterior margin convex towards the base; cell half the length; costal and subcostal widely separated; first subcostal at two fifths before end of cell, second at one third, third from the end, bifid, very slightly touching second near its base, fifth also from end; discocellular erectly-recurved, upper radial from above the middle, lower radial close to lower end; middle median close to end of cell, lower at two fifths: hind wing very short, quadrate, prolonged hindward, apex produced, exterior margin convexly-angular below the middle; cell more than half the length, disposed along the middle; costal vein bent down and touching subcostal near its base; two subcostals from end of cell; discocellular concave, the radial and two upper medians on a footstalk nearly half beyond end of the cell; lower median near end of cell. Body short; thorax stout; palpi ascending a little above vertex, rather slender, very laxly squamous, third joint short; antennæ long, stout, filiform; legs stout, fore and middle tibiæ laxly clothed; hind legs smooth; middle and hind spurs very long, slender, unequal.

CHANDICA QUADRIPENNIS, n. sp.

Male. Fore wing purplish-ochreous, reddish at the base; with a narrow yellow dentated-bordered costal and outer marginal band, the latter including the cilia; marginal band traversed by a row of purple-brown points; crossing the wing is a slender straight antemedial and a zigzag postmedial yellowish line; the inner border of the costal and outer band suffused with purplish-cinereous. Hind wing very pale ochreous-red, with the costal and abdominal border, including the cilia, whitish. Thorax ochreous-red; abdomen paler; collar, head, palpi, base of antennæ, fore and middle legs above yellow; collar, front, and tip of palpi edged with ochreous-red; bands on fore tibiæ purple-brown; antennæ ochreous.

Expanse 1¼ inch.

Hab. Darjiling. In coll. Dr. Staudinger.

A specimen of this species, from Borneo, is in the British Museum collection.

Family LITHOSIIDÆ.

Genus RŒSELIA, *Hübner.*

RŒSELIA LATIVITTATA, n. sp.

Male. White. Fore wing with an outwardly-oblique transverse broad medial pale brown band, the inner edge of which curves from base of the costa to middle of the posterior margin, the outer edge being angulated outward from lower end of the cell, and there merging into a similar-coloured outer marginal band; between the two bands is a transverse discal and a submarginal zigzag series of blackish points, of which the upper are dilated; cilia brown; raised tufts of scales within the cell black-tipped. Hind wing with the veins brownish, the apical border also slightly brown-tinged. Abdomen with slight brownish-speckled bands; palpi beneath and bands on legs brown.

Expanse $\frac{8}{10}$ inch.

Hab. Darjiling. In coll. Dr. Staudinger.

RŒSELIA INSCRIPTA, n. sp.

Male and female. Cinereous-white. Fore wing crossed by an excurved distinct black antemedial line, an irregular zigzag postmedial and a sinuous submarginal line, the latter with a broad black patch near its upper end and at its lower end; the inner transverse line is preceded by a parallel brown diffused line, the postmedial line by two recurved diffused brown outer lines, and the submarginal by a less-defined line; marginal slender line black; the raised tufts of scales also brown-tipped. Hind wing pale brownish-cinereous. Body brownish-cinereous; thorax whitish, with black bands and dorsal tufts; palpi and legs brownish; antennæ of male bipectinated.

Expanse, $\male$ $\frac{8}{10}$, $\female$ $\frac{9}{10}$ inch.

Hab. Darjiling. In coll. Dr. Staudinger and F. Moore.

RŒSELIA DENTICULATA, n. sp.

Cinereous-white. *Male:* Fore wing very sparsely brown-scaled; with a blackish basal costal patch, an inwardly-oblique antemedial black angulated line, which is dilated to the raised scales forming the orbicular and reniform spots, a less-distinct postmedial denticulated line, a submarginal line, and a marginal row of points; the area between the antemedial and postmedial lines more or less black-speckled: hind wing whitish. Thorax and head whitish; abdomen, palpi, and legs above brown. *Female:* Fore wing whiter; markings less defined: hind wing whitish-cinereous.

Expanse, $\male$ $\female$ $\frac{8}{10}$ inch.

Hab. Calcutta. In coll. Dr. Staudinger and F. Moore.

RŒSELIA SIKKIMA, n. sp.

Brownish-cinereous. *Male:* Fore wing with some confluent ochreous-brown streaks on the costa, an antemedial and postmedial transverse denticulated line, the medial area being brown-streaked; followed by a less distinct submarginal line, and marginal points; orbicular and reniform raised spots brown and prominent. Abdomen and fore legs with brown bands. *Female:* Paler brownish-cinereous: fore wing with less-defined and paler brown markings.

Expanse $\frac{8}{10}$ inch.

Hab. Darjiling. In coll. Dr. Staudinger and F. Moore.

x

Family DREPANULIDÆ.

Genus DREPANA, *Schrank.*

DREPANA PRUNICOLOR, n. sp.

Dark purplish-grey. Fore wing crossed by an inwardly-oblique yellowish antemedial line, which is bent wavily inward to the costa from end of the middle median; a brighter yellow postmedial similar bent line, followed by a submarginal paler line, the apical end of which is also yellow; a pure white dot at lower end of the cell : hind wing with a pale subbasal, medial, and submarginal pale yellowish line.

Expanse 1¼ inch.

Hab. Darjiling. In coll. Dr. Staudinger.

Family HYPOGRAMMIDÆ.

Genus SELEPA, *Moore.*

SELEPA ROBUSTA, n. sp.

Female. Cinereous-brown. Fore wing crossed by a slender angulated excurved antemedial pale-bordered blackish line, and a prominent diffused black curved postmedial line, which is slightly sinuous only at its posterior end; a black streak along the two lower median veins and the submedian vein between the two lines; outer area of the wing pale cinereous, traversed by three brown sinuous fasciæ, which are dilated at the costal end; marginal line black, slender; a short black and white streak running from base of the cell to the costa. Hind wing and abdomen paler; cilia whitish. Thorax, head, palpi, and bands on fore legs and middle legs cinereous-brown; body beneath and legs cinereous-white.

Expanse 1⁹⁄₁₀ inch.

Hab. Darjiling. In coll. Dr. Staudinger.

SELEPA ÆNESCENS, n. sp.

Fore wing ænescent brownish-ochreous; with a black-speckled bordered whitish streak at the base; crossed by two undulated antemedial and two outwardly-irregular recurved sinuous postmedial erect black-speckled lines; the outer lines with a contiguous less-defined parallel line, beyond which is an indistinct submarginal narrow sinuous fascia, and a prominent row of marginal points; between the middle lines is a large black-speckled orbicular and reniform mark, below which is a cluster of black speckles. Hind wing pale ænescent-brown; cilia pale ænescent-ochreous. Thorax, head, palpi beneath and legs ænescent-ochreous; middle of thorax and palpi above dark brown; fore legs blackish; abdomen and legs brownish.

Expanse 1⁹⁄₁₀ inch.

Hab. Darjiling. In coll. Dr. Staudinger.

Genus SYMITHA, *Walker.*

SYMITHA FASCIOSA, n. sp.

Female. Fore wing brownish-cinereous; crossed by a brown subbasal and two ante-medial slightly outwardly-oblique fasciæ, an erect medial slender sinuous line, and an irregular recurved zigzag discal line, the latter being outwardly bordered by a contiguous brown shade, and followed by a submarginal dark cinereous sinuous fascia; a pale-bordered brown lunule at end of the cell. Hind wing pale cinereous-brown. Thorax, head, palpi, fore tibia and tarsi above cinereous, with brown bands; abdomen and legs pale cinereous-brown.

Expanse 1 inch.

Hab. Darjiling. In coll. Dr. Staudinger.

SYMITHA SINUOSA, n. sp.

Female. Fore wing cinereous; crossed by a subbasal, medial, and a postmedial slender black erect zigzag line, the latter being dilated at the costal end, followed by a submarginal narrow sinuous fascia and marginal pointed line; the intervening spaces traversed by indis-tinctly-defined olivaceous wavy fasciæ; two clusters of blackish scales below the cell between the middle line, and a short streak below base of the cell. Hind wing very pale cinereous-brown, the veins and outer border brown; cilia edged with whitish-cinereous. Thorax and palpi speckled with blackish scales; fore legs with blackish bands.

Expanse ⅓ inch.

Hab. Darjiling. In coll. Dr. Staudinger.

SYMITHA LILACINA, n. sp.

Male. Fore wing lilacine-grey; crossed by a slender indistinct pale-bordered blackish erect antemedial line, an irregular recurved zigzag postmedial line, followed by indistinct pale fasciæ. Hind wing and abdomen pale cinereous-brown. Thorax grey; middle of thorax, front of collar, head, palpi, and bands on fore legs brown.

Expanse 1 9/10 inch.

Hab. Darjiling. In coll. Dr. Staudinger.

Family CATEPHIIDÆ.

Genus GYRTONA, *Walker.*

GYRTONA PUSILLA, n. sp.

Fore wing pale purplish brownish-ochreous; crossed by two indistinct erect medial wavy purple-brown lines, between which is a blackish-speckled fascia; beyond is a slight cell-spot, two postmedial outwardly-oblique less-distinct similar wavy lines, and two submar-ginal lines; basal area also with indistinct wavy lines. Hind wing brownish-cinereous, semidiaphanous, with the veins and outer border broadly dusky brown; cilia cinereous. Body, palpi, and legs brownish-ochreous.

Expanse 6/10 inch.

Hab. Calcutta. In coll. Dr. Staudinger and F. Moore.

Family POAPHILIDÆ.

Genus ARASADA, *Moore.*

Arasada, Moore, Lep. Ceylon, iii. 188.

ARASADA FASCIOSA, n. sp.   (Plate VIII. fig. 21.)

Ochreous-red. Fore wing crossed with inwardly-oblique red and broken yellow fasciæ from below the costal border ; a large pure white spot with black edge at end of the cell : hind wing with more regularly defined red and narrower subbasal, discal, and outer yellow lunular fasciæ ; a black dot at end of cell. Costal edge of fore wing pure white. Both wings with marginal black points, and the red fasciæ clouded with dusky brown interspersed with a few minute black scales. Collar pure white ; fore legs brownish above. Underside pale brownish-ochreous, with an oblique discal and submarginal brownish fascia. Body and legs beneath brownish-ochreous.

Expanse 1⅔ inch.

Hab. Darjiling. In coll. Dr. Staudinger.

ARASADA FRATERCULATA, n. sp.

Pale brownish-ochreous. Wings with a few scattered minute black scales. Fore wing with a very small dentate white spot at end of the cell ; traces of a brownish diffused recurved discal line, and a more-defined recurved submarginal series of pale inner-bordered black dentate spots, and a marginal row of spots : hind wing with a slight black-scaled spot at end of the cell, a black-scaled pale outer-bordered discal wavy line, a dentated submarginal slight fascia, and marginal spots. Body speckled with black scales ; head, tip of palpi, and fore legs dark brown.

Expanse ⁷⁄₁₆ inch.

Hab. Calcutta (November). In coll. Dr. Staudinger.

Much resembles the Ceylonese *A. pyraliformis.*

ARASADA RUPTIFASCIA, n. sp.

Pale purplish cinereous-ochreous. Wings with a very few minute black scales upon the basal area. Fore wing with a very indistinct blackish inwardly-oblique recurved wavy antemedial and postmedial line, and more-distinct blackish-scaled submarginal fascia, which latter is dentate anteriorly and diffused posteriorly ; marginal points black ; cell-spot small, indistinct, whitish-centred : hind wing with a prominent black-speckled white-centred slightly wavy duplex discal line, and a broad black-speckled outer fascia, which latter is interrupted across its middle ; marginal points black. Cilia posteriorly and of the abdominal margin whitish. Body slightly black-scaled ; head, palpi, and fore legs above brown.

Expanse ⁷⁄₁₆ inch.

Hab. Calcutta. In coll. Dr. Staudinger.

Family HYPENIDÆ.

Genus RIVULA, *Gn.*

RIVULA TRILINEATA, n. sp.

Allied to *R. pallida.* Larger. Ochreous-brown. Fore wing with three equidistant inwardly-oblique slender pale-bordered brown lines, each being bent inward to the costa and ending in a white point; some white points also before the apex, and a row of black-tipped white points along the exterior margin; an indistinct brown-speckled reniform mark at end of the cell. Hind wing paler towards the base. Thorax and head ochreous-brown; palpi, abdomen, and legs paler.

Expanse $\frac{7}{10}$ inch.

Hab. Darjiling (*Atkinson*); Cherra Punji (*Austen*). In coll. Dr. Staudinger and F. Moore.

INDEX.

O

EXPLANATION OF THE PLATES.

Plate I.

Fig. 1. Dodona deodata, p. 3.
2, 3. Zophoessa atkinsonia, p. 2.
Figs. 4, 5. Debis serbonis, p. 2.
6, 7, 8. Adolias satropaces, p. 1.

Plate II.

Fig. 1. Pramila atkinsoni, p. 9.
2. Trilochana scolioides, p. 10.
3. Herimba atkinsoni, p. 21.
4. Chærotricha bipartita, ♀, p. 49.
5. Artaxa venosa, ♀, p. 50.
6. ——— dispersa, ♀, p. 50.
7. Drepana flava, p. 84.
8. ——— postica, p. 84.
9. ——— biocularis, p. 84.
10. Carbisa venosa, ♂, p. 41.
11. Pimprana atkinsoni, p. 42.
12. Euchætes sikkimensis, p. 39.
13. Nayaca florescens, ♂, p. 43.
Fig. 14. Pangora distorta, ♀, p. 42.
15. Harapa testacea, ♀, p. 47.
16. Artaxa basalis, ♂, p. 51.
17. Daplasa irrorata, ♂, p. 52.
18. Pida lativitta, p. 49.
19. Arbudas bicolor, ♂, p. 20.
20. Adrepsa stilbioides, ♀, p. 38.
21. Caragola costalis, ♂, p. 46.
22. Gazalina transversa, ♀, p. 47.
23. Chærotricha quadrangularis, ♀, p. 50.
24. Euproctis variegata, ♂, p. 48.
25. ——— semivitta, ♂, p. 48.
26. Spilarctia obliquivitta, ♂, p. 40.

Plate III.

Fig. 1. Baryaza cervina, ♂, p. 45.
2. Dasychira albescens, ♂, p. 59.
3. ——— perdix, ♂, p. 58.
4. Porthetria umbrina, ♂, p. 55.
5. Lymantria grisea, ♀, p. 55.
6. Mahoba plagiodotata, ♂, p. 52.
7. Conthcyla thoracica, p. 74.
8. Miresa cuprea, p. 74.
9. Beara castanea, ♂, p. 71.
10. Danaka pyraliformis, ♂, p. 71.
11. Locharna strigipennis, p. 53.
12. Ramesa apicalis, p. 68.
Fig. 13. Hoplitis strigata, p. 68.
14. Spatalia gemmifera, ♂, p. 62.
15. Niganda strigifascia, ♂, p. 63.
16. Cadrusia virescens, ♀, p. 54.
17. Eutricha flavosignata, ♂, p. 77.
18. Mustilia hepatica, ♂, p. 82.
19. Gastropacha torrida, ♂, p. 76.
20. Mahanta quadrilinea, p. 78.
21. Trichiura khasiana, p. 82.
22. Scopelodes vulpina, p. 73.
23. Setora divergens, p. 75.
24. Narosa rufotessellata, p. 73.

EXPLANATION OF THE PLATES.

Plate IV.

Fig. 1. Risoba vialis, p. 91.
2. Pitrasa variegata, p. 94.
3. —— vitellina, p. 94.
4. Kerala multipunctata, p. 93.
5. Tyeracona transversa, p. 95.
6. Borolia furcifera, p. 98.
7. Calymera picta, p. 104.
8. Mamestra decorata, p. 111.
9. Motama cidarioides, p. 110.
10. —— aurata, p. 110.
11. —— decorata, p. 110.
12. Luperina pardaria, p. 114.
13. Apamea denticulosa, p. 109.

Fig. 14. Baorisa hieroglyphica, p. 133.
15. Pachetra heterocampa, p. 115.
16. Chandata partita, p. 114.
17. Dimya sinuata, p. 122.
18. Rauaja fasciata, p. 121.
19. Agrotis placida, p. 117.
20. Hadena hastata, p. 130.
21. —— constellata, p. 130.
22. Dryobata leucosticta, p. 129.
23. Dianthecia calamistrata, p. 124.
24. Nikara castanea, p. 126.
25. Euplexia sinuata, p. 125.
26. Hyada grisea, p. 130.

Plate V.

Fig. 1. Coarica fasciata, p. 153.
2. Thalatta fasciosa, p. 154.
3. Tambana catocalina, p. 156.
4. Blenina pannosa, p. 157.
5. —— quinaria, p. 158.
6. Amrella angulipennis, p. 158.
7. Donda thoracica, p. 161.
8. Erebcia costipannosa, p. 166.
9. Poaphila pallens, p. 172.
10. —— uniformis, p. 172.
11. —— oculata, p. 172.
12. Sypna pannosa, p. 168.
13. Catocala tapestrina, p. 166.

Fig. 14. Sadarsa longipennis, p. 165.
15. Mestleta acontioides, p. 179.
16. Durdara lobata, p. 177.
17. Sonagara strigosa, p. 180.
18. Anophia perdicipennis, p. 162.
19. Zarima dentifera, p. 162.
20. Dordura apicalis, p. 170.
21. Capnodes pallens, p. 176.
22. Poaphila quadrilineata, p. 172.
23. Sypna floccosa, p. 167.
24. —— plana, p. 167.
25. —— albovittata, p. 168.

Plate VI.

Fig. 1. Cultripalpa trifasciata, p. 184.
2. Zethes amynoides, p. 181.
3. Acharya crassicornis, p. 185.
4. Egnasia morosa, p. 185.
5. Phurys similis, p. 174.
6. —— fasciosa, p. 174.
7. Iluza duplexa, p. 174.
8. Raparna ochreipennis, p. 178.
9. Selenis reticulata, p. 178.
10. —— obscura, p. 178.
11. Ophiodes adusta, p. 169.
12. —— indistincta, p. 169.
13. Harmatelia basalis, p. 183.
14. Ophiusa falcata, p. 171.

Fig. 15. Iluza transversa, p. 174.
16. Corcobara angulipennis, p. 186.
17. Cephena costata, p. 196.
18. Herminia duplexa, p. 193.
19. Bocana marginata, p. 195.
20. Thyridospila fasciata, p. 181.
21. Bocana picta, p. 195.
22. Asthala silenusalis, p. 197.
23. Harita rectilinea, p. 187.
24. Apanda dentilineata, p. 187.
25. Bibacta truncata, p. 198.
26. Avitta fasciosa, p. 194.
27. Tochara obliqua, p. 175.

EXPLANATION OF THE PLATES.

Plate VII.

Fig. 1. Locastra lativitta, ♂, p. 199.
2. Pannucha basalis, ♂, p. 200.
3. Scopocera sinuosa, ♀, p. 203.
4. ——— variegata, ♀, p. 203.
5. Pyralis assamica, ♂, p. 205.
6. Patania semivialis, p. 209.
7. Cymoriza inextricata, ♂, p. 210.
8. ——— rivularis, p. 210.
9. Glyphodes chilka, ♂, p. 216.
10. Cirrhochrista bryozalis, ♀, p. 226.
11. Haritala recurrens, ♂, p. 215.
12. Microsca lobulata, p. 214.
13. Camadena vespertilionis, p. 214.
14. Samea quinquigera, p. 207.
15. Polythlipta peragrata, ♂, p. 216.

Fig. 16. Botyodes fraterna, ♂, p. 221.
17. Haritala auroralis, ♂, p. 215.
18. Hapalia dorsivittata, p. 223.
19. ——— flavofasciata, p. 223.
20. ——— fasciata, p. 223.
21. Pitama lativitta, p. 217.
22. Microsca fasciata, p. 213.
23. Charema imbecilis, p. 219.
24. Circobotys limbata, ♂, p. 220.
25. Polythlipta distorta, ♂, p. 215.
26. Botyodes leopardalis, ♂, p. 221.
27. Hapalia robusta, p. 222.
28. ——— kasmirica, p. 222.
29. Pionea nobilis, p. 224.
30. Parbattia vialis, p. 225.

Plate VIII.

Fig. 1. Decetia pallida, ♂, p. 232.
2. Norcia flava, ♀, p. 233.
3. Auzea reticulata, ♀, p. 233.
4. Cimicodes sanguiflua, ♀, p. 233.
5. ——— flava, ♀, p. 233.
6. Marcala flavifusata, ♂, p. 231.
7. Caustoloma acutipennis, p. 230.
8. Anisodes lunulosa, p. 250.
9. Menophra vialis, ♂, p. 238.
10. Remodes lineosa, ♂, p. 270.
11. ——— abnormis, ♂, p. 271.
12. Brabira pallida, ♂, p. 271.
13. Runeca ferrilineata, p. 252.
14. Macaria deletaria, ♂, p. 261.
15. Agathia divaricata, p. 250.

Fig. 16. Alcis vagans, p. 242.
17. Trygodes ferrifera, ♀, p. 255.
18. Emodesa sinuosa, p. 256.
19. Stegania purpurascens, p. 261.
20. Epione adustata, ♂, p. 229.
21. Arasada fasciosa, p. 290.
22. Bardanes plicata, ♂, p. 251.
23. Menophra deficiens, ♂, p. 237.
24. Onychia violacea, p. 279.
25. Collix flavofasciata, p. 270.
26. Carige lunulineata, p. 263.
27. Menophra torridaria, ♂, p. 237.
28. Eupithecia griseipennis, ♀, p. 269.
29. Metabraxas falcipennis, ♂, p. 266.
30. Lobophora sikkima, ♂, p. 272.

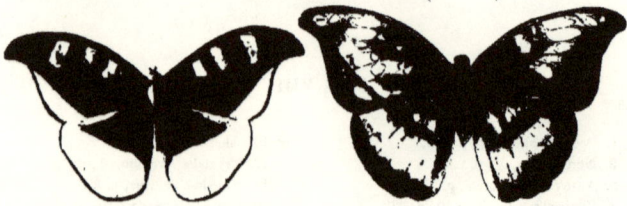

R.M.& A.Searit del. et lith. Minters. Bros. imp.

1. Pramila atkinsoni 2. Irilochana scolioides 3. Lemmoa atkinsoni 4. Chærotricha bipartita
5. Artaxa venosa 6. Artaxa dispersa 7. Drepana flava 8. Limpana postica 9. Limpana sp. 10. Carbsa venosa 11. Pimprana atkinsoni 12. Euonetes sikkimensis 13. Nayaca florescens
14. Pangora distorta 15. Harapa testacea 16. Artaxa basalis 17. Dappasa irrorata 18. Pramila sp.
19. Arbudas bicolor 20. Adrepsa subloides 21. Caragola costalis 22. Gazauna transversa
23. Chærotricha angularis 24. Euproctis variegata 25. Euproctis sp. 26. Euproctis sp.

1. Boryaza cervina 2. Dasychira alenoma 3. Dre_aina perdix 4. Leucoma molybra
5. Lymantria grisea. 6. Maheba plagiodonta 7. Taragnia the morea 8. Xanea nigra
9. Beara castanea 10. Danaka pyralmorphu. 11. Leharra surripunina. 12. Somera spatula
13. Hoplitis singata. 14. Spatalia gemmifera. 15. Nacuda singatalica 16. Delesina virescens
17. Eutricha flavosignata 18. Mustilia hepatica 19. Gastropacha torrida 20. Mahatara quinbalia
21. Trichiura khasiana. 22. Scopelodes vulgaris 23. Natara amergena. 24. Narosa caldesella

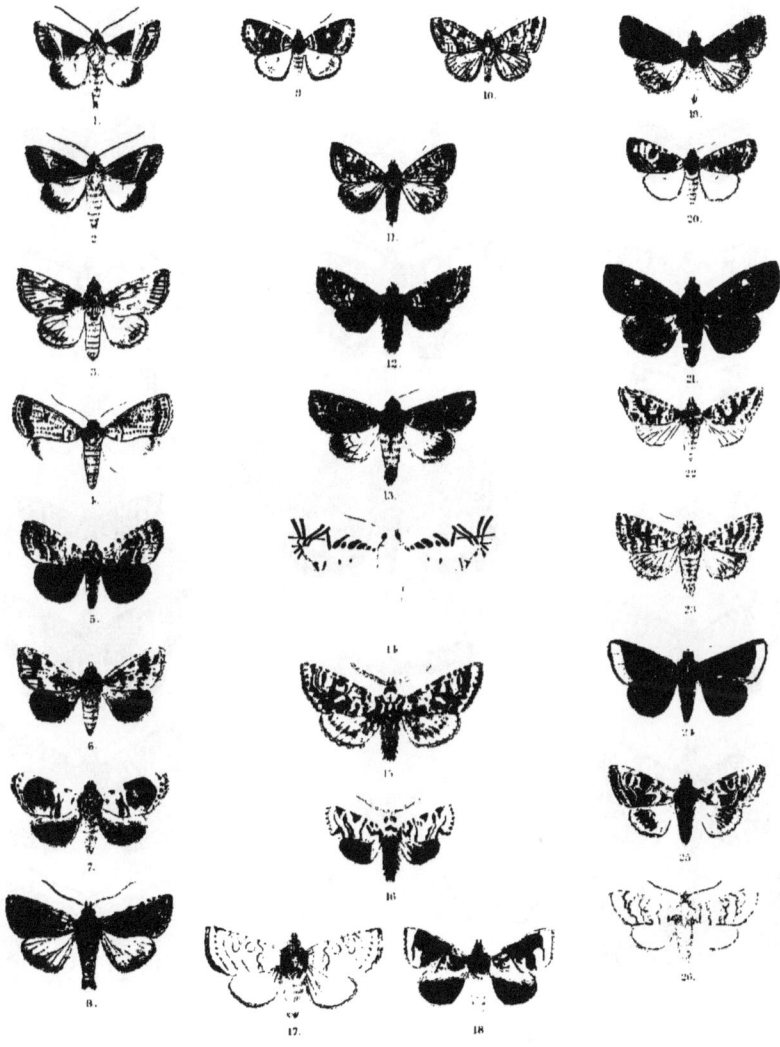

1, Ossova fuscosa   2, Thaluta fasciosa   3, Tachbaru catochurus   4, Bienura paruosa
5, Bienura quinoria   6, Anurefa angulipennis   7, Dorida thornicica   8, Encira conspiraosa
9, Poaphila pallens 10, Poaphila uniformis 11, Poaphila oculata 12, Sypna pannosa
13, Catocala tapestina 14, Sadarsa longipennis 15, Mestleta acontoides 16, Durdura lobata
17, Sonagara strigosa 18, Anophia perdicipennis 19, Zarima dentifera 20,) rdura apicalis
21, Caprodes pallens 22, Poaphila quadrilineata 23, Sypna floccosa 24, Sypna plana 25, Sypna rubovittata

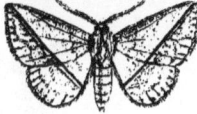

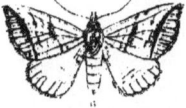
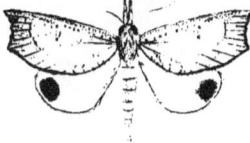

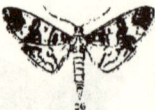

F. C. Moore, del. et lith.   Mintern Bros. imp.

1, Locastra lativitta. ♂   2, Paranucha cussanea ♂   3, Scoparera sinuosa ♀   4, ♀ variegata,♂   5, Pyralis kasmiralis ♂
6 Patana seminalis   7, Cymoriza trextricata.♂   8, C nivularis ♀   8, hypnodes bubo.♂   10, Cateclysta frigoralis. ♀
11, Haritala recurrens ♂   12, Microssa lobulata.♂   13, Cannaera vespertionis ♀   14, Samea quinculgera ♀. Pteridata petalbata ♀
18, Botyodes fraterna.♂   17, Haritana auroralis,♀   19, Hapalia dorsivittata   19. H. flav.lacer'. 20. H fascian.
21, Pitana lativitta   22, Microssa fasciata   23 Charema trinecinis.♂   24, Circobotys lineata.♂   25 Hapalopta distincta
26, Botyodes leopardalis.♂   27, Hapala robusta   28. H kasmirica   29, Pionea tuccinis   30. Parthchtia vians

P C Moore del et lith.　　　　　　　　　　　　　　　　　　　　Mintern Bros. imp.

1, Decetia pallida . 2,Noreia flava　3.Auzea reticulata　4,Cimicodes sanguiflua　5, C flava,♀.
6, Marcala flavifusata　7,Caustoloma acutipennis. 8 Anisodes iunulosa　9,Menophra vialis　10.Remodes lineosa
11, R abnormis　12,Brabira pallida　13,Runica ferrilineata　14, Macaria deletaria. 15, Agathia divaricata
16, Alcis vagans　17 Trygodes ferrifera　18,Emodesa sinuosa　19. Stegania purpurascens　20. Epione adustata
21, Arasada fasciosa. 22,Bardanes plicata. 23.Menophra deficiens　24. Onychia violacea　25. Collix flavofasciata
26, Carige lunulineata　27, Menophra torridaria. 28,Eupithecia griseipennis. 29, Metabraxas falcipennis　30,Loboplhora sikkima .

www.ingramcontent.com/pod-product-compliance
Lightning Source LLC
Chambersburg PA
CBHW021501210326
41599CB00012B/1086